DESERTIFICATION IN DEVELOPED COUNTRIES

DESERTIFICATION IN DEVELOPED COUNTRIES

INTERNATIONAL SYMPOSIUM AND WORKSHOP ON DESERTIFICATION IN DEVELOPED COUNTRIES: WHY CAN'T WE CONTROL IT?

Edited by

DAVID A. MOUAT AND CHARLES F. HUTCHINSON

U.S. Environmental Protection Agency,
National Health and Environmental Effects Laboratory,
Corvallis, OR U.S.A.

Sponsored by
U.S. Bureau of Land Management and Environmental Protection Agency

in collaboration with:

U.S. Forest Service
Desert Research Institute
University of Arizona
The Nature Conservancy
USDA Agricultural Research Service
Soil Conservation Service
Instituto de Agroquimica y Tecnologia de Alimentos

Supported by

International Arid Lands Consortium
European Society for Soil Conservation

Reprinted from Environmental Monitoring and Assessment, Volume 37, Nos. 1–3, 1995.

KLUWER ACADEMIC PUBLISHERS
DORDRECHT / BOSTON / LONDON

A C.I.P. Catalogue record for this book is available from the Library of Congress

ISBN-13: 978-94-010-7231-1 e-ISBN-13: 978-94-009-1635-7
DOI: 10.1007/978-94-009-1635-7

Published by Kluwer Academic Publishers,
P.O. Box 17, 3300 AA Dordrecht, The Netherlands.

Kluwer Academic Publishers incorporates
the publishing programmes of
D. Reidel, Martinus Nijhoff, Dr W. Junk and MTP Press.

Sold and distributed in the U.S.A. and Canada
by Kluwer Academic Publishers,
101 Philip Drive, Norwell, MA 02061, U.S.A.

In all other countries, sold and distributed
by Kluwer Academic Publishers Group,
P.O. Box 322, 3300 AH Dordrecht, The Netherlands.

Cover design:

The Goat Well drop structure, San Simon Valley, Arizona, USA. Photo taken by C.F. Hutchinson, 1995. The San Simon Valley in southeastern Arizona is an example of land degradation in the United States. The channel of the San Simon was downcut from 3 to 10 m over a reach of 100 km between 1893 and 1916. Headcutting extended up many tributaries, such as this one. Several factors appear to have converged between 1885 and 1895 to cause this large-scale erosion event, including: (a) overgrazing, (b) intense wagon and animal traffic to nearby mines, (c) construction of a railroad and drainage ditch along the main axis of the valley, and (d) severe drought. A series of water and sediment control structures have been built in the valley that have dramatically slowed - and in many cases reversed - erosion. For example, headward erosion at this site was halted in 1940 by construction of a concrete drop structure from which the photograph was taken. Although erosion has been abated at this site, the isolated surfaces on either side of the gully show no signs of recovery after 55 years. However, for the valley at-large conditions are greatly improved, particularly in the most productive, lower-lying areas.

Printed on acid-free paper

TABLE OF CONTENTS

TABLE OF CONTENTS

INTRODUCTION *

DAVID A. MOUAT, CHARLES F. HUTCHINSON and BEAUMONT C. MCCLURE**
*Editors Special Issue on Desertification in Developed Countries, ** Chairman, Symposium and Workshop Steering Committee*

This special issue of *Environmental Monitoring and Assessment* is devoted to a selection of peer-reviewed papers presented at the International Symposium and Workshop on Desertification in Developed Countries: Why Can't We Control It? The symposium and workshop were held in Tucson, Arizona, October 24–29, 1994, with the last day being devoted to six workshops, each of which discussed and made recommendations on one of the following desertification-related topics: common indicators, stressors, socioeconomic factors, innovative approaches, consistent problems, and a proposed symposium and workshop to be held in 1997.

Desertification (sometimes referred to as land degradation) is a phenomenon occurring in arid, semiarid, and dry subhumid areas. Resulting from various factors, including climatic and other natural stresses and human activities, desertification plagues all regions of the world. One sixth of the world's population is threatened by the effects of desertification, and 76% and 73% of the drylands of North America and Africa respectively are degraded (UNEP, 1992). The destruction of the land's productive capacity brought on by desertification costs the world more than US$42 billion each year (1990 estimate from UNEP, 1992).

The Earth Summit held in Rio de Janeiro in June 1992 generated renewed interest in efforts to combat desertification. The following December, the United Nations General Assembly passed resolution 47/188, which established an "Intergovernmental Negotiating Committee for the Elaboration of an International Convention to Combat Desertification in Those Countries Experiencing Serious Drought and/or Desertification, Particularly in Africa." The Committee adopted a Convention in June 1994 (UN, 1994). The Convention was opened for signature in October 1994 and has since been signed by more than 100 countries. The symposium from which this special issue emanates is in part the beginning of an effort to assess, monitor, and mitigate problems of desertification. The Rio Summit was partially responsible for this effort.

The International Symposium and Workshop on Desertification was limited to a few participants (approximately 150 scientists and land managers from 15 countries) in order to concentrate on specific monitoring and assessment techniques and mitigative interventions. The symposium and workshop organizers felt that, although many lessons have been learned regarding desertification throughout the world, we should at first concentrate our efforts on a few countries to distill

* The opinions expressed in this publication do not necessarily represent any policy or position of the sponsors of the symposium and workshop.

Environmental Monitoring and Assessment **37:** 1–4, 1995.

those lessons into a transferable format. Between now and 1997, when another symposium is planned, we will continue to examine efforts to combat desertification by encouraging regional workshops. These workshops will focus on local problems, solutions, and success stories.

Desertification has many effects, including soil and vegetation degradation and social and economic losses. Soil degradation includes displacement of soil material by water and wind erosion and internal soil deterioration by such physical and chemical processes as salinization, waterlogging, and nutrient depletion (UNEP, 1992). Vegetation degradation may include loss of cover and biomass and compositional changes such as the replacement of native by exotic species. Schlesinger and others (1990) examined desertification through changes in ecosystem structure and function and within the context of spatial and temporal distribution of soil resources relative to vegetation. They found that in undisturbed productive grasslands, soil resources such as water and nitrogen have a relatively uniform distribution. But when these grasslands become desertified, the distribution of these resources becomes increasingly heterogeneous in time and space. Typically this process also results in changes in land and water use and settlement patterns and demographics. Economic productivity also declines (Reining, 1978).

Desertification can be assessed and monitored through an analysis of its physical, biological, social, and economic indicators, which include nutrient availability measured through surface soil condition, litter accumulation, and organic carbon; water budget measured by such indicators as runoff and erosion, infiltration, rainfall regime, and water quality; energy balance assessed through measurements of biomass, photosynthetic efficiency, and surface temperature; and biological diversity, including the ratio of native to exotic species, species composition, and indicator species. Social and economic indicators might include population pressures, inappropriate government policies such as certain types of agricultural subsidies, unsuitable farming practices, and overgrazing. The United Nations Environment Programme (UNEP, 1992) stated that the only globally comprehensive data base of characteristics of desertification involves soil degradation, which is caused by overgrazing, deforestation, and unsustainable agricultural practices. The World Atlas of Desertification (UNEP, 1992) shows that soil degradation is a problem that is just as common in developed countries as in less developed ones.

Mitigation measures in developed countries seem to stem largely from changes in governmental policies. In addition, practices affecting large-scale agricultural and grazing enterprises may be transferred from developer to user. Some of the papers included in this special issue discuss mitigating measures for combatting desertification in developed countries. On the surface there would appear to be few lessons that administrators or resource managers from developing countries might take away from the practices of developed countries. Aside from the physical and biological characteristics of the lands they share, these classes of countries greatly differ in (1) the vigor, size, and integration of their economies; (2) the level and amount of technical support available to resource managers; and (3) the effectiveness of institutions in influencing their behavior. It seems clear, however, that opportunities abound for two-way communication on desertification control.

First, despite some of the economic advantages enjoyed by developed countries, resource managers are constrained by costs. Excluding irrigated agriculture and mining, most resource exploitation in arid and semiarid regions operates on a fairly thin margin of profitability. Thus, it is difficult to justify, attract, or maintain large capital investment in arid and semiarid regions; there are too many other areas and opportunities in which returns are potentially higher. This argues strongly for efficiency, whether the setting is a developed or a developing country. If a resource manager cannot justify an intervention in economic terms, it will surely fail.

Second, the multitude of problems that constitute desertification has no single solution. The technical "magic bullet" does not exist, and, more to the point, it is unrealistic to expect that it ever will. Physical, biologic, economic, and social factors vary so widely and occur in such a bewildering array of combinations that they conspire to defeat most formula-based approaches for dealing with the issues. In short, like developing countries, developed countries do not have all the answers and must continue to pursue them through programs of focused interdisciplinary action.

Third, desertification control (and related resource development) is perhaps more a social undertaking than a technical or economic one. Thus, using the lesson learned from developing countries, the need for bottom-up, grassroots, and community-based participatory development emerges as a dominant theme. Development must be planned, implemented, and assessed by those it is intended to help. It cannot be achieved through decree, regardless of social value or scientific merit.

This special issue on desertification in developed countries includes 24 peer-reviewed papers. More than 65 papers were presented at the Symposium and Workshop on Desertification in Developed Countries. These papers focused on (1) the social, economic, political, and institutional factors that have resulted in successful interventions; (2) desertification assessment and systematic observations (monitoring); and (3) techniques that have been tried to halt or reverse desertification processes. The papers in this special issue reflect this range of interest.

We wish to thank everyone who has contributed to this special issue: the authors of the oral and poster papers presented at the symposium, the many reviewers who gave considerably of their time and advice, and especially Mr. Herbert K. McGinty, Bureau of Land Management National Training Center, who spent many long hours editing the manuscripts.

References

Reining, P. (editor). 1978. *Handbook on desertification indicators*, American Association for the Advancement of Science, Washington.

Schlesinger, W.H., J.F. Reynolds, G.I. Cunningham, L.F. Huenneke, W.M. Jarrell, R.A. Virginia, and W.G. Whitford. 1990. Biological feedbacks in global desertification. *Science* 247: 1043–1048.

United Nations. 1994. *Earth Summit: Convention on desertification*, New York. United Nations Conference on Environment and Development, Rio de Janeiro, 3–14 June 1992.

United Nations Environment Programme (UNEP). 1992. *World atlas of desertification*, Edward Arnold, London.

EFFECT OF MEDITERRANEAN SHRUB ON WATER EROSION CONTROL

V. ANDREU, J. L. RUBIO and R. CERNÍ
Centro de Investigaciones sobre Desertificación-CIDE Jaime Roig, 11. 46010-Valencia, Spain

Abstract. In the Spanish Mediterranean environment, scrub vegetation occupies a greater area than does forest. The impact of wildfire on the scrub vegetation and recovery afterward affects a number of other processes, including water erosion. While recovered vegetation considerably influences soil protection and erosion control, this function has scarcely been studied. This study discusses the behavior and architecture of recovering (or regenerating) typical Mediterranean shrub vegetation and the subsequent impact on soil protection. The study compared two protective forage species (*Medicago arborea L.* and *Psoralea bituminosa L.*). The research was performed in field conditions on a set of four experimental plots. A control plot was maintained with no vegetation cover. Runoff and soil loss by water erosion between 1989 and 1992 were studied on each of these plots. The natural vegetation was found to have a more significant protective effect (69.2% decrease in soil loss) than the other species tested. Soil loss on the *Medicago* plot decreased by 41.7%, and soil loss on the *Psoralea* plot decreased by 29.3%. That the *Psoralea* was only recently planted must be considered in evaluating its protective effects.

1. Introduction

The Mediterranean environment is characterized not only by a unique climate but also by the physiognomy of its vegetation, its land use patterns, and often the general appearance of its landscape. One of the most common phenomena of these areas is the incidence of soil erosion and wildfires. For the past centuries the combined action of humans and fire has produced a typical pattern of the Mediterranean landscape (Naveh, 1975). In many cases a repeated occurrence of fires and the recovery processes following them produce “perturbation dependent systems” (Naveh, 1990; Vogel, 1980) that characterize the present Mediterranean forest and shrublands. But in recent years the number of wildfires has increased causing the manual destruction of around 200,000 ha of Mediterranean forest and scrub as well as a concomitant increase in erosion (Sanroque and Rubio, 1982).

This increase in fire frequency with a reduction in the recovery periods could drastically change (with matorral increase) or even eliminate vegetation. The subsequent reduction of vegetation cover might intensify the erosive processes and cause severe structural breakdown of the soil (Al’Benskii and Nikitin, 1967; Brandt *et al.*, 1972; Bridge *et al.*, 1983; Chartres and Mücher, 1989).

Although it covers a greater area than forests in the Mediterranean region (Tomaselli, 1977), the matorral (or maquis) vegetation has received little research attention, especially of its role in ameliorating erosion-induced desertification. The objective of this research was to study under field conditions the effect of a typical

Environmental Monitoring and Assessment **37:** 5–15, 1995.

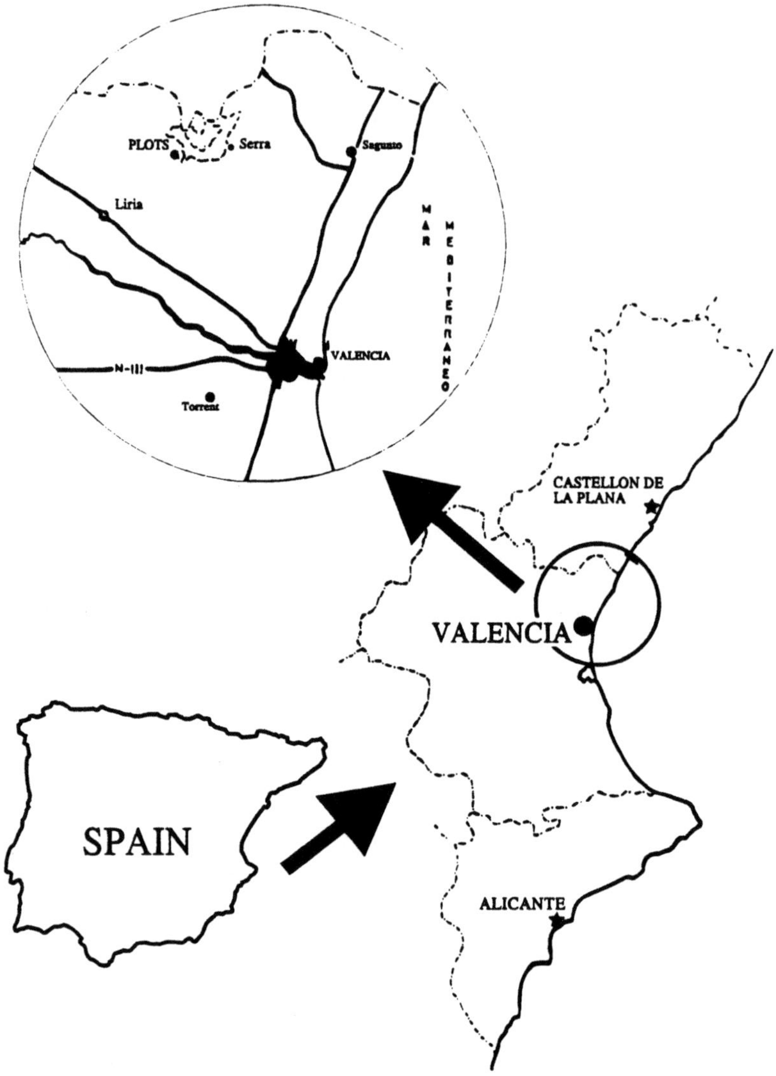

Fig. 1. Geographical location of the studied area.

Mediterranean shrub association (*Rosmarinus ericion*) on water erosion processes, comparing two forage species with different biotypes (*Medicago arborea L.* and *Psoralea bituminosa L.*). In addition, the research investigated the development of plants and the effects of such development on erosion parameters from 1989 to 1992.

2. Materials and Methods

The experimental plots are located near Porta-Coeli (Valencia, Spain), as shown in Figure 1, on a southwest-facing hillside that was forested before being consumed by fire in 1982. The area's lithology consists of Muschelkalk materials of clayey-sandy marl and conglomerate with shallow Calcic Xerochrept soils (Andreu *et al.*, 1991). The study area contains frequent outcrops of consolidated conglomerate material with abundant stoniness ($\simeq$ 40%) (Rubio *et al.*, 1990; Andreu *et al.*, 1991). Some physical and chemical characteristics of these soils are shown in Table I.

Climatically, the area belongs to the dry ombroclimate stage of the thermo-mediterranean climate according to Thornthwaite's system (Thornthwaite, 1948). The study area's annual precipitation is approximately 425 mm, and the mean annual temperature is 17 °C. The thermopluviometric features of the plots have been measured by a system of sensors connected to a recorder allowing data to be recorded in real time.

To establish the vegetal composition and the distribution of species in the study area, vegetation was inventoried at 15 sites within the same area as the plots (320 m^2). The dominant vegetation type corresponds to the *Rosmarinus ericion* association (Stübing, 1985), in which the most abundant species include *Rosmarinus officinalis L.*, *Thymus vulgaris L.*, *Stipa tenacissima L.* and *Chamaerops humilis L.* (Andreu *et al.*, 1994).

The design for runoff determination consisted of four 40 × 8 m plots in a slightly concave hillside at an elevation of 220 m, with a mean slope of 20%. The plots lay on a slope with representative soil, vegetation, and landscape characteristics of the area. Each plot was delimited by long, hydrophobically coated bricks following the direction of the slope in a rectangular pattern. At the lower end of the plots 2 m wide collectors directed runoff and sediment into 2000 l tanks (Figure 2).

One of the plots was left in a "natural" condition following a wildfire occurring 12 years ago. This plot had about the same composition as that of the mean of the inventoried plots. The shrub cover of this natural plot had had a vigorous growth (with a mean height of 85 cm and a ground cover of approximately 65% of the soil surface). Interspersed among the shrubs were numerous *Pinus* seedlings, showing a general trend toward the previous *Pinus halepensis* dominated cover.

In a second plot 90 seedlings of *Medicago arborea L.* were planted in 1989. The *Medicago* were planted in groups of three, approximately 20 cm apart. Transverse fences have been favored with this kind of planting to evaluate how well plants restrain runoff and soil removal.

In the same manner as the *Medicago* plot, 90 seedlings of *Psoralea bituminosa L.* were planted in a third plot in 1990. The delay of a year in their planting was due to a failure in a previous attempt to establish seedlings of *Atriplex nummularia L.* in the plot, and to difficulties in the dispersal of these seedings. Both species were selected to evaluate their effects on soil retention. The *Medicago* is a woody plant normally growing about 2 m high, whereas the *Psoralea* is a herbaceous plant

TABLE I

Soil physical and chemical characteristics in the different plots

	Natur	Medic	Psor	Bare	Particle size	Plot	
					⌀ (mm)	Bare	Natur
Humidity (%)	2.75	3.00	2.95	3.95	0.0–0.002	6.14	7.66
Str. St. (%)	34.75	30.75	28.79	22.85	0.002–0.005	6.16	7.52
pH	8.1	8.1	8.00	8.00	0.005–0.02	18.55	20.77
E.C. ($S\ m^{-1}$	0.37	0.55	0.70	0.81	0.02–0.05	22.15	22.48
$CaCO_3$ (%)	32.30	33.85	32.07	36.67	0.05–0.10	13.60	12.60
Org. Mat. (%)	6.26	5.94	5.04	4.92	0.10–0.25	12.09	10.44
Min. N ($mol_c\ kg^{-1}$)	0.45	2.84	3.33	4.77	0.25–0.50	10.51	9.04
Avail. P ($mol_c\ kg^{-1}$)	0.11	0.13	0.21	0.17	0.50–1.0	5.30	5.46
CEC ($cmol_c\ kg^{-1}$)	22.11	23.97	22.53	20.90	1.0–2.0	5.50	4.03
Ca ($mol_c\ kg^{-1}$)	20.18	21.84	20.43	18.54	Gravel (% in weight)		
Mg ($mol_c\ kg^{-1}$)	1.33	1.51	1.42	1.71	>2.0	31.57	42.42
Na ($mol_c\ kg^{-1}$)	0.07	0.07	0.07	0.07			
K ($mol_c\ kg^{-1}$)	0.58	0.51	0.55	0.53			

Str. St.: structural stability; E.C.: electric conductivity; Org. Matt.: organic matter; Min. N: mineral nitrogen; Avail. P: available phosphorous; CEC: cation exchange capacity; Natur: natural vegetation; Medic: *Medicago*; Psor: *Psoralia*.

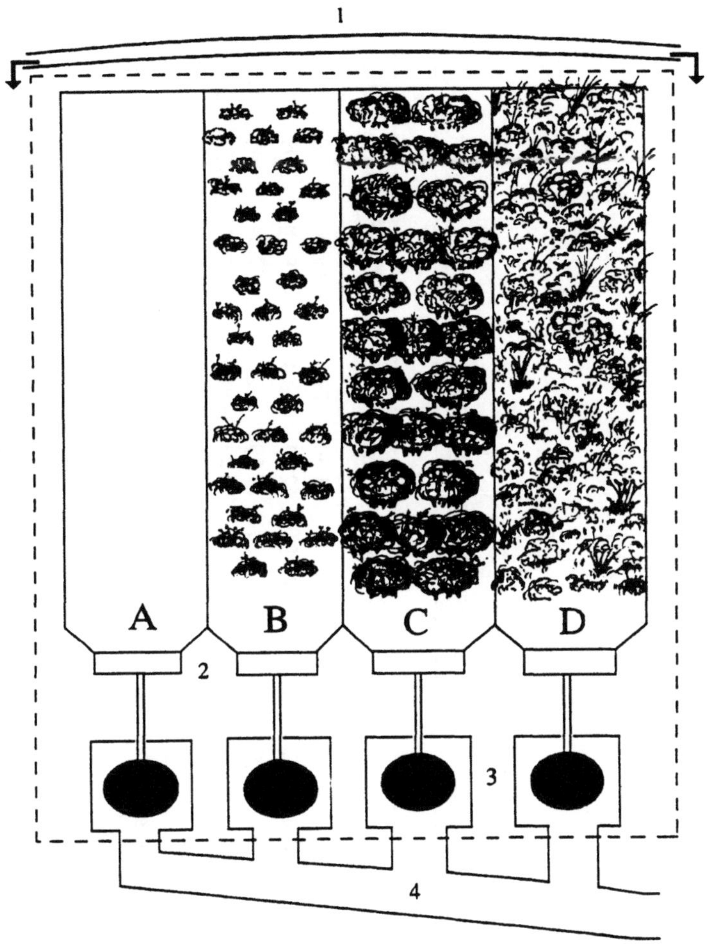

Fig. 2. Layout of the experimental plots. A – Bare; B – *Psoralea*; C – *Medicago*; D – Natural. 1 – Drainage channel; 2 – Collectors; 3 – Deposits; 4 – Deposits drainage channel.

rarely higher than 1 m and highly capable of colonizing and developing a more creeping-type structure.

The last, and control, plot was maintained bare of vegetation by manual removal of the vegetal cover and treating the soil with herbicides.

For each plot the following characteristics were measured: vegetation height, cover, stem diameter and angle, leaf area index, compaction of soil in the root area, and other traits. Between 1988 and 1992, 45 erosive rain events occurred, from which runoff and sediment discharge were collected on all plots.

TABLE II

Mean annual values of some vegetation characteristics on the erosion plots

	1989			1990			1991			1992		
	Natur	Medic	Psor	Natur	Medic	Psor	Natur	Medic	Psor	Natur	Medic	Psor
Cover (%)	65.60	40.80	–	66.20	41.10	16.73	66.70	41.30	18.64	67.95	41.50	18.96
ø Stem (cm)	17.65	2.80	–	17.83	2.87	0.40	18.91	2.93	0.45	19.10	3.03	0.51
Leaf area index	50.00	23.05	–	50.90	25.00	12.66	51.30	27.50	14.21	52.64	29.40	15.50
Stem angle	52.79	73.25	–	53.15	73.40	33.02	53.00	75.42	38.70	53.01	76.32	39.43
Shear strength* ($N\ m^{-2}$)	6.20	4.58	–	6.43	5.05	4.35	6.67	5.75	5.00	6.02	6.80	5.90

Natur: Natural vegetation; Medic: *Medicago arborea*; Psor: *Psoralea bituminosa*; * Shear strength measured in the soil root mass.

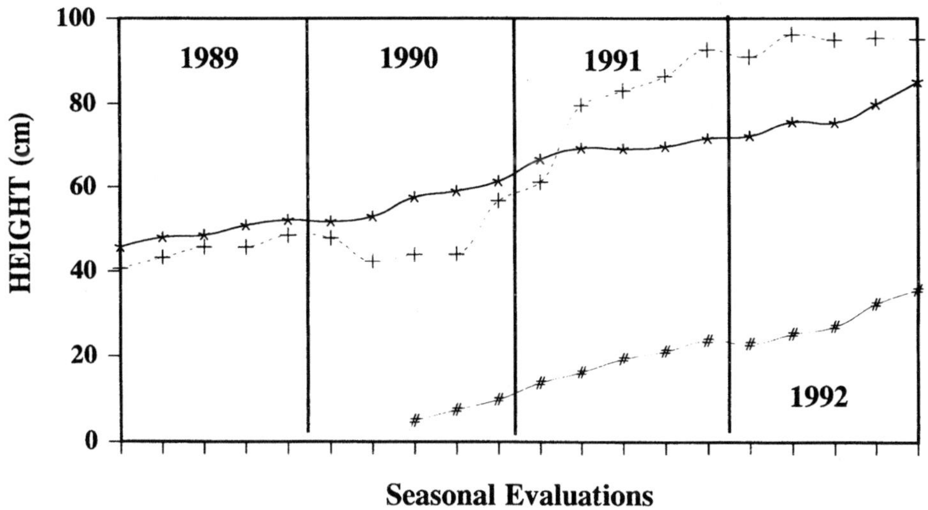

Fig. 3. Vegetation development in height (* = Natural; + = *Medicago*; # = *Psoralea*).

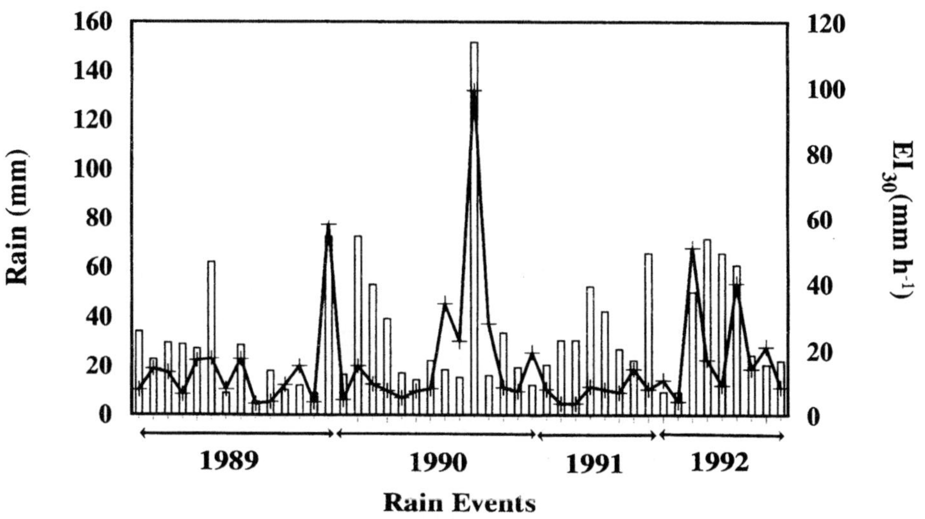

Fig. 4. Quantity of rain and EI_{30} of the erosive events during 1989–1992.

3. Results and Discussion

Figure 3 compares yearly changes in the average height of the two introduced shrub species with natural vegetation. Table II gives values for percent cover, stem diameter and angle, leaf area index, and soil shear strength in the root zone for the natural vegetation and the two introduced species.

The total amount and the intensity of rainfall of the erosive rain events (with production of runoff) during the studied period are shown in Figure 4. The runoff

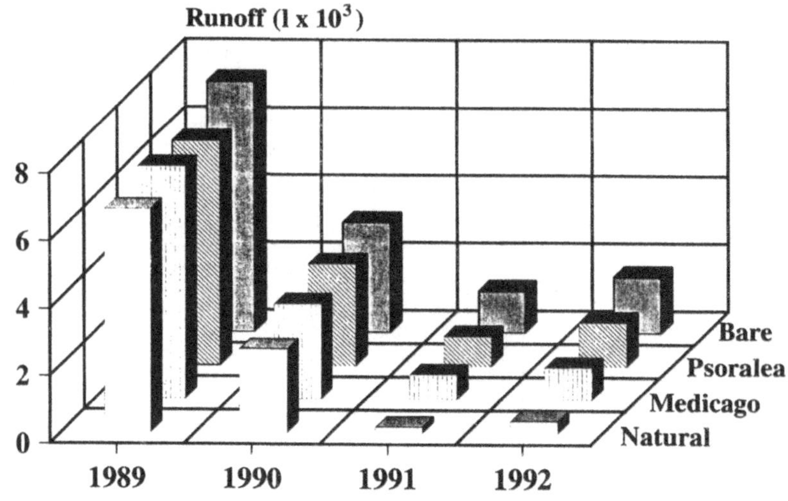

Fig. 5. Total annual runoff collected in the plots.

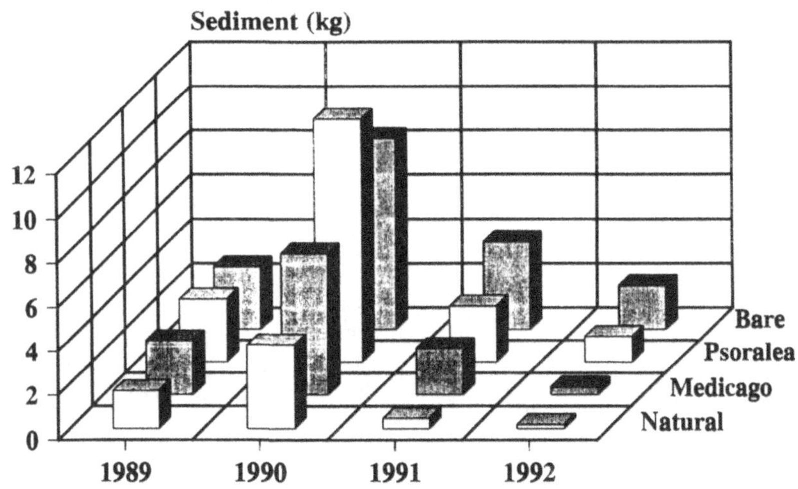

Fig. 6. Total soil losses collected in the plots.

and sediment collected annually from each plot during 1989–1992 are shown in Figures 5 and 6 respectively.

The vegetation growth was steady in all cases, but individuals of the same species, mainly forage plants, differed markedly. *Medicago arborea L.* was among the fastest growing of the species, reaching the mean height value of the natural vegetation in 1991 and even surpassing it by 10.5% in 1992. But the natural species offered substantially greater protection in intercepting rainfall than the planted species, as is reflected by the leaf area index (Table II), which is twice the planted species value.

Since being planted in 1990, the *Psoralea* showed a slow but steady development for a species with slow and limited growth. As shown in Table II, the planted individuals had a higher canopy cover than did the natural vegetation.

The planted species had a pronounced influence on soil structural properties, especially immediately near the plants themselves. Marked differences in structural stability and organic matter content were noted (Table I) and were likely due to the differences in biomass production and its incorporation into the soil matrix.

The highly irregular distribution and type of rainfall can be divided into two periods: 1989–1990 and 1991–1992. Greater and more intense precipitation fell in 1989–1990 (Figure 4), whereas the second period experienced fewer erosive rains (of low intensity). In fact, during the first period, EI_{30} reached 99 mm/h, whereas in the period of 1991–1992 the maximum intensities didn't exceed 50 mm/h. In the same manner, during 1991–1992, the mean duration of rains increased considerably. (The average duration of rains was 561 minutes during the first period and 3629 minutes during the second.) This observation is reflected by a decrease in rainfall energy producing less runoff, soil detachment, and transport during 1991–1992.

In a pattern similar to that of rainfall, the most important sediment and runoff production episodes appeared during 1989–1990, reaching a runoff volume of more than 2000 l per plot (in the episodes of 9 November 1989 and 9 September 1990).

The amount of sediment collected was not excessively high compared to other Mediterranean soilscapes, even in the high EI_{30} episodes, mainly because of high soil stability, high soil organic matter content (Table I), and the protective action of the surface stoniness.

When the yearly data are compared, the control plot (bare of vegetation) clearly shows more significant rates of soil loss and runoff production than the other plots (Figures 5 and 6), except for *Psoralea* in 1990, which represents the highest amount of soil lost (see Figure 6). But the *Psoralea* was planted in 1990, and the soil disturbed by the planting procedures, even if minimal, could have resulted in the removal of soil particles by runoff.

As compared to the bare soil, soil loss during the 4-year period decreased by 41.7 percent for the *Medicago* plots, 29.3% for the *Psoralea* plots, and 69.2% for the natural vegetation. König (1992) reported similar results with *Medicago* and other leguminous shrubs as contour edges.

Medicago performs better than *Psoralea* for erosion control mainly because of *Medicago*'s faster growth and development, which produces an effective barrier to rain impact and runoff. Despite having a better architecture for erosion control than *Medicago*, *Psoralea* has a more open canopy and therefore develops less protection.

Plots with natural vegetation produce lower runoff and soil loss than do the plots with *Medicago* and *Psoralea*, but the plots dominated by the introduced shrubs (especially *Medicago*) have effectively protected the soil by favoring high soil structural stability, increasing organic matter content, and promoting soil shear strength in the zone under the canopy (Tables I and II).

4. Conclusion

The loss of vegetation cover increases soil removal through sediment transport by runoff. This study found these losses to be 4.29 times higher for sediment weight and 2.59 times higher for runoff in bare plots than in the plots with natural vegetation. In 1991, 3.98 kg of sediments were lost on the bare plots, whereas only 0.46 kg were lost on plots supporting natural vegetation.

The use of plant species to protect and retain soil is an effective measure against erosive processes. But even with the positive results reported in this study, planted species protect the soil less than does natural vegetation. A summary example of this observation using the two planted species, *Medicago arborea L.* and *Psoralea bituminosa* follows. During the study period (1989–1992) the total runoff collected was 9,631.14 l on the plot with natural vegetation, 11,395.14 l on the *Medicago* plot, 11,817.17 l on the *Psoralea* plot, and 13,552.38 l on the bare plot.

On degraded Mediterranean landscapes, especially with steep slopes, planting species of the described types can be one of the first actions to undertake to diminish erosion impacts and allow and favor the progressive development of an indigenous plant cover. Such management can regenerate the whole ecosystem.

References

Al'Benskii, A.V. and P.D. Nikitin. 1967. *Handbook of afforestation and soil amelioration*, Israel Program for Scientific Translation, pp. 187–276.

Andreu, V., R. Cerní and J.L. Rubio. 1991. The Porta-Coeli experimental sites, in M. Sala and J.L. Rubio (Eds.), *Excursion guide book* (ESSC Workshop, Valencia-Barcelona 1991), Universidad Internacional Menendez y Pelayo, Valencia, pp. 2–14.

Andreu, V., J.L. Rubio, J. Forteza, and R. Cerní. 1994. Long term effects of forest fires on soil erosion and nutrient losses, in M. Sala and J.L. Rubio (Eds.), *Soil erosion as a consequence of forest fires*, Geoforma Ediciones, Logroño, pp. 79–90.

Aschmann, H. 1973. Distribution and peculiarity of Mediterranean ecosystems, in F. di Castri and H.A. Mooney (Eds.), *Mediterranean type ecosystems*, Springer-Verlag, Heidelberg, pp. 11–20.

Bailey, H.P. 1979. Semi-arid climates: Their definition and distribution, in A.E. Hall, G.H. Cannell and H.W. Lawton (Eds.), *Agriculture in semi-arid environments*, Springer-Verlag, Berlin, pp. 73–96.

Brandt, G.H., E.S. Conyers, M.P. Ettinger, F.J. Lowes, J.W. Mighton, and J.W. Pollack. 1972. *An economic analysis of erosion and sediment control methods for watersheds undergoing urbanization*, The Dow Chemical Co., Midland, Texas.

Bridge, B.J., J.J. Mott, and R.C. Hartigan. 1983. The formation of degraded areas in the dry savanna woodlands of northern Australia. *Australian Journal of Soil Research* **21**: 91–104.

Chartres, C.J. and H.J. Mücher. 1989. The effects of fire on the surface properties and seed germination in two shallow monoliths from a rangeland soil subjected to simulated raindrop impact and water erosion. *Earth Surface Processes* **14**: 407–417.

FAO-Unesco. 1988. *Soil map of the World 1:5.000.000*. Revised legend. FAO, Roma.

König, D. 1992. The potential of agroforestry methods for erosion control in Rwanda. *Soil Technology* **5**: 167–176.

Lal, R. 1988. *Soil erosion research methods*, Soil and Water Conservation Society-International Society of Soil Science, Ankeny, Iowa.

Naveh, Z. 1975. The evolutionary significance of fire in the Mediterranean region. *Vegetatio* **29**(3): 199–208.

Naveh, Z. 1990. The role of fire and its management in the conservation of the mediterranean ecosystems and landscapes, in UIMP (Eds.), *El papel del fuego en los ecosistemas mediterráneos*, Universidad Internacional Menendez y Pelayo, Valencia, pp. 37–52.

Rubio, J.L., V. Andreu, and R. Cerní. 1990. Degradación del suelo por erosión hídrica: Diseño experimental y resultados preliminares, in J. Albaladejo, M.A. Stocking and E. Díaz (Eds.), *Degradación y regeneración del suelo en condiciones ambientales mediterráneas*, Instituto de Agroquimica y Technologia de Alimentos, Consejo Superior de Investigaciones Cientificas, Madrid, pp. 215–235.

Sanroque, P. and J.L. Rubio. 1982. *El suelo y los incendios forestales*, Diputación de Valencia, Valencia, pp. 57–63.

Stübing, G. 1985. *Estudio fitosociológico de los matorrales termófilos valencianos*, University of Valencia, Valencia, pp. 127–144.

Thornthwaite, C.W. 1948. An approach towards a rational classification of climate. *Geography Revue* **30**: 55–94.

Tomaselli, R. 1977. Degradation of the Mediterranean maquis, in Unesco (Eds.), *Mediterranean forest and maquis: Ecology, conservation and management*, Unesco, Paris, pp. 33–72.

Vogel, R.J. 1980. The ecological factors that produce perturbation dependent ecosystems, in J. Cairns (Eds.), *The recovery process in damaged ecosystems*, Ann Arbor Science Publishers, Ann Arbor, Michigan, pp. 63–94.

SECONDARY DESERTIFICATION DUE TO SALINIZATION OF INTENSIVELY IRRIGATED LANDS: THE ISRAELI EXPERIENCE

A. BANIN and A. FISH

Department of Soil and Water Sciences and The Seagram Center for Soil and Water Sciences, The Hebrew University, P.O. Box 12, Rehovot 76100, Israel

Abstract. Secondary salinization of intensively irrigated lands is an increasingly alarming redesertification process experienced in many irrigated regions of the developed countries. The major cause is a profound interference in the geochemical/salt balances of irrigated regions. A case-in-point is the recent salinization of the Yizre'el Valley, a 20,000 ha intensively irrigated region in Israel. The extremely intensive and advanced agroecosystem developed in the region since the 1940s included pumping and importing irrigation water by the National Water Carrier, large-scale reclamation and reuse of municipal sewage water, winter flood impoundment in reservoirs for summer irrigation, and cloud seeding to enhance rainfall. Modern irrigation methods were applied, including sprinkler, trickle, moving-line, and center-pivot systems. Water use efficiency at any level was very high. Nevertheless, large-scale salinization of regional water resources and many fields had developed in the mid-1980s. Reconstructing and evaluating the water and salt balances of the Yizre'el Valley (using Cl as the representative salt constituent) shows that as water use in the valley increased to about 60 million m^3 per year, the importing of soluble salts by water totaled 15,000 tons of Cl per year. Recirculated salt – salt picked up by impounded surface water and applied to fields – increased significantly and in the late 1980s amounted to more than 9,000 tons Cl per year. The source of recirculated salts was the accumulated salts in soils and in the shallow aquifer in the valley, which were leached by floodwater or drained or infiltrated into reservoirs, grossly and adversely affecting water quality. Analysis of the Yizre'el Valley's case points to the utmost importance of maintaining the geochemical balances in addition to increasing irrigation efficiency. An irrigated region may achieve geochemical balance by the following means: limiting the extent of irrigated areas, developing a well-maintained drainage system that drains tail-water and salinized shallow-aquifer water, and devoting a significant portion of water for regional leaching. The sustained long-term productivity of irrigated lands in arid zones crucially depends on correctly managing water and soil resources. Regional management of irrigated lands to prevent secondary desertification will be aimed at carefully balancing the undisputed benefits of irrigation with the long-term (on time scales of 10 to 100 years) detrimental processes set in motion when irrigation is introduced to arid and semiarid zone soils.

1. Introduction and Background

1.1. General

Arid and semiarid lands have been rapidly irrigated since the 1950s, more than doubling in area (Jensen *et al.*, 1990). Generally, agricultural production increases in these irrigated zones only during a relatively short period (20–40 years). A series of decline processes immediately begins with the initiation of large-scale irrigation, causing slow but steady yield declines on regional scales.

One of the most serious decline processes is the development of secondary salinization, often followed by sodification of soils. This form of desertification

Environmental Monitoring and Assessment **37:** 17–37, 1995.

is grossly affecting semiarid and arid regions around the latitudes of 30°N and 30°S. In the long run these processes lower the productivity of fertile lands and the quality of local groundwater. Such processes have to be controlled to ensure the continued viability and productivity of irrigated lands. Such control requires close monitoring of the salt balances in single fields as well as in whole regions, with the objective of avoiding long-term salt accumulation in the root zone and in regional water resources.

The naturally limited leaching of arid lands, coupled with increased input of salts by irrigation and urbanization and elevated groundwater levels, limit water drainage and removal of salts, and lead to a slow increase (on time scales of 10–100 years) of the salinity of soils and local water resources. Many reports in recent years have documented the observations of encroaching secondary salinization in arid zones following the introduction of intensive large-scale irrigated agriculture in developed and developing countries (Westcot, 1988; Oosterbaan, 1989; Jensen *et al.*, 1990).

Increased salinity harms plants by increasing osmotic pressure and decreasing water availability. Toxic effects of specific ions on plants are another stress factor. Increased concentration of exchangeable sodium in soils caused by the increased relative concentration of Na^+ in salinized soil solutions (i.e. elevated SAR in the soil) is causing soil-structure destabilization, decreased hydraulic conductivity, waterlogging, crust formation, and increased susceptibility to erosion.

1.2. Balances of Soluble Salts and Other Chemical Constituents

The salt balance concept was introduced by Scofield (1940). Originally, it evolved from the recognition of the need to maintain the soil in the root zone in the irrigated field at a desired salinity level that would not cause crop stress during the growing season. Crop stress, in turn, could lead to yield decline. The simplest representation of the salt balance, basically a mass conservation equation, is as follows:

$$\mathrm{IN} - \mathrm{OU} = \mathrm{AC} \tag{1}$$

Where IN and OU are the quantities of soluble salts (or ionic component(s)) added and removed, respectively, from the root zone in a given area during a given period of time, and AC is the accumulation in the root zone. A “Salt Balance Index” may be defined as: SBI = OU/IN. Three situations can then be identified:

$$\text{Stable salinity state}\quad \mathrm{IN} = \mathrm{OU};\quad \mathrm{AC} = 0;\quad \mathrm{SBI} = 1 \tag{2}$$

$$\text{State of leaching}\quad \mathrm{IN} < \mathrm{OU};\quad \mathrm{AC} < 0;\quad \mathrm{SBI} > 1 \tag{3}$$

$$\text{State of salinization}\quad \mathrm{IN} > \mathrm{OU};\quad \mathrm{AC} > 0;\quad \mathrm{SBI} < 1 \tag{4}$$

Customarily the balance is calculated annually for either total salt loads or representative chemical constituents.

1.3. Salt Balance Equations

Three cases can be formulated for regional salt balances. In general, they all follow the format presented in equation (1).

(a) Non-disturbed region:

$$S_{rw} + S_{we} + S_{eo} + S_{gw} + S_{pr} = S_{dw} + S_{ro} + S_{cp} + S_{pu} \pm \Delta S_a \pm \Delta S_s \quad (5a)$$

(b) Agricultural, irrigated region:

$$S_{rw} + S_{we} + S_{eo} + S_{gw} + S_{pr} + S_{iw} + S_{fr} + S_{ps} + S_{am} + S_{fl} = S_{dw} + S_{ro} + S_{cp} + S_{pu} \pm \Delta S_a \pm \Delta S_s \quad (5b)$$

(c) Agricultural and urbanized region:

$$S_{rw} + S_{we} + S_{eo} + S_{gw} + S_{pr} + S_{iw} + S_{fr} + S_{ps} + S_{am} + S_{fl} + S_{sw} + S_{ga} + S_{at} = S_{dw} + S_{ro} + S_{cp} + S_{pu} \pm \Delta S_a \pm \Delta S_s \quad (5c)$$

where the various S terms represent constituent input or output (mass per year). The terms are

Inputs

S_{rw} = rainwater
S_{we} = rock and soil weathering
S_{eo} = eolian (long-range)
S_{gw} = groundwater penetration
S_{pr} = plant residues
S_{iw} = irrigation water
S_{fr} = fertilizers
S_{ps} = pesticides
S_{am} = soil amendments (manure, gypsum)
S_{fl} = feedlots
S_{sw} = sewage (domestic and industrial)
S_{ga} = garbage
S_{at} = atmospheric fallout (short range)

Outputs

S_{dw} = drainage water
S_{ro} = runoff water
S_{cp} = chemical precipitation
S_{pu} = plant uptake

Accumulation

S_a = adsorbed constituents
S_s = soluble constituents

Among the input terms in the salt balance equations, S_{ra} (rain contribution) is usually low. But in the coastal plain of Israel significant NaCl concentrations may be found in rainwater (Herut, 1992). The major source of salts in coastal regions (up to 100 km from the sea) consists of cyclical salts, i.e. salts buoyed from the sea as spray or aerosols and introduced into clouds. Chloride and sodium are therefore the major constituents of salts in the rain under these conditions. Furthermore, calculations show that during soil formation in desert zones, salt added to the top layer of the soil by rain may cause high salinization of this soil layer due to very limited leaching. S_{fr} (fertilizer contribution) is usually small, except for cases of heavy fertilization and limited leaching. S_{we} (soil dissolution) is also usually small but may be important in the top layer of the soil during rain. Under normal irrigated agriculture conditions in arid regions the major term is thus S_{iw}, the salt load added by irrigation water (good irrigation water, applied at a rate of 7000 m^3/ha/yr and with TDS = ~700 mg/l adds about 5 tons of salts per hectare per year). Under improper management, where the groundwater level is high and water rises and evaporates from the soil surface, the term S_{gw} may become increasingly important or even the major one, and salt loads added to the soil from this source may become very large.

1.4. Managing the Field Salt Balance

In the past, and particularly according to the approach developed by the U.S. Department of Agriculture Salinity Laboratory in Riverside, California, during the 1930s and 1940s (Richards, 1954), reliance was mostly on S_{dw} to maintain a salt balance, i.e. keep input=output and accumulation=zero. Their methodology to control salt balance in the irrigated field defined the leaching requirement, i.e. the excessive fractional amount of irrigation water, above the consumptive use of the crop, added to the irrigation water to maintain long-term zero accumulation of soluble salts in the root zone.

The optimal-situation balanced equation was written as

$$(W_i * C_i)/A = (W_d * C_d)/A \qquad (6)$$

where W is applied water volume, C is concentration of salt(s), and A is the irrigated area. Subscripts i and d represent irrigation water and drainage water, respectively. The leaching fraction (LF) was then computed by

$$LF = W_d/W_i = C_i/C_d \qquad (7)$$

C_d was selected according to crop sensitivity to salt stress in the attempt to minimize yield reduction.

As experience accumulated in irrigated regions, researchers recognized that the leaching requirement approach might have to be modified to reduce excessive return flows and drainage water burdens on the regional scale (Bouwer, 1969; Van Schilfgaarde *et al.*, 1974). Furthermore, they recognized that on the regional scale and on medium-time scales other sources of soluble salts would have to be addressed, including inputs from urban and other anthropogenic sources, floodwater, soil and rock weathering, and groundwater penetration.

It was also found that chemical constituents might become serious environmental hazards as they accumulate in drainage water from irrigated fields. One group of constituents consists of potentially toxic elements from natural-geochemical sources, such as selenium, undergoing enhanced leaching by irrigation (Tanji and Hanson, 1990). Another group of constituents such as nitrates may result from such excessive agricultural inputs as mismanaged nitrogen fertilization.

1.5. Case studies

Carter (1975) discussed regional salt balances, considering the full complement of contributors to the regional salt balance. Bower *et al.* (1969) analyzed the soluble-salt balance of the Coachella Valley in California and correlated it to expanding drainage systems. Kaddah and Rhoades (1976) analyzed the salt balance of California's Imperial Valley for 1944–1973, documenting the increase in the average salinity of the regional irrigation water, as measured by the electrical conductivity and the SBIs of the major ionic constituents. Over a 30-year period (1944–1973), the annual regional WBI (water balance index = WBI = water output/input) ranged from 0.31 to 0.43 with averages of 0.40 and 0.36 for the periods from 1944 to 1954 and from 1955 to 1973, respectively. The overall regional SBI, estimated from weighted averages of the electrical conductivity of regional influent and effluent waters, varied between 0.90 and 1.17 for the first period and between 1.01 and 1.23 for the second. The increase in the SBI during the period 1955–1973 is attributed to increases in the area of subsurface drained land coupled with increased leaching. Detailed analyses of ionic components in the influent and effluent waters during 1973 showed SBI < 1.0 for HCO_3^-, $SO_4^=$ and Ca^{++} (0.50, 0.90 and 0.73, respectively). These findings were attributed to precipitation of $CaCO_3$ and $CaSO_4$ in the soil. SBI > 1 was observed for Cl^-, $Na^+ + K^+$ and Mg^{++} (2.21, 1.56 and 1.13, respectively). It was sugggested that this partly resulted from groundwater contributions. The influent water is sulfatic in nature ($Cl/SO_4 = 0.52$). Generally, the experience in the Imperial Valley was that overall stable salinity state was maintained (overall SBI $\cong$ 1.0) but the regional cost in water was quite significant, amounting to more than one third of the diverted water (WBI = 0.36–0.40).

Faci *et al.* (1985) studied the water and salt balance of an irrigated region fed by the Ebro River in Spain. They found that in the 2-year period studied the region was undergoing net leaching (SBI > 1), which was adding considerable salt load to the Ebro. The major components in the regional drainage water were Ca^{++} and $SO_4^{=}$, resulting from dissolution of $CaSO_4$ in the soils. The regional WBI was also quite high (0.53–0.58). Excessive leaching in this case resulted mainly in soil dissolution.

Lennaerts *et al.* (1988) studied the salt balance of a pilot irrigated area in Egypt. They modelled the effect of groundwater pumping on water quality and forecast the long-term effect of recycling the regional groundwater through the agricultural system on the chemical composition of the soil and the aquifer itself.

Reporting on the characteristics of agricultural drainage water in Bahrain, Raveendran and Madany (1991) found large-scale salinization of the shallow groundwater.

Rhoades and Loveday (1990) reviewed the practices for monitoring and controlling salinity in irrigated lands, outlining the current approach to leaching fraction and salt balance in fields from the viewpoints of crop response in the single field at any given season as well as in relation to regional management of salinity. In this context, Rhoades (1989) considered the optimal strategy for reuse and disposal of saline irrigation drainage water for irrigation. He concluded that one can obtain more flexibility in using the total water supply by isolating drainage water with its excessive salt loads from higher quality water.

Wescot (1988) recently reviewed the issue of reuse and disposal of high-salinity drainage water from irrigated regions. He pointed out that in many regions in the world drainage water from irrigated fields threatens to salinize and pollute the water resources (rivers, aquifers) and regional soils. In reviewing options for drainage water management, he did not consider explicitly the need to maintain the salt balance of the irrigated region. But it becomes clear that many of the management options considered offer only temporary relief from the problem by discharging the salt burden downstream or to future users.

We can no longer deny the need to establish and monitor the full chemical balance of the irrigated region as a basis for its proper management. Such actions should particularly constitute a basic component of the reconstruction and remediation of deteriorating irrigation systems suffering from decades of neglect in the management of regional salt balances.

Under the Israeli conditions, with limited amounts and high costs of water for irrigation and with the presence of winter rain causing effective leaching, the Riverside Leaching Requirement was not strictly applied. The Israeli experience showed furthermore that a more complete approach to the system should involve considering water quality in aquifers and not just those of the single field. Where salt inputs from agricultural return flows and leaching water reach the aquifer, the long-range effects of this process may lower the aquifer's water quality. In Israel's coastal plain the shallow Pleistocene aquifer is being salinized at an increasing rate in recent years (Hydrological Service, 1994). As a result, water quality for irrigation and other uses is deteriorating. One possibility considered both by the Riverside school (Rhoades and Loveday, 1990; Van Schilfgaarde *et al.*, 1974) and in Israel is that of causing salt precipitation in insoluble forms in or below the root zone (the term S_{cp} in Eq. 5). When salt components such as $SO_4^{=}$, $CO_3^{=}$, and Ca^{++} are present in the water, their precipitation can be an efficient controlling factor of the salt balance. But control of the more soluble components (Cl, Na) is difficult, and large-scale desalination is now being considered as the only long-term remedy for removing excessive salts from aquifers and maintaining their long-term quality.

1.6. Water Use Efficiency and Salt Balance

Increased water use efficiency in irrigation projects is a goal of the management and reconstructive activity of irrigation systems in many countries. The Field Water Use Efficiency is defined as $E_f = W_m/W_f$, where W_m is the volume of water needed and made available for evapotranspiration by the crop to avoid undesirable water stress in the plants during the growing cycle, and W_f is the volume of water furnished to the field (ICID, 1978). One should be aware, however, that increased field water use efficiency, i.e. an increased fraction of evapotranspired water out of the furnished water, may result in decreased salt leaching, leading eventually to regional salinization (Rhoades, 1984). Regional salinization is particularly possible and dangerous in regions where natural leaching by rainfall is limited. The proper management of irrigated regions at all scales requires continuous monitoring and management of salt balances in addition to water management.

2. The Salinization of the Yizre'el Valley

2.1. Yizre'el Valley Setting and Environments

The Yizre'el Valley lies in northern Israel, separating the Shomron from the Galilee Mountains (Figure 1). Shaped like a triangle, this 400 km^2 valley has a base length of 50 km near Karmel Mountain and an apex in the Nazareth Mountains. It is divided into the eastern and western sections by a low northeast-southwest anticlinar range. The valley is drained into the Mediterranean Sea by the Qishon River. The river drains a total area of 685 km^2, bordered by the Tabor and Nazareth Mountains to the north, the Gilboa Mountains to the east, and the Ramot Menashe/Karmel Mountains to the south and west. A dam built in 1953–1954 at Kefar-Barukh formed an artificial lake that divided the basin into an upper part (470 km^2) and a lower part (215 km^2). Annual precipitation ranges from 500 to 600 mm, decreasing from west

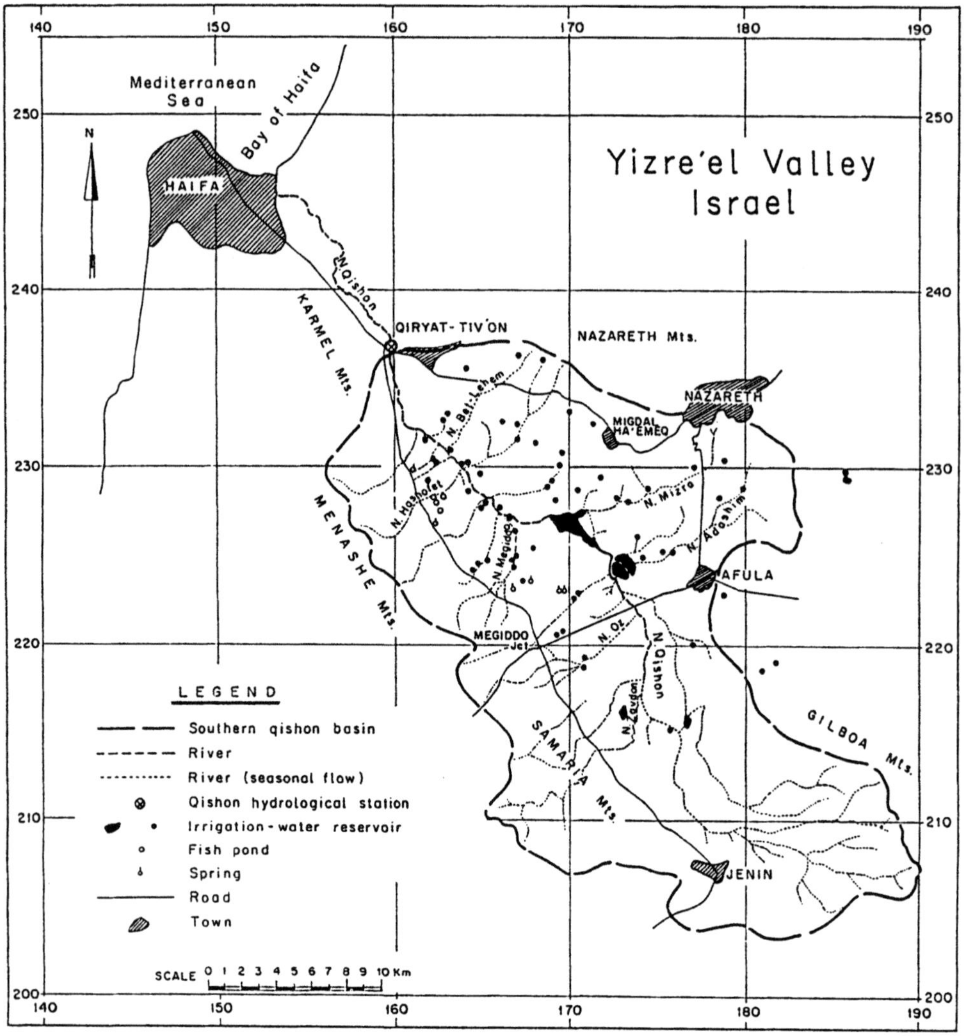

Fig. 1. Location map and hydrological features of Yizre'el Valley, Israel.

to east. The average annual temperature is 19–21 °C with monthly average extremes of 10–12 °C in January and 26–28 °C in August. The average annual evaporation rate is 5–6 mm/day (~2000 mm/yr), peaking in June-August (7–8 mm/day).

The Yizre'el Valley was formed in the Pleistocene (2–3 million years B.P.) during the formation of the Jordan Rift and the Galilee and Karmel anticlines. The valley's formation involved considerable faulting. During the Holocene heavy erosion from the surrounding anticlinal areas filled the valley with thick layers of clastic materials ranging from breccia and gravel (coarse fragments) to clays (fine grains) (Yatir, 1973).

The soils of the region are typically clayey vertisols rich in montmorillonite. They were formed from alluvial materials derived from mountain region soils mostly developed on hard calcareous rocks. The soils are deep and chemically fertile, contain some calcite, and have low to medium organic matter content. Soil structure is friable in the top layer, turning into blocky-prizmatic in deeper horizons. Certain areas show typical secondary hydromorphic nature, developed where the alluvial soil was seasonally (or permanently) flooded. The hydromorphic soils have gone through chemical reduction processes and tend to contain in their deep horizons gley layers resulting from the presence of reduced iron and manganese oxides. They are particularly compact and plastic, manifesting low hydraulic permeability at the deeper horizons (Ravikovitch, 1992).

2.2. Settlements and Agricultural Activity in the Yizre'el Valley

After the establishment of the State of Israel, new cities and rural settlements were established, and the population of the Yizre'el Valley rapidly grew, as did the intensity of agricultural production. Small-scale irrigation began in 1938, and the irrigated area underwent major expansion during the 1950s and 1960s.

Over the years an extremely intensive and advanced agroecosystem developed in the region. Major developments included pumping and importing Lake Kinneret water by the National Water Carrier (NWC), large-scale reclaiming and reusing of municipal sewage water, impounding winter floods for summer irrigation, and cloud seeding to enhance rainfall. Modern irrigation methods were applied, including sprinkler, trickle, moving-line, and center-pivot systems. Water use efficiency at any level was very high. Nevertheless, large-scale salinization of regional water resources and many fields developed in the mid-1980s.

2.2.1. *Irrigation Development and Salinization*

The development of irrigation in the valley can be divided into four stages (Table I). At the fully developed stage, in the 1980s, an area of 14,000 ha was irrigated in the basin. The area consisted of 8,500 ha in its upper part and 5,500 ha in its lower part (as divided by the Kefar-Barukh Dam). Mostly field crops were cultivated, mainly cotton, corn, and fodder, as well as citrus and avocados. Irrigation water was supplied from various sources, including Lake Kinneret water pumped by the National Water Carrier; groundwater from local and western Galilee wells; and reclaimed sewage effluents from local farming settlements, valley cities, and Haifa – a major city in the vicinity (Figure 2). Floodwater is impounded in many reservoirs built during the 1970s across the valley and used for summer irrigation. In 1990, some 60 million m^3 of water were used for irrigation, and the average water duty was 450–500 mm per year. Crops were watered mostly by sprinkling (including moving lines) and drip irrigation, and water was conveyed by high-pressure pumping and a well-maintained piping conveyance system. As a result, field water use efficiency was high, and conveyance water losses, common in other

TABLE I

Evolution of the regional average SBI for Yizre'el Valley during different stages in its development

Period	Description, Key Dates	Av. SBI (± Std)
A. 1945–1954 "Non-disturbed"	Limited interference in regional leaching; Limited irrigation; 1953/4 – Construction of of K. Barukh dam and reservoir.	1.02 (± 0.48)
B. 1955–1964 "K. Barukh Reservoir"	Increasing irrigation using imported water. 1964 – Lake Kinneret water imported by National Water Carrier.	0.49 (± 0.14)
C. 1965–1974 "National Water Carrier"	Continued increase in irrigated area using imported water and impounded floodwater. 1972–1982 – Large-scale construction of reservoirs throughout the valley.	0.34 (± 0.12)
D. 1975–1990 "Full regional development" (Floodwater impoundment; Reclaimed sewage water use)	Stable water supply. 1980s – Cotton becomes the major crop. 1984 – Maaleh-Qishon complex becomes operational. Replacement of Lake Kinneret water by Haifa effluents. 1985–1987 – Major salinization outbreak in K. Barukh Reservoir. Salinity in reservoir higher than imported fresh water and Haifa effluents.	0.26 (± 0.12)

irrigated regions, were minimal. But water may have seeped from the many earth reservoirs built in the 1970s, particularly during spring and early summer before it could be used for irrigating summer crops.

Salinity was detected in the valley soils in the 1950s (Ravikovitch *et al.*, 1960), but it was limited in extent (about 450 ha) and was detected mostly in the hydromorphic soils and at deep soil horizons.

In the southern part of the valley a major agronomic problem in that period was waterlogging, which prevented timely cultivation in the spring. This problem was solved by installing shallow drainage systems (0.8–1.0 m deep) (Shalhevet, 1965).

In the mid-1980s the Kefar-Barukh reservoir suffered a major outbreak of salinization. The Cl concentration reached 700 mg/l, preventing the use of the water for irrigating most crops in the area. Soil surveys conducted later found 7450 ha suffering from salinity hazard. Drainage systems were installed on 4100 ha, partly solving the problem. The remaining 3350 ha are in various stages of salinization

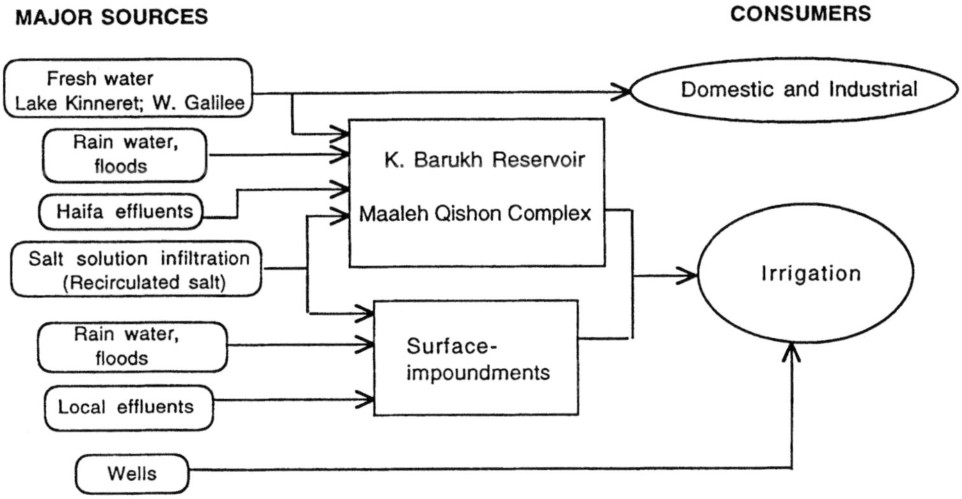

Fig. 2. Water and salt flow scheme for Yizre'el Valley.

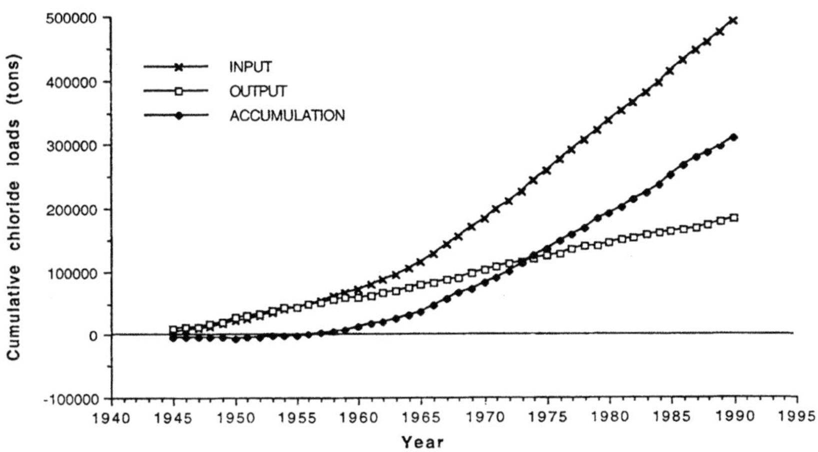

Fig. 3. Cumulative regional input, output and accumulation of chloride salts from 1945 to 1990. Note input-slope change accompanying the beginning of importation of Lake Kinneret water by the National Water Carrier in 1964 and accumulation-slope changes due to K. Barukh reservoir construction (1953) and Lake Kinneret water importation (1964).

and are not drained. Irrigation water salinity is generally high, particularly for the impounded floodwater, which constitutes a major portion of the irrigation water.

The reasons for the increased salinization of soil and water resources are still debated (Gafni and Salingar, 1992; Adar *et al.*, 1992). In the following section, we will discuss the salt balance of Yizre'el Valley as it has changed over the years and suggest a plausible explanation for the salinization process.

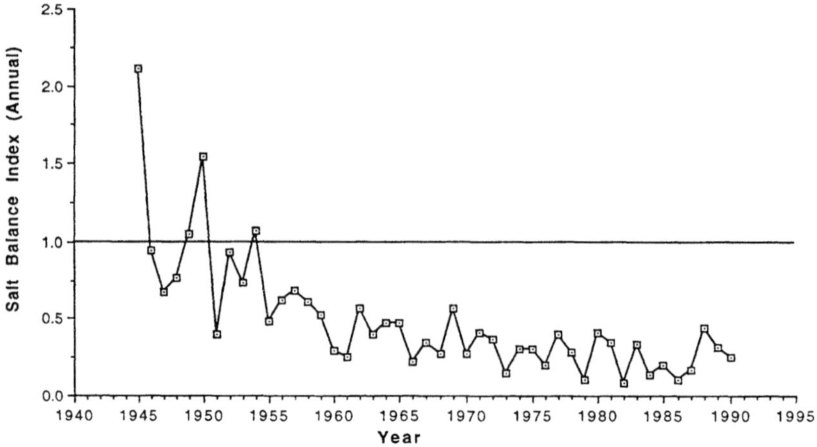

Fig. 4. Annual salt balance index (SBI) of the Yizre'el Valley from 1945 to 1990.

2.3. SALT BALANCE OF YIZRE'EL VALLEY

Figure 2 presents a scheme of the water and salt flows in Yizre'el Valley. The salt (chloride) balance of the valley for 1945–1990 was calculated using water input and chloride concentration data as described in the Appendix.

Cumulative regional input, output, and accumulation of Cl^- from 1945 to 1990 are presented in Figure 3. The change in the annual SBI during the same period is shown in Figure 4. Average annual SBI values for the different stages in the development of the valley's irrigation scheme are shown in Table I. During the initial period (1945–1954; Table I), accumulation was negative or zero due to limited imports of irrigation water. The construction of the K. Baruckh dam and reservoir in 1953–1954 and somewhat increased imports of water to the region initiated accumulation in the mid-1950s since the Qishon output had not increased. Further increase in water imports, mainly from Lake Kinneret by the National Water Carrier, significantly changed the input-curve slope in the mid-1960s. Whereas at the initial period SBI $\cong$ 1 and the valley was practically balanced with respect to salt input by rain and output by the Qishon River flows, increased importation of irrigation water and salts with no concurrent increase in outputs resulted in the initiation of regional salinization signified by steadily decreasing SBI. The low value of SBI may be contrasted with that maintained in Imperial Valley (California) as reported by Kaddah and Rhoades (1976) and cited above.

Detailed data and a breakdown of the water and salt inputs and outputs for representative years at the different stages described in Table I are given in Table II and discussed below.

In 1953 regional irrigation development was beginning. This year may be taken as a reference for the natural background input/output relationships of the valley. At this stage the main input term was rainwater, and accumulation (AC) was relatively small. The region was in a steady state with the multiannual AC $\cong$ 0. Continued

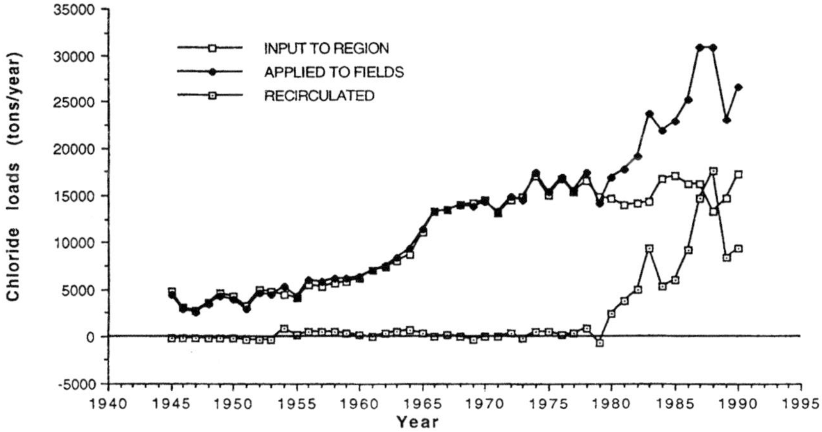

Fig. 5. Annual regional chloride salts inputs, amounts applied to fields and amount recirculated. Note onset of salt recirculation attributed to large-scale construction of reservoirs in the 1970s.

development of irrigation using imported water (mainly National Water Carrier supplies from Lake Kinneret) roughly doubled the chloride inputs in 1965. But the removal rate did not change, and regional accumulation of chlorides was enhanced (AC > 0), amounting to 5800 tons, about 50% of the annual input.

In the 1970s a large number of reservoirs were built throughout the valley to impound winter floodwater for summer irrigation (mainly of cotton). The increased water supply from the reservoirs did not change the *net* input of chloride into the valley but initiated a major process of salt recirculation into the top layer of the soils. The effect was seen in the year of 1983, when somewhat increased importing of National Water Carrier water resulted in increased Cl input. Again, output from the valley did not increase enough to cope with the elevated inputs, and accumulation was therefore high, amounting to about 66% of the input. Chloride applied to fields, however, increased significantly, and a major fraction (~40%) of the applied chloride was contributed by recirculated salts. At this stage, estimated regional accumulation (since 1945) exceeded 220,000 tons of Cl, and annual recirculated salt was 4.25% of this amount.

In 1984 the Qishon complex went into operation, delivering (for irrigation of the valley) secondarily treated effluents from Haifa, a city of 250,000, 35 km west of the valley's center. Supply and capture of local waste water and its reuse for irrigation also increased as the hills surrounding the valley became urbanized. (Nazareth, Migdal Haemek, Kiryat Tiveon, and Afula are the major cities in the region.) This stage is represented by the data for 1990. Annual net input of chlorides was then about three to four times the natural background (1953), but removal increased only by about 20%. Accumulation increased to about 70–80% of the annual input. The volume of impounded floodwater increased as more floodwater reservoirs were built, and total impounding capacity reached 50 million m^3. At

TABLE II

Yizre'el Valley: Regional water and chloride salt inputs, outputs, and balances in selected years (see details in Appendix)

Year				1953	1965	1983	1990
Rainwater			(10^6 m^3)	402.7	443.4	482.0	402.0
Water supplies	Wells		(10^6 m^3)	1.8	2.0	2.1	1.8
	Surface-impounded		(10^6 m^3)	0.0	2.4	21.6	20.5
	Mekorot	Fresh water	(10^6 m^3)	12.9	29.0	45.4	36.4
		K. Baruch reservoir (floods)	(10^6 m^3)	0.0	8.0	7.5	2.5
		Haifa effluents	(10^6 m^3)	0.0	0.0	0.0	15.0
		Total Mekorot	(10^6 m^3)	12.9	37.0	53.0	53.9
Water consumption by sector	Agriculture	Well water	(10^6 m^3)	1.8	2.0	2.1	1.8
		Fresh water (imported by Mekorot)	(10^6 m^3)	10.2	23.8	32.3	20.6
		K. Baruch reservoir (floods)	(10^6 m^3)	0.0	8.0	7.5	2.5
		Haifa effluents	(10^6 m^3)	0.0	0.0	0.0	15.0
		Surface-impounded	(10^6 m^3)	0.0	2.4	21.6	20.5
		Total agriculture	(10^6 m^3)	12.0	36.1	63.5	60.3
	Domestic		(10^6 m^3)	2.6	4.7	10.6	12.0
	Industrial		(10^6 m^3)	0.1	0.5	2.6	3.9
Total water supply			(10^6 m^3)	14.7	41.3	76.7	76.2
Qishon water outflow			(10^6 m^3)	13.6	36.0	27.7	8.9
Chloride concentration	Rain water		(mg/l)	7	7	7	7
	Wells, Yizre'el Valley		(mg/l)	123	225	407	426
	Mekorot fresh water		(mg/l)	133	261	221	224
	K. Baruch water supply/surface-impounded		(mg/l)	133	230	317	393
	Haifa effluents		(mg/l)	0	0	0	370
	Qishon river (winter)		(mg/l)	236	139	146	448

Table II Continued

Year			1953	1965	1983	1990
Chloride loads	Rain	(ton/yr)	2819	3104	3374	2814
	Fresh water for irrigation (imported by Mekorot)	(ton/yr)	1357	6216	7129	4603
	Domestic and industrial water	(ton/yr)	359	1358	2905	3544
	Haifa effluents	(ton/yr)	0	0	0	5550
	Yizre'el wells	(ton/yr)	221	439	872	786
	Surface-impounded water	(ton/yr)	0	544	6849	8046
	K. Baruch water supply (including fresh water and floods)	(ton/yr)	1357	7326	12614	14948
Input of salt to the region		(ton/yr)	4756	11117	14280	17297
Output of salt from the region (Qishon River)		(ton/yr)	3528	5252	4855	4385
Accumulated salt (AC) (Annual)		(ton/yr)	1228	5865	9426	12912
Salt applied to fields		(ton/yr)	4397	11412	23709	26594
Recirculated salt (REC)		(ton/yr)	–359	295	9429	9298
Regional accumulation (REG.AC) (Since 1945)		(tons)	–2362	36176	221818	308096
Accumulated/input		(%)	–	52.8	66.0	74.6
Recirculated/applied (Annual)		(%)	–	2.6	39.8	35.0
Recirculated/input		(%)	–	2.7	66.0	53.8
AC/REG.AC		(%)	–	16.2	4.2	4.2
REC/REG.AC		(%)	–	0.8	4.3	3.0
Delta Crec		(mg/l)	–29.9	8.2	148.4	154.1
Water out/input (WBI)			0.033	0.076	0.052	0.019
Salt out/input (SBI)			0.742	0.472	0.340	0.254

TABLE III
Integrated water and salt loads and balances in Yizre'el Valley (1945–1990)

	Total	Annual average (± std.)
Water (10^6 m^3)		
Rainfall	17,121	372.0 (± 91.0)
Imported water	1,618	35.2 (± 17.7)
Irrigation water	1,709	37.2 (± 20.4)
Qishon outflow	830	18.0 (± 19.6)
Water Balance Index (WBI = output/input)		0.039(± 0.035)
Salt (Chloride, tons)		
Inputs	489,799	10,648 (± 5,099)
Outputs	181,702	3,950 (± 1,777)
Accumulation	308,096	6,698 (± 5,217)
Salt Balance Index (SBI = output/input)		0.49 (± 0.38)
Weighted average CI concentration in Qishon (mg/l)		
Observed		379 (± 202)
Calculated for SBI = 1		1,548 (± 1,930)

this stage the average concentration of Cl in the impounded water had increased significantly. Water in some of the reservoirs was salinized, reaching concentrations of 900–1000 mg Cl/l. Chloride in surface-impounded water does not constitute *net* regional input of salts and thus does not aggravate the regional salt balance. But this chloride does represent recirculation of previously accumulated salts and gravely affects water *quality* and use. The change in the annual amount of recirculated salts during 1945–1990 is shown in Figure 5. In 1990 the recirculated chloride load was estimated at about 9300 tons, adding on the average 9300/60.3 = 154 mg Cl/l to the total agricultural water supply.

Table III presents an overall summary of the water and salt loads and balances of the valley between 1945 and 1990.

3. Summary and Conclusions

Our study shows that the increased input of salts since the 1950s and practically constant output by the Qishon River are the cause of significant accumulation of soluble chloride salts in the Yizre'el Valley. These salts reside mostly in the topsoil, subsoil, and a shallow aquifer at depths of 5–10 m throughout the valley. The irrigation water reservoirs in the valley, built mainly in low-lying areas and near seasonal rivers and drainageways, received salinized floodwater as well as slowly seeping (or diffusing) salt fluxes, which contributed a large amount of recirculated salts to the valley's irrigation water. The lack of an extensive drainage network (now being expanded) prevented efficient annual leaching of the soils during winter. This condition was aggravated by the impounding and reuse of saline floodwaters and saline water pumped from the shallow aquifer and by highly efficient irrigation, i.e. by applying only a very small leaching fraction during the summer irrigation period. The combination of all these factors has resulted in (a) net accumulation of soluble salts in the surface-slab of the valley and (b) recirculation of these salts by means of the surface-impounded water onto the fields in ever-increasing amounts.

Remedying the situation in the Yizre'el Valley and rebalancing its salt balance will require increased allocation of water for leaching of salts to the Mediterranean. Because of the low hydraulic permeability of the valley's clayey soils, well-managed leaching may be needed during the summer in addition to the winter leaching by rains.

Allocating irrigation water to leaching during the summer may mean a decrease in irrigated areas since the valley's water quota is fixed. But to ensure long-term, sustainable irrigated agriculture this may be an essential and appropriate price to pay.

The lesson to be learned from this case study is that the maintenance of long-range optimal salt balance in irrigated regions depends on minimizing salt loads by minimizing water additions and leaching out excess soluble salts to prevent regional salinization. Utmost attention must be paid to the long-term effects of irrigation practices on aquifer water qualities and regional soil salinization. The geochemical balance in irrigated regions has to be preserved by leaching and removing salts. Proper drainage systems have to be maintained, and enough excess water has to be allocated for regional leaching to ensure the removal of excess salts from fields and from the whole region and the disposal of these salts in the sea.

Acknowledgements

In the course of collecting and researching the data for the present study, we have been helped by many people who shared data and observations with us. It is a pleasure to thank them all: Dr. Y. Eren, Mr. R. Boaz, Mr. A. Guttman, Ms. D. Kaplan, Mr. A. Moran, Mr. Y. Reuvenii, Dr. A. Ben-Zvi, Dr. M. Ben-Zvi, Prof. E. Adar, Prof. M. Revhon, Dr. Y. Salingar, Mr. D. Shkedii, and Mr. A. Livneh. We also thank Mr. S. Tal for the expert preparation of the map and Ms. S. Salomon for word processing of the manuscript. This study was supported in part by a grant from the Hebrew University.

References

Adar, E.M., A.S. Issar, S. Sorek, and I. Gev. 1992. Modelling of flow pattern in a shallow aquifer affected by reservoirs. *Transport in Porous Media* **8**: 1–20.

Bouwer, H. 1969. Salt balance, irrigation efficiency, and drainage design. *American Society of Civil Engineers. Irrigation and Drainage Division. Journal* **100**: 153–170.

Bower, C.A., J.R. Spencer, and L.O. Weeks. 1969. Salt and water balance, Coachella Valley, California. *American Society of Civil Engineers. Irrigation and Drainage Division. Journal* **100**: 55–64.

Carter, D.L. 1975. Problems of salinity in agriculture. In *Plants in saline environments*, Springer-Verlag, Berlin, pp. 25–35.

Faci, J., R. Aragues, F. Alberto, D. Quilez, J. Machin, and J.L. Arrue. 1985. Water and salt balance in an irrigated area of the Ebro River (Spain). *Irrigation Science* **6**: 29–37.

Gafni, A. and Y. Salingar. 1992. Hydraulic and salinity regime of two affected sites in the Yizre'el Valley, Israel. *Hydrological Processes* **6**: 55–65.

Herut, B. 1992. The chemical composition and sources of dissolved salts in rainwater in Israel. Ph.D. thesis, Hebrew University, Jerusalem (in Hebrew).

Hydrological Service. 1994. *The evolution of exploitation and state of groundwater resources in Israel* (until fall of 1993), Hydrological Service, Jerusalem.

ICID/International Commission on Irrigation and Drainage. 1978. Standards for the calculation of irrigation efficiencies. *ICID Bulletin* **27**: 91–101.

Jensen, M.E., W.R. Rangeley, and P.I. Dieleman. 1990. Irrigation trends in world agriculture. In B.A. Stewart and D.R. Nielsen (Eds.), *Irrigation of agricultural crops*, Agronomy Monograph No. 30, American Society of Agronomy, Madison, Wisconsin, pp. 31–67.

Kaddah, M.T. and J.D. Rhoades. 1976. Salt and water balance in Imperial Valley, California. *Soil Science Society of America Journal* **40**: 93–100.

Lennaerts, A.B.M., F.A.R. Attia, and S.J. de Jong. 1988. Forecasting the suitability of pumped groundwater for irrigation in the Nile Valley. *Agricultural Water Management* **14**: 525–535.

Oosterbaan, R.I. 1989. *Effectiveness and environmental impacts of irrigation projects: A review.* Annual Report, 1988, International Institute for Land Reclamation and Improvement/ILRI, Wageningen, The Netherlands, pp. 18–33.

Raveendan, E., and I.M. Madany. 1991. Characteristics of agricultural drainage water in Bahrain. *Science of the Total Environment* **104**: 239–247.

Ravikovitch, S. 1992. *The soils of Israel*, Hakibbutz Hameuchad Publishing House, Tel-Aviv, (Hebrew with English Summary).

Ravikovitch, S., H. Koyumdjisky, and Y. Dan. 1960. *Soils of western and central Valley of Yizre'el.* Bulletin No. 64, Agriculture Research Station, Beit-Dagan, Israel, (Hebrew with English Summary).

Rhoades, J.D. 1984. Salt problems from increased irrigation efficiency. *Journal of Irrigation and Drainage* **111**: 218–229.

Rhoades, J.D. 1989. Intercepting, isolating and reusing drainage waters for irrigation to conserve water and protect water quality. *Agricultural Water Management* **16**: 37–52.

Rhoades, J.D. and J. Loveday. 1990. Salinity in irrigated agriculture. In B.A. Stewart and D.R. Nielsen (Eds.), *Irrigation of agricultural crops*, Agronomy Monograph No. 30, American Society of Agronomy, Madison, Wisconsin, pp. 1089–1142.

Richards, L.A. 1954. *Diagnosis and improvement of saline and alkali soils*. Agricultural Handbook 60, U.S. Department of Agriculture, Washington.
Scofield, C.S. 1940. Salt balance in irrigated areas. *Journal of Agricultural Research* **61**: 17–39.
Shalhevet J. 1965. Drainage of fine-textured soils - Hazorea Experiment. Final Report, Volcani Institute of Agricultural Research, Department of Soils and Water, Rehovot, Israel, (Hebrew with English summary).
Tanji K.K. and B.R. Hanson. 1990. Drainage and return flows in relation to irrigation management. In B.A. Stewart and D.R. Nielsen (Eds.), *Irrigation of agricultural crops*, Agronomy Monograph No. 30, American Society of Agronomy, Madison, Wisconsin, pp. 1057–1087.
Van Schilfegaarde, J., L. Bernstein, J.D. Rhoades, and S.L. Rawlins. 1974. Irrigation management for salt control. *American Society of Civil Engineers. Irrigation and Drainage Division. Journal* **100**: 321–328.
Westcot, D.W. 1988. Reuse and disposal of higher salinity subsurface drainage water – A review. *Agricultural Water Management.* **14**: 483–511.
Yatir (Vishkin), Y. 1973. The geological structure and hydrogeochemistry as a basis to the understanding of the hydrological regime in the Yizre'el Valley. M.Sc. thesis, Hebrew University of Jerusalem, Israel (in Hebrew).

Appendix

A. Construction of the Annual Chloride-Salt Balance of Yizre'el Valley 1945 to 1990

Construction of the data base needed for calculating the chloride salt balance of the Yizre'el Valley was complicated because of changes of geographical region boundaries according to which reporting was done over the years, changes in supply patterns by Mekorot (The Israel Water Supply Co.) and incomplete records, especially for the earlier part of the studied period. After cross-checking the records of water supply and consumption by sector published by the Water Commission (WCAC) and by the Joint Planning Authority (PAASR), we developed a coherent data base, using the sources listed in Table A-I. We then applied the data base in calculating the chloride salt balance.

Salt (chloride) balance is calculated for an imaginary regional slab whose areal borders are those of the southern part of the Qishon River basin (Figure 1) and whose depth from the soil surface to the bottom of the upper shallow aquifer is 5–10 m. The only surface output conduit is the Qishon River, and we tacitly assume that it is the only output from the regional surface slab. We neglected chloride leaks to deeper horizons and chloride output from the region as underground drainage flows through the western opening of the valley.

We also assumed that all chloride inputs are carried by rain and water supplies. Regional inputs of chlorides as dry salts for agricultural consumption, domestic consumption (direct food consumption, detergents and domestic water softening) and for industrial uses (food processing, particularly meat processing for Kashrut, industrial water softening, textile industry, high-tech industry, etc.) are thus neglected on the input side.

TABLE A-I
Summary of data sources

Item	Water volume	Cl concentration
Input		
Rainwater	1945–1990 – Meteorological Survey/Hydrological Survey.	Herut (1992); Average for Lower Galilee.
Well water	1945 – Historical record, *Encyclopaedia Hebraica.* 1956, 1959 and 1962–1990: From Water Commission Allocation and Consumption (WCAC) Yearbooks. Other years – estimated interpolations.	Yizre'el Valley, Cell 380, HS(94) p. 83.
Surface-Impounded Water		
(a) Private reservoirs (b) Kefar Barukh reservoir	WCAC Yearbooks and reservoir construction records since 1962. Estimates from Yatir (Vishkin) 1973, Shaham *et al.* (1988), Qishon flows, QCMP (1985–1990).	Function fitted to Cl concentration in Kefar Barukh reservoir (Shaham *et al.*, 1988); Average concentration function and actual weighted average of Cl concentrations recorded in valley reservoirs from 1982 to 1990 (PMR; QCMP).
Fresh Water (Mekorot)	Same as for well water above.	1945–1964 – Cl concentration in Western Galilee Plant, Cell 360, HS(94) p. 83. 1965–1969 – Mixture of Western Galilee/Lake Kinneret waters: 50/50; 40/60; 30/70; 20/80; 10/90 for the 5 years respectively. HS(94) p. 83; 167–168. 1970–1990 – Cl concentration in Lake Kinneret, HS(94) pp. 167–168, annual average.
Haifa sewage effluents	QCMP (1985–1990).	QCMP (1985–1990).
Output		
Qishon River	Monthly records of Hydrological Survey (HS (71) and other records), at station 8146, coordinates E159.38, N236.30 on Qishon River at the quarry, near the Amakim Jct.	Computed from a synthetic recharge-concentration function based on Hydrological Survey records at Station 8146 (in preparation).

Fresh water is defined as water supplies from wells (Western Galilee and deep regional wells) and from Lake Kinneret by the National Water Carrier as distinguished from surface-impounded water.

Imported chloride salts (INPUT) were computed as the sum of products of annual water inputs times their respective chloride concentration. Included were rain; private well waters; Haifa sewage effluents; and pumped fresh water by Mekorot from deep groundwater sources at the margins of the region, from the Western Galilee Regional Plant and from Lake Kinneret by the National Water Carrier (274 cm pipe). Chloride concentrations in the fresh water sources were obtained from water quality records compiled by the Hydrological Service of Israel (1994).

Chloride salts applied to soils were computed by summing up contributions from rain, fresh water (less domestic and industrial use), surface-impounded water and imported Haifa sewage effluents. Contributions from local sewage effluents were part of the surface-impounded contribution.

Recirculated chloride salt was computed as Applied minus Imported. This yearly addition of chloride to soils above the imported amount results from salt fluxes into impounded surface water in the many reservoirs scattered over the valley.

B. List of Data Sources

1. Center for Environmental Engineering and Water Resources (1985–1990). Qishon Complex Monitoring Project, Annual Reports No. 1 to 6. Faculty of Civil Engineering, Technion, Haifa (QCMP (year)).
2. *Encyclopaedia Hebraica* (1956). Massada, Tel-Aviv, Israel (in Hebrew), Vol. VI, pp. 829–830.
3. Hydrological Service (1971). *Hydrological Year-Book of Israel.* Summary of records to October 1967. Hydrological Service, Jerusalem, Israel (Hebrew + English Summary; 4 maps) 270 + XIII p. (HS(71)).
4. Hydrological Service (1994). *The evolution of exploitation and state of groundwater resources in Israel* (until Fall of 1993). Hydrological Service, Jerusalem, Israel. ISSN-0793-1093 (Hebrew; Graphs and Tables) 171 p. (HS(94)).
5. Palgei Miim – Reservoir Survey: Water quality and quantities, 1988, 1990, 1991, 1992 (PMR).
6. Planning Authority for Agriculture, Settlement and Rural Areas (Annual Reports). Agricultural planning and agricultural production. Joint Planning Authority, Ministry of Agriculture, State of Israel and Department of Settlement, Jewish Agency, Hakirya, Tel-Aviv, Israel (PAASR (year)).
7. Shaham, G., Z. Shein, Y. Yarmut, Z. Soffer (1988). Changes in the water supply system of Yizre'el Valley for solving drainage problems and salt removal. Report to Palgei Miim, Emek Yizre'el (in Hebrew) 24 p.
8. Water Commission (annual reports). Survey of water consumption in Israel. Division of Allocation and Permits, Water Commission, Ministry of Agriculture, Hakirya, Israel (WCAC (year)).

SURFACE DISTURBANCES: THEIR ROLE IN ACCELERATING DESERTIFICATION

JAYNE BELNAP
National Biological Service, 2282 S. West Resource Blvd., Moab, Utah 84532, USA

Abstract. Maintaining soil stability and normal water and nutrient cycles in desert systems is critical to avoiding desertification. These particular ecosystem processes are threatened by trampling of livestock and people, and by off-road vehicle use. Soil compaction and disruption of cryptobiotic soil surfaces (composed of cyanobacteria, lichens, and mosses) can result in decreased water availability to vascular plants through decreased water infiltration and increased albedo with possible decreased precipitation. Surface disturbance may also cause accelerated soil loss through wind and water erosion and decreased diversity and abundance of soil biota. In addition, nutrient cycles can be altered through lowered nitrogen and carbon inputs and slowed decomposition of soil organic matter, resulting in lower nutrient levels in associated vascular plants. Some cold desert systems may be especially susceptible to these disruptions due to the paucity of surface-rooting vascular plants for soil stabilization, fewer nitrogen-fixing higher plants, and lower soil temperatures, which slow nutrient cycles. Desert soils may recover slowly from surface disturbances, resulting in increased vulnerability to desertification. Recovery from compaction and decreased soil stability is estimated to take several hundred years. Re-establishment rates for soil bacterial and fungal populations are not known. The nitrogen fixation capability of soil requires at least 50 years to recover. Recovery of crusts can be hampered by large amounts of moving sediment, and re-establishment can be extremely difficult in some areas. Given the sensitivity of these resources and slow recovery times, desertification threatens million of hectares of semiarid lands in the United States.

1. Introduction

Most arid and semiarid lands of the United States have been heavily affected by human use since the late 1800s. Historically, most of this impact has been from livestock grazing near areas with forage and water. More recently, substantial increases in off-road vehicle use and hiking have greatly expanded direct and indirect human impacts both spatially and temporally. The combination of recreational use and livestock grazing is resulting in unprecedented levels of local and regional disturbances, accelerating desertification processes.

Many resources in arid ecosystems are sensitive to compressional and shear forces associated with current uses. Soils in arid regions are often highly erodible, and soil formation extremely slow, taking 5,000 to 10,000 years or more (Dregne, 1983a). Low levels of organic matter, large particle size ranges, and shallow soil freezing depths generally found in these soils result in compactible soils with slow dilation rates. Compaction of soils results in less water infiltration and less locally available water (Webb, 1983; Wilshire, 1983), which in turn influence soil biota activity, N cycle dynamics (Torbert and Wood, 1992), vascular plant vigor and reproduction (Crawford, 1979; Skujins, 1984), and decomposition rates of soil

Environmental Monitoring and Assessment **37:** 39–57, 1995.

organic matter (West, 1981). Compaction reduces soil aggregates and pore space, which are important for soil stability and infiltration and as microenvironments for soil biota (Dregne, 1983a; Stolzy and Norman, 1961).

Surface disturbance also affects surface cyanobacterial-lichen soil crusts. In the western United States these crusts are important for increased soil stability, water infiltration, and fertility of soils (Harper and Marble, 1988; Johansen, 1993; Metting, 1991; Belnap and Gardner, 1993). Absence of these crusts can lead to increased erosion, resulting in loss of organic matter, fine soil particles, nutrients, and microbial populations in soils (Harper and Marble, 1988; Schimel *et al.*, 1985).

Normal nutrient cycles in these semiarid regions can also be disrupted by soil surface disturbance. Nitrogen is often limiting in desert systems (Zak and Whitford, 1988). Cyanobacterial-lichen soil crusts have been shown to be the dominant source of nitrogen in a cold-desert pinyon-juniper and a grassland ecosystem in southern Utah (Evans and Ehlringer, 1993; Evans and Belnap, unpublished data). Experiments have demonstrated that all types of surface disturbance tested dramatically decreased nitrogenase activity in these crusts (Belnap *et al.*, 1994). Plants growing on undisturbed sites consistently show higher N content when compared to adjacent disturbed sites (Belnap and Harper, in press; Harper and Pendleton, 1993).

Cyanobacterial-lichen soil crusts are an important source of fixed carbon for these sparsely vegetated areas (Beymer and Klopatek, 1991). In addition, soil disturbance can alter soil food webs and thereby affect nutrient availability in these systems (Ingham *et al.*, 1989). Disruptions of soil food webs can reverberate throughout the ecosystem, affecting macro-floral and faunal components (Hendrix *et al.*, 1992; Coleman *et al.*, 1992). Soil surface disturbance can also affect plant community composition and architecture. Changes in these critical habitat components have been shown to affect invertebrate and vertebrate populations (MacMahon, 1987). Since preventing desertification depends on maintaining stability and fertility of soils, as well as diversity of processes and species in ecosystems, impacts to sensitive soil surfaces can accelerate the desertification process.

2. Materials and Methods

2.1. Study Sites

Arches National Park: Arches is located 16 km north of Moab, Utah, at 1370 m elevation. Annual rainfall is approximately 200 mm. A heavily trampled area was compared with an adjacent, relatively undisturbed area. Both sites are on Arches loamy fine sand with similar depth, slope, inclination, and exposure. The dominant vegetation is *Coleogyne ramosissima* (blackbrush). Plant interspaces in the untrampled area are covered with well-developed cyanobacterial-lichen crust.

Behind-the-Rocks: Situated on Rizo fine sandy loam at 1400 m elevation, this site lies 15 km south of Moab, Utah. Annual rainfall is approximately 200 mm. This site is dominated by *C. ramosissima*. Undisturbed soils are covered by a cyanobacterial crust similar to that at the Arches untrampled site.

Natural Bridges National Monument: Located 73 miles southwest of Moab at 1980 m elevation, these two site have annual rainfall of approximately 300 mm. Both sites are on Rizo fine sandy loam of similar depth, slope, inclination, and exposure, dominated by *Pinus edulis* and *Juniperus utahensis*. A trampled site was compared to an area 200 m away that had received little trampling.

Canyonlands National Park: Located 31 miles south of Moab at 1513 m elevation, these study sites are in adjacent grasslands. Both have sandy loam soils (Mido and Begay) with similar soil depth, slope, and exposure. Both sites are dominated by the grasses *Stipa hymenoides* and *S. comata* (Kleiner and Harper, 1977). Annual rainfall averages 180 mm.

2.2. Analyses

Five randomly located transect lines were placed in each area at the Arches study site. Ground and vegetation cover were estimated with twenty 0.25 m^2 quadrats per transect using a nested frequency quadrat frame, Daubenmire cover classes, and density counts. All perennial plants were mapped. For shrubs intersecting the transect line, distance from nearest neighbor and hummock height were estimated. Hummock height was estimated as the height at which a rod held parallel to the surface from the base of the shrub intersected a rod held vertically at the lowest point in the interspace between the considered shrub and its nearest neighbor.

Plant tissue elemental content was determined from composite samples comprising at least five individuals of each species. Analysis was conducted by the Brigham Young University Soil Laboratory. Nitrogenase activity in cyanobacterial crusts was determined by methods outlined in Belnap (1992, 1994). Soil biota were sampled at 0–10 cm, the depth at which maximum bacterial, fungal, and nematode biomass was found. Nematodes, total and active bacteria, microarthropods, and total and active fungal biomass were evaluated following methods outlined in Ingham *et al.* (1989).

Physical soil characteristics were determined for cores 0–10 cm deep. Soils were analyzed by the Brigham Young University Soils Laboratory. Soil depth was determined with a metal probe. Compaction was measured with a recording penetrometer and bulk density measurements. Soil temperatures were taken midday on five occasions during June and July 1994. Measurements were taken with a 1 mm wide probe inserted just under the soil surface. Results were analyzed using analysis of variance (ANOVA), Duncan's multiple range test, and unpaired t-tests.

TABLE I

Soil physical characteristics of disturbed and undisturbed areas at Arches National Park

	Undisturbed		Disturbed		Significant
	Mean	Std. Dev.	Mean	Std. Dev.	differences
pH	8.30	0.00	8.20	0.00	$p < 0.03$
% Sand	82.23	0.80	85.76	0.70	$p < 0.05$
% Clay	8.83	0.50	7.24	0.80	
% Silt	8.96	0.02	7.00	1.50	
% OM	0.55	0.14	0.40	0.30	
ppm P	34.04	0.14	44.38	0.08	$p < 0.01$
ppm Zn	0.40	0.10	0.50	0.14	
ppm Fe	19.90	0.26	26.50	0.13	
ppmMn	3.20	0.17	4.30	0.08	
ppm Cu	0.20	0.00	0.20	0.00	
ppm N	591.70	47.10	558.30	23.60	
ppm K	90.00	0.71	110.00	14.10	
ppm Ca	4390.00	99.00	3445.00	7.10	$p < 0.02$
ppm Mg	70.00	0.00	70.00	0.00	
ppm Na	90.00	7.10	110.00	14.14	
0–3 cm BD	104.9	4.4	114.9	10.0	$p < 0.007$
3–6 cm BD	99.5	4.9	108.1	5.6	$p < 0.001$

3. Results and Discussion

3.1. Soil Physical Characteristics

Soils at the trampled and untrampled sites in Arches were very similar (Table I). Percent clay, silt, and organic matter were low at both sites and did not differ significantly. No statistically significant differences were found in zinc, iron, manganese, copper, total nitrogen, potassium, magnesium, or sodium levels. At the two sites pH and percent sand differed slightly but significantly, and differences in phosphorus and calcium were greater and significant.

Bulk density and soil porosity were compared in the two areas. Bulk density was found to be significantly higher at 0–3 and 3–6 cm depths in the trampled area than in the untrampled area. Soil porosity did not significantly differ at 0–3 cm depth between the two areas but was significantly lower at 3–6 cm depth in the trampled area than in the untrampled area. Increasing bulk density often results in lower microbial populations and lower activity levels (Torbert and Wood, 1992). In addition, increased bulk density generally results in less water infiltration and increased runoff (Webb and Wilshire, 1980; Loope and Gifford, 1972). Decreased

TABLE II

Biological characteristics at Arches National Park in disturbed and undisturbed areas. All differences are statistically significant at $p < 0.05$, except for frequency of litter, which did not significantly differ

	Undisturbed		Disturbed	
	Mean	Std. Dev.	Mean	Std. Dev.
Percent cover				
Cyano	45.50	7.20	78.20	14.10
Lichen	2.00	1.20	0.20	0.30
Collema	4.40	2.40	0.60	0.70
Moss	32.70	5.50	4.40	4.00
Litter	31.80	5.00	8.60	3.20
Bare Ground	0.06	0.13	8.36	12.20
Exotics	0.06	0.13	1.40	1.30
Grass	3.56	1.48	2.38	0.95
Shrub	29.81	16.48	29.53	9.99
Frequency				
Lichen	0.36	0.15	0.02	0.02
Collema	0.76	0.05	0.14	0.15
Moss	0.95	0.04	0.31	0.24
Bare Ground	0.03	0.05	0.88	0.06
Gravel	0.08	0.15	0.50	0.40
Grass	0.64	0.17	0.39	0.13
Shrubs	0.81	0.25	0.49	0.21
Community				
# of Shrubs	7.20	1.24	3.20	0.58
Festuca Weight, g	0.80	0.16	1.68	0.33
# of Shrub Species	2.30	0.50	1.40	0.55
Interspace, cm	43.00	4.95	41.31	5.09
Hummock Height, cm	11.11	1.26	25.19	1.97
# of Exotics	0.30	0.50	1.40	0.55
# of Plant Species	8.00	1.20	10.00	1.40

water availability locally can lead to increased water stress for plants, thereby accelerating desertification processes (Dregne, 1983b).

3.2. Soil Surface Cover and Cyanobacterial-Lichen Soil Crusts

At the Arches site, soil surfaces of the trampled and untrampled sites were compared (Table II). Cover of plant litter, lichens, mosses, and cyanobacteria was significantly

Fig. 1. Scanning electron micrograph of dry cyanobacterial sheaths winding through sandy soils from Moab, Utah. Note firm attachment of sheath material to the individual sand grains, even though sheath material is dry (magnification × 90).

higher in the untrampled area. The untrampled area had 32% litter cover, contrasted with only 8% litter cover in the trampled area. Six species of lichens and four species of cyanobacteria were found in the untrampled area, whereas only one species of cyanobacteria was found in the trampled area. Lichens and mosses combined covered 39% of the surface in the untrampled area but only 4% in trampled area. Lichen and moss species had a combined frequency four times greater in the untrampled than the trampled area. Trampled interspace surfaces were generally flat, whereas untrampled interspace surfaces supported cyanobacterial pedicels up to 7 cm high.

Litter and cryptobiotic crust cover and roughened microtopography play important roles in protecting soil surfaces from wind and water erosion and in enhancing water infiltration. Combining litter, vascular plant, lichens, and moss cover, the untrampled area had virtually no surfaces exposed to wind or water erosion, whereas most of the interspaces in the trampled area had little if any protection. Surfaces trampled flat promote sheet erosion. Untrampled, pedicelled surfaces create microcatchments for water and reduce the velocity of surface water flows, thus encouraging suspended sediments to settle out (Harper and Marble, 1988).

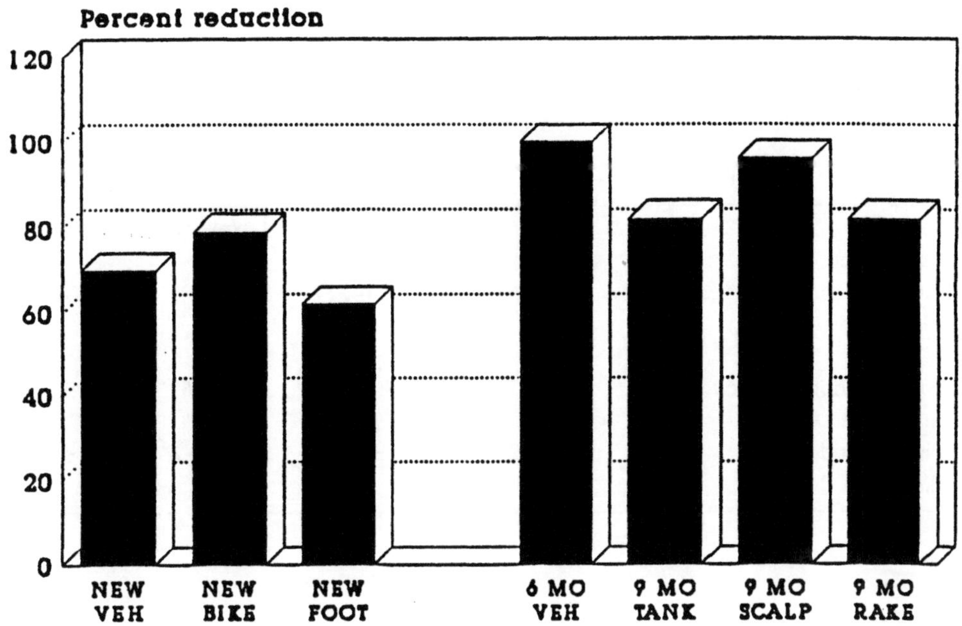

Fig. 2. Percent reduction of nitrogenase activity after soil surface disturbances were experimentally applied. All differences from the control were significant. Bicycle, motor vehicle, and footprint treatments were applied near Moab, Utah. Tank, rake, and scalped treatments were applied at Dugway Proving Ground, Utah (Belnap *et al.*, 1994).

Consequently, trampling can greatly accelerate desertification processes through increased soil loss and water runoff (Dregne, 1983b).

As cyanobacteria move through the soil, they leave behind polysaccharide material that firmly adheres to soil particles, even when dry or no longer associated with a living cyanobacterium (Figure 1). Sheath material can absorb up to eight times its weight in water, increasing the water-holding capacity of the soil (Brock, 1975; Campbell, 1979; Campbell *et al.*, 1989). In addition, nutrient-rich clay particles adhere to sheath material, thus increasing the fertility of soils (Belnap and Gardner, 1993; Belnap and Harper, in press). Soil crusts can also affect higher plant seedling establishment and survival. Experiments with fine and coarsely textured soils demonstrated that seedling establishment was higher for forbs and grasses in crusted areas than uncrusted areas. Survival over a 3-year period was enhanced in two species and unaffected in one species. Overall plant survival was three times higher in the crusted plots (Belnap, 1994). Other studies have reported similar enhancement of seedling germination and establishment in crusted areas when compared to uncrusted surfaces (Harper and Marble, 1988; St. Clair *et al.*, 1986; Lesica and Shelley, 1992).

Surface disturbance has been shown to negatively affect the cohesion and coverage of cyanobacterial crusts, since crustal components are brittle when dry and

TABLE III

Active and total bacterial and fungal biomass at Arches National Park and Natural Bridges National Monument. AF = Active fungi; TF = Total fungi; AB = Active bacteria; TB = Total bacteria; TF/B = Total fungi/bacteria. Asterisk (*) denotes statistical differences at $p < 0.05$

	AF	TF	AB	TB	TF/B
Arches undisturbed	0.3 ± 0.2	14.5 ± 5.5	2.3 ± 1.0*	3.78 ± 0.82*	4.0 ± 1.4*
Arches disturbed	0.4 ± 0.5	11.6 ± 4.9	1.2 ± 0.7*	5.35 ± 1.6*	2.5 ± 1.7*
Natural Bridges undisturbed	0.59 ± 0.67*	25.2 ± 7.7	0.8 ± 0.35*	3.4 ± 0.6*	7.5 ± 2.4*
Natural Bridges disturbed	0	31.5 ± 24.5	0.12 ± 0.14*	14.2 ± 4.8*	3.1 ± 3.6*

easily crushed (Belnap, 1993; Campbell *et al.*, 1989; Harper and Marble, 1988). Surface disturbance also reduces the depth to which abandoned cyanobacterial sheath accumulates, thereby reducing resistance to water erosion. At many disturbed sites, sheath material is often not observed below 1 mm depth, in contrast to 7 cm thick crusts in untrampled areas. Consequently, impacts to surface crusts can reduce soil stability, soil fertility, and soil moisture retention, thus accelerating desertification in impacted areas (Dregne, 1983b).

Cyanobacteria and cyanobacterial components of soil lichens fix atmospheric nitrogen (Belnap, 1992; Skujins and Klubek, 1978; West and Skujins, 1977; Terry and Burns, 1987). Studies have shown that nitrogen fixed by soil cyanobacteria is available to neighboring vascular plants under laboratory conditions (Mayland *et al.*, 1966; Mayland and McIntosh, 1966). In some desert systems, these crusts have been found to be the dominant source of this often-limiting element (Evans and Ehlringer, 1993).

Nitrogenase activity was dramatically reduced in soil crusts regardless of the type of experimentally applied disturbance (Figure 2; Belnap *et al.*, 1994; Belnap, unpublished data). Disturbance by human feet, mountain bikes, four-wheel drive trucks, tracked vehicles (tanks), and shallow and deep raking all resulted in an immediate 40–80% reduction in nitrogenase activity. Measurements taken 6–9 months after treatment showed that nitrogenase activity had dropped still lower in the disturbed areas, with treatments showing an 80–100% reduction (Belnap, *et al.*, 1994). At the Arches trampled site, no nitrogenase activity could be detected, while the undisturbed site showed normal nitrogenase activity. Because disturbance generally increases nitrogen losses through enhanced denitrification and accelerated soil erosion (Peterjohn and Schlesinger, 1990), less nitrogen input often results in

TABLE IV-A

Nematode populations at Arches National Park and Natural Bridges National Monument. Total numbers, bacterial feeders, and fungal feeders were significantly higher in the undisturbed area ($p < 0.05$). Numbers of root feeders did not significantly differ

	Bacterial feeders	Fungal feeders	Root feeders	Total individuals
Nematode functional groups				
Arches undisturbed	505.0 ± 287.0*	97.0 ± 142.0*	35.0 ± 38.0	651.0 ± 646.0*
Arches disturbed	370.0 ± 287.0*	54.0 ± 127.0*	67.0 ± 54.0	493.0 ± 360.0*
Natural Bridges undisturbed	36.12 ± 33.56*	30.09 ± 44.0*	4.08 ± 9.5	70.44 ± 82.56*
Natural Bridges disturbed	5.02 ± 2.85*	2.81 ± 2.71*	0.07 ± 2.52	8.56 ± 4.9*

TABLE IV-B

Microarthropods: Numbers of microarthropods in a never grazed and a previously grazed site in Canyonlands National Park. These samples were not statistically analyzed

	Shrub total species	Grass total species	Shrub canopy	Shrub interspace	Grass canopy	Grass interspace	Total individuals
Microarthropods							
Undisturbed	15	11	44	7	14	10	72
Disturbed	11	7	18	1	5	7	34

less total nitrogen for the ecosystem. Reduced fertility of systems is one of the most definitive and problematic aspects of desertification (Dregne, 1983b).

3.3. Soil Food Webs

Biomass of active bacteria, total bacteria, active fungi, and total fungi, as well as nematode numbers, were measured in paired trampled and untrampled areas in Arches National Park and Natural Bridges National Monument. Results obtained from the two areas were similar. At Arches the biomass of active bacteria in the untrampled area exceeded by more than three times that in the trampled area (Table III). Total bacterial biomass was higher in the trampled area, whereas active and total fungal biomass were not significantly affected by disturbance. The ratio of total fungi to bacteria had significantly shifted in the trampled area by a factor of two, with relatively more bacteria present. Nematode populations were also significantly affected (Table IVa). The number of bacterial and fungal feeding nematodes in the untrampled area was twice that found in the trampled area. Numbers of plant root feeders in the trampled and untrampled area did not significantly differ. Twice as many nematodes were found in the untrampled area as in the trampled area.

Soil biota were also analyzed at Natural Bridges National Monument (Tables III and IVa) with results similar to those from Arches. Biomass of active bacteria and fungi was six to seven times higher in the untrampled area than in the trampled area. Total bacterial biomass was three times higher in the trampled area, while total fungal biomass did not significantly differ. The ratio of fungi to bacteria was found to have significantly shifted in the trampled area by a factor of three, with relatively more bacteria in the trampled area. Nematode numbers were significantly affected as well, with numbers of bacterial and fungal feeders 7–10 times higher in the untrampled area. Plant root feeders were not significantly affected. Total nematode numbers were almost seven times greater in the untrampled area than in the trampled area.

Microarthropod populations were compared between trampled and untrampled areas in Canyonlands National Park. Although the trampled site had not been grazed for 30 years, numbers of individuals of each species were still higher in the untrampled area, both under plants and in plant interspaces, than in the disturbed area (Table IVb). Total individuals were more than twice as abundant (44 vs. 18) under shrubs of the untrampled than the trampled area. In shrub interspaces, individuals were seven times more abundant (1 vs. 7) in the untrampled area than in the trampled area. In the grass community, the same pattern was observed. Individuals were almost three times as abundant in the untrampled area under grass plants (14 vs. 5) than the trampled area. Microarthropod individuals were 1.4 times more abundant (10 vs. 7) in the untrampled interspace between grasses than in trampled interspaces. Total numbers of individuals found in the untrampled area were more than double those observed in the trampled area (72 vs. 34). Fewer species were found in trampled than untrampled areas in both shrub (11 vs. 15 species) and grass (7 vs. 11 species) communities. The results represent a decline in species richness of 26% and 36% respectively. Four species found in the trampled area were not found in the untrampled area, and six species from the untrampled area were not found in the trampled area.

Soil food webs can profoundly influence nutrient availability in ecosystems, and changes in soil biota populations can result in slower decomposition rates and fewer nutrients available to plants, thereby hastening the desertification process. Nutrient uptake rates can be affected by the ability of bacteria to solubilize or chelate elements (Lange, 1974). Symbiotic fungi can increase a plant's ability to exploit more soil volume and increase the spatial distribution of carbon in soils. Carbon- and nitrogen-fixing soil microbiota contribute to site fertility. Larger

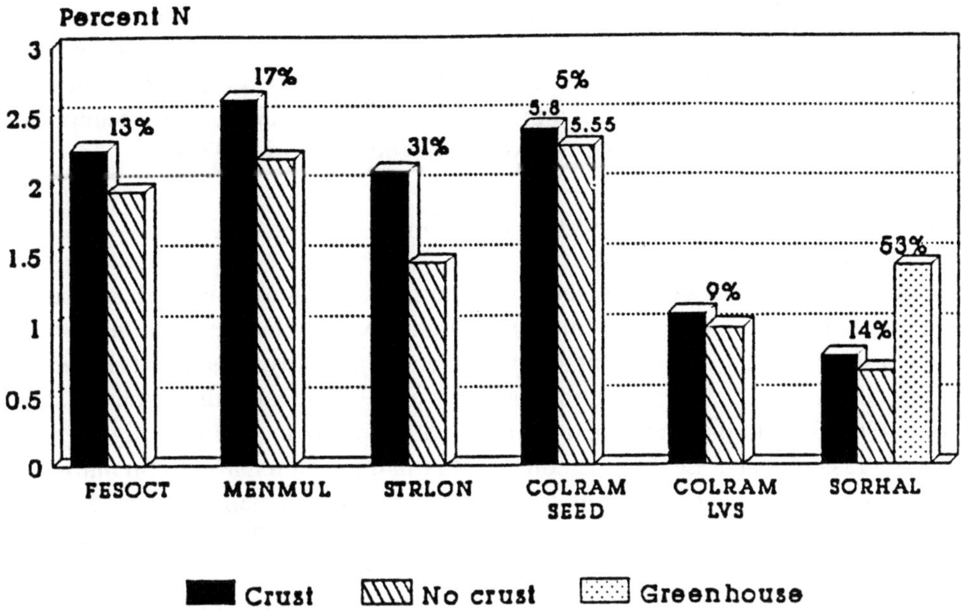

Fig. 3. Nitrogen levels in vascular plant tissue from sites around the state of Utah, USA. Plants were collected from adjacent crusted and uncrusted soils. All differences were statistically significant at $p < 0.05$. FESOCT = *Festuca octoflora* (annual grass); MENMUL = *Mentzelia multiflora* (annual forb); STRLON = *Streptanthella longirostris* (perennial forb); COLRAM = *Coleogyne ramosissima* seedlings and leaves (perennial shrub); SORHAL = *Sorghum halepense* (annual grass).

microbial populations mean greater retention of N, P, and other nutrients bound up in microbial biomass (Skujins, 1984). Bacteria are known to secrete IAA and other plant growth regulators. Both bacteria and fungi secrete polysaccharide material that acts to aggregate soil particles (Lange, 1974). The presence of bacteria and fungi can also affect vascular plant root morphology (Skujins, 1984).

Ratios of microbes vary in different systems. Shrub-dominated systems generally have relatively higher fungal numbers to decompose more recalcitrant woody litter. Grass systems have relatively more bacteria. Disturbance often shifts these ratios. Both the blackbrush community at Arches and the pinyon-juniper community at Natural Bridges resulted in stimulation of bacterial activity and a depression in fungal activity. Because different soil biota process litter differently, such a shift in soil food web structure or relative activity levels can significantly lower litter decomposition rates and nutrient availability to systems and thus increase desertification rates.

Predators, such as nematodes and microarthropods, are also an essential part of the soil food web. These organisms play important roles in decomposition cycles by shredding litter, mixing soil layers, and carrying litter underground. They act as dispersers of other soil microorganisms. They can modify soil pH and contribute N, P, and other nutrients to soils through fecal material and release of nutrients bound up in microbial biomass. They also influence soil food web structure through preferential predation on other soil organisms. Lowering microarthropod numbers and altering population structure can significantly lower decomposition rates and again accelerate the desertification process (Zak and Whitford, 1988; Santos and Whitford, 1981; Seastedt and Crossley, 1984; Kitchell *et al.*, 1979).

3.4. Nutrient levels in vascular plants

Concentrations of nitrogen and other macronutrients in annual, biennial, and perennial plants were found to be significantly higher when they grew on undisturbed crusted surfaces than on adjacent trampled areas at all study sites tested (Figure 3). At Arches, leaf tissue N in the untrampled area was 9% higher in the perennial shrub *Coleogyne*, 31% higher in the perennial forb *Streptanthella*, and 13% higher in the annual grass *Festuca*. Leaf tissue content of N in the biennial *Mentzelia multiflora* was 17% higher in plants growing in a crusted area than in plants from an adjacent dune at a nearby site in Arches (Belnap and Harper, 1995). The same results were obtained at other field sites around Utah and in greenhouse experiments, where levels of nitrogen in sorghum and rape were higher in pots with cyanobacteria than in pots without cyanobacteria. Dry weight of plants in pots with cyanobacteria were up to four times greater than in pots without cyanobacteria (Harper and Pendleton, 1993). Dry weight of *Festuca* plants in the untrampled area was twice the *Festuca* dry weight of the trampled area (Belnap and Harper, 1995).

Maintaining normal nitrogen cycles is critical to maintaining the fertility of semiarid soils and preventing desertification (Dregne, 1983b). After water, nitrogen is generally considered the element most likely to limit primary productivity in semiarid and arid ecosystems (Ettershank *et al.*, 1978; Romney *et al.*, 1978).

3.5. Vascular plant community structure

Plant community composition and structure were examined at the Arches trampled site and compared to a nearby untrampled area (Table II). Many significant differences were observed. Fewer, larger individuals of the dominant shrub *Coleogyne ramosissima* were found in the trampled area than in the untrampled area. Shrubs dominated the interspace of the untrampled area, whereas annuals or perennial bulbs (with above-ground parts only in the spring) dominated the interspace in the trampled area. Exotic grasses contributed significantly to grass cover in the trampled area, whereas the untrampled area had only a few exotic plant individuals.

Interspaces between *Coleogyne* plants were significantly larger and hummocks under the shrubs four times higher in the trampled area than in the untrampled area. Consequently more soil is exposed to erosion in the trampled area. In addition, small animals face crossing an average of 4.8 m between *Coleogyne* plants in the trampled area, compared to 2.0 m in the untrampled area, resulting in higher exposure to predation. The trampled area also had only small annual plants in the

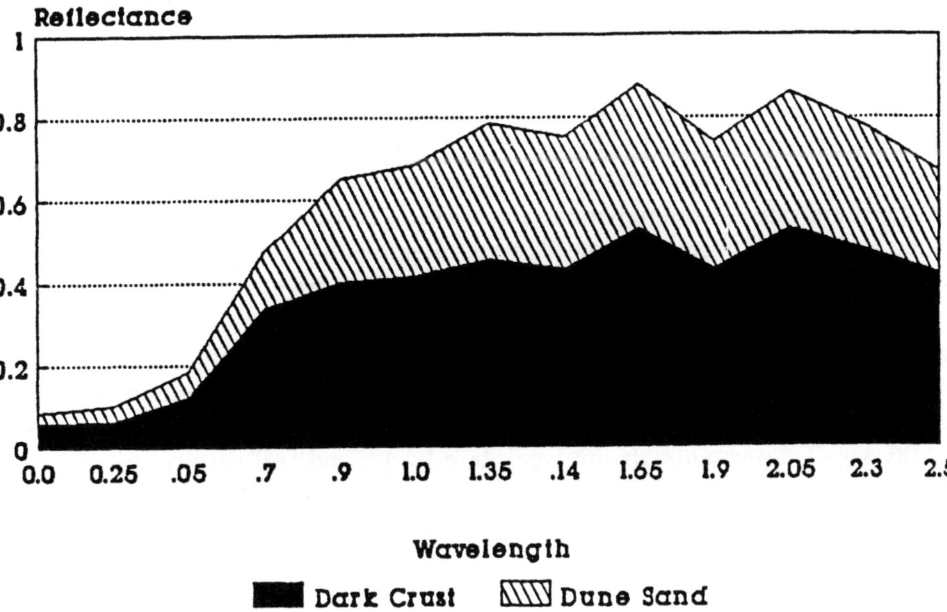

Fig. 4. Reflectance of bare sand and *Microcoleus*-dominated soil surfaces. Measurements were taken using a laboratory spectrophotometer. (From Gregg Swayze and Roger Clark (U.S. Geological Survey, Denver).

interspaces, instead of perennials such as *Opuntia polyacantha*, which might offer more protective cover.

Changes in plant community architecture have been shown to affect populations of invertebrates and vertebrates living in these communities (MacMahon, 1987). As a result, on semiarid and arid landscapes trampling may have many as yet undocumented effects that increase susceptibility to desertification.

3.6. SURFACE ALBEDO AND TEMPERATURE

Albedo is also of concern in semiarid and arid systems. When trampled surfaces were compared to untrampled surfaces, there was up to a 50% increase in reflectance from 0.25 to 2.5μ (Figure 4). This represents a change in the surface energy flux of approximately 40 watts/m^2 (R. Clark, U.S. Geological Survey, Denver, personal communication). Extensive trampled areas, combined with lack of urban areas to offset this energy loss, could lead to changes in regional climate patterns in many semiarid regions (Sagan *et al.*, 1979).

Concomitant with albedo changes, trampled and untrampled surfaces significantly differed in surface temperatures. Mid-day temperatures were taken in June and July 1994. Air temperatures were found to average 39 °C (S.D. 0.25 °C), bare sand 52 °C (S.D. 0.5 °C), and dark crusted surfaces 62 °C (S.D. 1.0 °C). These differences were all significant. In the winter, surface temperatures of well-developed crusts were up to 14 °C higher than ambient air temperature.

Surface temperatures can be very important in desert systems. Nitrogenase activity is heavily temperature dependent, with lower temperatures resulting in lowered activity levels (Rychert *et al.*, 1978). Altered soil temperatures can be expected to affect microbial activity, plant nutrient uptake rates, and soil water evaporation rates. Soil temperatures have been shown to affect seed germination time and seedling growth rates for vascular plants. Timing of these events is often critical in deserts, and relatively small delays can reduce species fitness and seedling establishment, which may eventually affect community structure (Bush and Van Auken, 1991). Food and other resources are often partitioned among ants, arthropods, and small mammals on the basis of surface temperature-controlled foraging times (Doyen and Tschinkel, 1974; Crawford, 1991; Wallwork, 1982). Many small desert animals are weak burrowers, and soil surface microclimates are of great importance to their survival (Larmuth, 1978). Consequently, altering surface temperatures can affect nutrient availability and community structure for many desert organisms, thus increasing susceptibility to desertification.

3.7. Recovery from Surface Disturbance

Some research has been conducted to estimate recovery rates of soil disturbances in arid systems. Recovery of soils from compaction has been estimated to require 100–130 years in the Great Basin Desert (Knapp, 1992) and 80–140 years in the Mojave Desert (Webb and Wilshire, 1980). Recovery rates of soil biota are expected to be slow, given the slow recovery of soil porosity, moisture-dependent microbial activity in a dry environment, limited dispersal ability, naturally low microbial populations, and large disturbance areas.

Much more is known about the recovery rates of cyanobacterial-lichen soil crusts. Recovery rates have been found to depend on the type and extent of disturbance, the availability of nearby inoculation material, and the temperature and moisture regimes that follow disturbance. Estimates of time for visually assessed recovery have varied from 5 to 100 years (Anderson *et al.*, 1982; Jeffries and Klopatek, 1987; Callison *et al.*, 1985; Cole, 1990). However, Belnap (1993) showed that many components of recovery cannot be assessed visually. Assuming linear recovery rates, recovery was estimated to be 35–65 years for cyanobacterial biomass, 45–85 years for lichen cover, and 250 years for moss cover in scalped 0.25 m^2 plots surrounded by well-developed crusts. Since recovery time depends on the presence of a nearby inoculant, larger disturbed areas will take longer to recover. Several studies have shown that inoculation can hasten the biological recovery of disturbed crusts (Tidemann *et al.*, 1980; Ashley and Rushforth, 1984; Belnap, 1993).

Nitrogenase activity recovery appears to be quite slow. In scalped areas, no nitrogenase activity was detectable after 9 years, and N content of soils was still much lower when compared to adjacent control plots. In areas disturbed by four-wheel drive vehicles, no recovery could be documented after 2 years (Belnap, unpublished data). With the use of isotopic ratios of N, soil and plant N and nitrogenase activity levels were found to be significantly lower in an area that livestock had not grazed for 30 years than in an area that livestock had never grazed. These data suggest that negative effects on nitrogen dynamics may persist in systems for extended periods after disturbance ceases. This slow recovery from disturbance indicates that this system is vulnerable to desertification.

Restoration of normal surface albedos and temperatures will depend on the restoration of cyanobacteria, lichens, and mosses. While cyanobacteria form a dark matrix in which other components are embedded, dark mosses and lichens contribute up to 40% of the cover in an undisturbed crust (Belnap, 1993). Consequently, recovery of surface albedo characteristics in severely disturbed areas could take up to 250 years for even very small areas.

Different deserts may show very different resistance and resilience to surface disturbances. Some characteristics of deserts suggest that these ecosystems probably evolved with low levels of soil surface disturbance by ungulates. These characteristics include the presence of cryptobiotic soil crust, the morphology and phenology of perennial bunch grasses found in these deserts, and the dominance of C_3 grasses (Mack and Thompson, 1982). Some deserts, such as those in the Colorado Plateau biogeographic province, may have also evolved with low levels of microdisturbance such as those created by soil and surface invertebrates, and thus depend more heavily on soil surface integrity for natural ecosystem functioning. As a result, these deserts may be harmed more by soil surface disturbances than deserts that evolved under higher levels of surface disturbance.

4. Conclusion

Thirty percent of the United States consists of semiarid or arid landscapes. Preventing desertification of these areas requires maintaining the sustainability and productivity of the ecosystems within these areas. Because most current land use practices involve extensive impacts to soil surfaces, we must understand the short- and long-term effects of such impacts on nutrient cycles and soil food webs in these ecosystems and how these processes accelerate desertification.

The data presented here show that nutrient cycles in deserts can be extremely sensitive to soil surface disturbances. Nitrogen cycles may be at the greatest risk from current land use practices. Input of available nitrogen from atmospheric and parent material sources is extremely low in arid regions, and these systems generally depend on biologically fixed nitrogen and nitrogen released from the decomposition of organic matter. The above data show that both of these input pathways are threatened by soil surface disturbances. Because nitrogen loss is a continuous process, understanding factors that decrease the input or availability of nitrogen is of critical importance to maintaining the sustainability of arid systems (Peterjohn and Schlesinger, 1990) and preventing desertification.

Preventing soil loss is another major challenge in preventing desertification of these areas. Many semiarid and arid areas rely on soil surface cyanobacteria for stability, and current land uses are affecting large areas of this resource (Belnap, personal observation). In addition, loss of diversity of processes and species results from current types of soil surface disturbance. The sensitivity of the semiarid and arid lands of the western United States to these impacts, combined with the low resiliency of these systems, indicate that they are at risk from desertification. These areas must be managed to conserve the fertility and stability of soil resources.

Acknowledgements

I wish to acknowledge the following people and organizations for their support of my research. Greg Swayze and Roger Clark provided albedo data; Dave Evans conducted the isotopic analyses; Nancy Stanton provided nematode analysis; Andy Moldenke provided microarthropod identifications; Danielle Tilford, Gary Gurtler, and Cindy Furman provided field and technical assistance; and Kimball Harper and Elaine Ingham provided helpful comments and discussion. This research was funded by the U.S. Army Corp of Engineers, Construction Engineering Laboratory, Champaign, Illinois; Department of Defense Legacy Program; and the National Park Service.

References

Anderson, D.C., K.T. Harper, and S.R. Rushforth. 1982. Recovery of cryptogamic soil crusts from grazing on Utah winter ranges. *Journal of Range Management* **35**: 355–359.

Ashley, J. and S.R. Rushforth. 1984. Growth of soil algae on topsoil and processed oil shale from the Uintah Basin, Utah, USA. *Reclamation and Revegetation Research* **3**: 49–63.

Belnap, J. 1992. *Effects of wet and dry pollutants on the physiology and elemental accumulation of cryptobiotic crusts and selected rock lichens on the Colorado Plateau.* Final Report, National Park Service, Washington Air Quality Division, December 1991. NPS/AQ-91, on file at Washington Office of Air Quality, National Park Service, Denver, Colorado.

Belnap, J. 1993. Recovery rates of cryptobiotic crusts: Inoculant use and assessment methods. *Great Basin Naturalist* **53**(1): 89–95.

Belnap, J. 1994. Potential value of cyanobacterial inoculation in revegetation efforts. In *Proceedings, ecology and management of annual rangelands*, edited by S.B. Monsen and S.G. Kitchen, General Technical Report INT-GTR-313, USDA Forest Service, Intermountain Research Station, Odgen, Utah, pp. 179–185.

Belnap, J. and J.S. Gardner. 1993. Soil microstructure in soils of the Colorado Plateau: The role of the cyanobacterium *Microcoleus vaginatus*. *Great Basin Naturalist* **53**(1): 40–47.

Belnap, J. and K.T. Harper. 1995. The influence of cryptobiotic soil crusts on elemental content of tissue in two desert seed plants. *Arid Soil Research and Rehabilitation* **9**: 107–115.

Belnap, J., K.T. Harper and S.D. Warren. 1994. Surface disturbance of cryptobiotic soil crusts: Nitrogenase activity, chlorophyll content, and chlorophyll degradation. *Arid Soil Research and Rehabilitation* **8**: 1–8.

Beymer, R.J. and J.M. Klopatek. 1991. Potential contribution of carbon by microphytic crusts in pinyon-juniper woodlands. *Arid Soil Research and Rehabilitation* **5**: 187–198.

Brock, T.D. 1975. Effect of water potential on a *Microcoleus* from a desert crust. *Journal of Phycology* **11**: 316–320.

Bush, J.K. and O.W. Van Auken. 1991. Importance of time of germination and soil depth on growth of *Prosopis glandulosa* seedling in the presence of a C_4 grass. *American Journal of Botany* **78**: 1732–1739.

Callison, J., J.D. Brotherson, and J.E. Bowns. 1985. The effects of fire on the blackbrush (*Coleogyne ramosissima*) community of southwest Utah. *Journal of Range Management* **38**: 535–538.

Campbell, S.E. 1979. Soil stabilization by a prokaryotic desert crust: Implications for Precambrian land biota. *Origins of Life* **9**: 335–348.

Campbell, S.E., J.S. Seeler, and S. Golubic. 1989. Desert crust formation and soil stabilization. *Arid Soil Research and Rehabilitation* **3**: 217–228.

Cole, D.N. 1990. Trampling disturbance and recovery of cryptogamic soil crusts in Grand Canyon National Park. *Great Basin Naturalist* **50**: 321–325.

Coleman, D.C., E.P. Odum, and D.A. Crossley Jr. 1992. Soil biology, soil ecology, and global change. *Biology and Fertility of Soils* **14**: 104–111.

Crawford, C.S. 1979. Desert detritivores: A review of life history patterns and trophic roles. *Journal of Arid Environments* **2**: 31–42.

Crawford, C.S. 1991. The community ecology of macroarthropod detritivores. In *Ecology of Desert Communities*, edited by G. Polis, University of Arizona Press, Tucson, Arizona, pp. 89–112.

Doyen, J.T. and W.F. Tschinkel. 1974. Population size, microgeographic distribution and habitat separation in some tenebrionid beetles. *Annals of the Entomological Society of America* **67**: 617–626.

Dregne, H.E. 1983a. Physical effects of off-road vehicle use. In *Environmental effects of off-road vehicles: impacts and management in arid regions*, edited by R.H. Webb and H.G. Wilshire, Springer-Verlag, New York, pp. 15–30.

Dregne, H.E. 1983b. *Desertification of arid lands*, Harwood Academic Publishers, New York, pp. 1–15.

Ettershank, G., J. Ettershank, M. Bryant, and W.G. Whitford. 1978. Effects of nitrogen fertilization on primary production in a Chihuahuan Desert ecosystem. *Journal of Arid Environments* **1**: 135–139.

Evans, D. and J.R. Ehlringer. 1993. Broken nitrogen cycles in arid lands: Evidence from ^{15}N of soils. *Oecologia* **94**: 314–317.

Harper, K.T. and J.R. Marble. 1988. A role for nonvascular plants in management of arid and semiarid rangeland. In *Vegetation science applications for rangeland analysis and management*, edited by P.T. Tueller, Kluwer Academic Publishers, Dordrecht, pp. 135–169.

Harper, K.T. and R.L. Pendleton. 1993. Cyanobacteria and cyanolichens: Can they enhance availability of essential minerals for higher plants? *Great Basin Naturalist* **53**(1): 59–72.

Hendrix, P.F., D.C. Coleman, and D.A. Crossley Jr. 1992. Using knowledge of soil nutrient cycling processes to design sustainable agriculture. *Journal of Sustainable Agriculture* **2**: 63–82.

Ingham, E.R., D.C. Coleman, and J.C. Moore. 1989. An analysis of food-web structure and function in a shortgrass prairie, a mountain meadow, and a lodgepole pine forest. *Biology and Fertility of Soils* **8**: 29–37.

Jeffries, D.L. and J.M. Klopatek. 1987. Effects of grazing on the vegetation of the blackbrush association. *Journal of Range Management* **40**: 390–392.

Johansen, J.R. 1993. Cryptogamic crusts of semiarid and arid lands of North America. *Journal of Phycology* **29**: 140–147.

Kitchell, J.F., R.V. O'Neill, D. Webb, G.W. Gallepp, S.M. Bartell, J.F. Koonce, and B.S. Ausmus. 1979. Consumer regulation of nutrient cycling. *BioScience* **29**: 28–34.

Kleiner, E.F. and K.T. Harper. 1977. Soil properties in relation to cryptogamic ground cover in Canyonlands National Park. *Journal of Range Management* **30**: 202–205.

Knapp, P. 1992. Soil loosening processes following the abandonment of two arid western Nevada townsites. *Great Basin Naturalist* **52**: 149–154.

Lange, W. 1974. Chelating agents and blue-green algae. *Canadian Journal of Microbiology* **20**: 1311–1321.
Larmuth, J. 1978. Temperatures beneath stones used as daytime retreats by desert animals. *Journal of Arid Environments* **1**: 35–40.
Lesica, P. and J.S. Shelley. 1992. Effects of cryptogamic soil crust on the population dynamics of *Arabis fecunda* (Brassicaceae). *American Midland Naturalist* **128**: 53–60.
Loope, W.L. and G.F. Gifford. 1972. Influence of a soil microfloral crust on select properties of soils under pinyon-juniper in southeastern Utah. *Journal of Soil and Water Conservation* **27**: 164–167.
Mack, R.N. and Thompson, J.N. 1982. Evolution in steppe with few large, hooved mammals. *The American Naturalist* **119**: 757–773.
MacMahon, J.A. 1987. Disturbed lands and ecological theory: An essay about a mutualistic association. In *Restoration ecology*, edited by W.R. Jordan, M. Gilpin, and J.D. Aber, Cambridge University Press, New York, pp. 221–238.
Mayland, H.F. and T.H. McIntosh. 1966. Availability of biologically fixed atmosphere nitrogen-15 to higher plants. *Nature* **209**: 421–422.
Mayland, H.F., T.H. McIntosh, and W.H. Fuller. 1966. Fixation of isotopic nitrogen in a semi-arid soil by algal crust organisms. *Soil Science of America Proceedings* **30**: 56–60.
Metting, B. 1991. Biological surface features of semiarid lands and deserts. In *Semiarid lands and deserts: Soil resource and reclamation*, edited by J. Skujins, Marcel Dekker, Inc., New York, pp. 257–293.
Peterjohn, W.T. and W.H. Schlesinger. 1990. Nitrogen loss from deserts in the southwestern United States. *Biogeochemistry* **10**: 67–79.
Romney, E.M., A. Wallace, and R.B. Hunter. 1978. Plant response to nitrogen fertilization in the northern Mohave Desert and its relationship to water manipulation. In *Nitrogen in desert ecosystems*, edited by N.E. West and J.J. Skujins, Dowden, Hutchinson & Ross, Stroudsburg, Pennsylvania, pp. 232–243.
Rychert, R., J. Skujins, D. Sorensen, and D. Porcella. 1978. Nitrogen fixation by lichens and free-living microorganisms in deserts. In *Nitrogen in desert ecosystems*, edited by N.E. West and J.J. Skujins, Dowden, Hutchinson & Ross, Stroudsburg, Pennsylvania, pp. 20–30.
Sagan, C., O.B. Toon, and J.B. Pollack. 1979. Anthropogenic albedo changes and the earth's climate. *Science* **206**: 1363–1368.
St. Clair, L.L., J.R. Johansen, and B.L. Webb. 1986. Rapid stabilization of fire-disturbed sites using a soil crust slurry: Inoculation studies. *Reclamation and Revegetation Research* **4**: 261–269.
Santos, P.F. and W.G. Whitford. 1981. The effects of microarthropods on litter decomposition in a Chihuahuan Desert ecosystem. *Ecology* **62**: 554–669.
Schimel, D.S., E.F. Kelly, C. Yonker, R. Aguilar, and R.D. Heil. 1985. Effects of erosional processes on nutrient cycling in semiarid landscapes. In *Planetary ecology*, edited by D.E. Caldwell, J.A. Brierley, and C.L. Brierley, Van Nostrand Reinhold, New York, pp. 571–580.
Seastedt, T.R. and D.A. Crossley, Jr. 1984. The influence of arthropods on ecosystems. *BioScience* **34**: 157–161.
Skujins, J. 1984. Microbial ecology of desert soils. *Advances in Microbial Ecology* **7**: 49–91.
Skujins, J. and B. Klubek. 1978. Nitrogen fixation and denitrification in arid soil cryptogamic crust microenvironments. In *Environmental Biogeochemistry and Geomicrobiology*, vol. 2., edited by W. E. Krumbein, Ann Arbor Publishers, Ann Arbor, Michigan, pp. 543–552.
Stolzy, G. and A.G. Norman. 1961. Factors limiting microbial activities in soil. *Archiv Mikrobiologie* **40**: 341–350.
Terry, R.E. and S.J. Burns. 1987. Nitrogen fixation in cryptogamic soils crusts as affected by disturbance. In *Proceedings, Pinyon-Juniper Conference*. Technical Report INT-215, USDA Forest Service, Intermountain Research Station, Odgen, Utah, pp. 369–372.
Tidemann, A.R., W. Lopushinsky, and H.J. Larsen Jr. 1980. Plant and soil responses to a commercial blue-green algae inoculant. *Soil Biology and Biochemistry* **12**: 471–475.
Torbert, H.A. and C.W. Wood. 1992. Effects of soil compaction and water-filled pore space on soil microbial activity and nitrogen losses. *Communications in Soil Science Plant Analysis* **23**: 1321–1331.
Wallwork, J.A. 1982. *Desert soil fauna*, Praeger Scientific Publishers, London.

Webb, R.H. 1983. Compaction of desert soils by off-road vehicles. In *Environmental effects of off-road vehicles: Impacts and management in arid regions*, edited by R.H. Webb and H.G. Wilshire, Springer-Verlag, New York, pp. 31–80.

Webb, R.H. and H.G. Wilshire. 1980. Recovery of soils and vegetation in a Mojave Desert ghost town, Nevada, USA. *Journal of Arid Environments* **3**: 291–303.

West, N.E. 1981. Nutrient cycling in desert ecosystems. In *Arid land ecosystems: Structure, functioning, and management*, vol. 2, edited by D.A. Goodall and R.A. Perry, Cambridge University Press, Cambridge, pp. 301–324.

West, N.E. and J. Skujins. 1977. The nitrogen cycle in North American cold-winter semi-desert ecosystems. *Oecologia Planta* **12**: 45–53.

Wilshire, H.G. 1983. The impact of vehicles on desert soil stabilizers. In *Environmental effects of off-road vehicles: Impacts and management in arid regions*, edited by R.H. Webb and H.G. Wilshire, Springer-Verlag, New York, pp. 31–50.

Zak, J. and W. Whitford. 1988. Interactions among soil biota in desert ecosystems. *Agriculture, Ecosystems, and Environment* **24**: 88–100.

A HIERARCHICAL APPROACH FOR DESERTIFICATION ASSESSMENT

GER BERGKAMP
Landscape and Environmental Research Group, University of Amsterdam, Nieuwe Prinsengracht 130, 1018 VZ Amsterdam, The Netherlands

Abstract. Environmental systems are complex and multi-scaled open systems. To understand land degradation, one has to consider the interactions between landscape patterns and environmental processes at different scales. Patterns and processes in the landscape are perceived to be organized in nested hierarchical structures. To study land degradation in this context, a coupled top-bottom/bottom-up approach was developed. The top-bottom landscape analysis is aimed at identifying landscape systems at different scales. The bottom-up analysis focuses on system dynamics at finer scales. Applied in a study in central Spain, the approach is aimed at understanding the functional differences between three types of degraded seminatural slopes at different scales. Six levels of organization were distinguished: ped-level, terracette-level, hummock-level, slope part-level, slope level, and watershed level. Properties that characterize these levels were selected for different disciplines. The bottom-up analysis focused on water movement at different spatial scales. Successfully applying this approach revealed the importance for land degradation of the close linkage between spatial patterns and hydrological processes at different spatial scales. Identifying constraining and dynamic indicators related to water conservation at different scales can be useful for assessing desertification.

1. Introduction

In arid and semiarid areas the natural environment is known to degrade at many scales. Most desertification-related research seems to focus on a single temporal and spatial scale. Little attention has been given to integrating patterns characterizing land degradation and environmental processes governing it at different scales. This lack of attention to the importance of this need is closely related to the linear concepts and reductionistic views many scientists continue to hold.

But reductionistic landscape theories aimed at precisely describing equilibrium conditions of environmental systems are gradually being replaced by combined holistic-reductionistic views. Recent work on the organization of complex, dynamic open systems has pointed to the limitations of reductionistic approaches in environmental sciences. (O'Neill *et al.*, 1986; Phillips, 1992; Renwick, 1992; Trofimov and Phillips, 1992). O'Neill *et al.* (1986) proposed a new theory of complexity and hierarchical organization of dynamic, open environmental systems. The theory provides a framework for analyzing hierarchically organized processes at different spatial and temporal scales and their linkage with tangible, spatial structures at different scales. This framework is much more suitable for studying land degradation patterns and processes than those based on traditional conservative systems views, which emphasize equilibrium concepts.

Environmental Monitoring and Assessment **37:** 59–78, 1995.

Only recently has hierarchy theory been applied in environmental studies that integrate processes and patterns at different spatial and temporal scales. Although this application has been restricted mainly to the field of ecology, a much wider use of its concepts seems feasible in studies of various disciplines in the context of land degradation.

This paper proposes a coupled top-bottom/bottom-up approach to studying land degradation, discussing how this approach was applied in a study of a degraded area in central Spain. Considering the degrading landscape as a multi-scaled, complex, open environmental system, the approach uses important elements of hierarchy theory and provides a means for selecting constraining and dynamic indicators of desertification. Before discussing the approach, this paper gives an overview of some hierarchy theory elements directly relevant to the approach and its application.

2. Theory

2.1. Development of Dynamic Systems Thinking

The development of the theory of open, dynamic systems began with works of Ludwig von Bertalanffy (1950, 1968) and Ilya Prigogine (1945). They postulated that open dynamic systems generate internal structure through the fluxes of energy, matter, and water and the dissipation of entropy. The nonequilibrium conditions of these systems become a source of order leading to so-called dissipative structures (Prigogine, 1978; Prigogine and Stengers, 1985).

Away from equilibrium conditions, the dynamics of these nonlinear open systems are extremely rich and often difficult to predict. Inspired by the early works of Bertalanffy, Weiss (1971) and Strahler (1952) developed dynamic systems thinking in biology and geomorphology respectively. Ecosystems are either thought of as sections of the landscape or as patterns of interactions (Tansley, 1935; Weiss, 1971). Weiss defined an ecosystem as patterns of connections between complex units in space and time so constituted that its subunits, by "systematic" cooperation through a network of interactions, preserve its internal configuration of structure and behavior and tend to restore it after nondestructive disturbances. As such, the ecosystem is a paradigm of the principles of interdependency, partly prestructured and partly in free systematic interaction (Weiss, 1971). For systems of this kind it is not trivial to derive the properties of the whole from the behavior of the parts (Simon, 1973).

During the 1950s and 1960s most concepts of system dynamics in both ecology and geomorphology perceived systems through Hamiltonian concepts in which one large canonical ensemble could describe the whole system. Imbedded in concepts of equifinality, climax vegetation, and dynamic equilibrium, systems thought of as being conservative instead of dissipative (Culling, 1987). Further developments during the 1970s and 1980s led to the construction of a theory for open systems with dissipative structures, which are necessarily hierarchically organized (Koestler 1967; Weiss, 1971; Pattee, 1973; Allen and Star, 1982; O'Neill *et al.*, 1986).

2.2. Landscape Research and Hierarchy Theory

2.2.1. *The Holon*

Within hierarchy theory a dynamic system is perceived as an open, dissipative system that through the process of self-organization can generate a hierarchically organized complexity (Bertalanffy, 1968; Koestler, 1967; Weiss, 1971; Allen and Star, 1982; O'Neill *et al.*, 1986). To describe a system's functioning constrained by a higher level system and at the same time constraining lower level subsystems, Koestler (1967) introduced the term *holon*. The holon is defined by a set of governing principles, within which it has an array of options of possible responses to environmental change (Koestler, 1978; Haigh, 1987). As the systems view is only a representation of the world, the observer is allowed to define holons and the hierarchical levels of organization according to the aim of the investigation (Allen and Star, 1982).

2.2.2. *Hierarchical Levels*

Generally two types of hierarchies are distinguished: functional hierarchies, which are process oriented (Levandowsky and White, 1977; Allen and Star, 1982; O'Neill *et al.*, 1986), and structural hierarchies, which are entity oriented (Salthe, 1985). Structural hierarchies are not necessarily time or space dependent. Rather they are used for separating objects from their context (Allen and Hoekstra, 1990). Patterns and processes cannot be understood by looking only at structures or only at the functional organization. An integration of structure and function is necessary. At the same time a selective loss of detail is needed to give rise to a hierarchical controlled organization (Pattee, 1973).

Hierarchical levels are differentiated by three principles:

(a) Specific levels of organization are linked to specific spatial and temporal scales, so that higher levels are linked to broader spatial and temporal scales.
(b) Rate differences of at least one order of magnitude exist between different levels, so that higher levels have lower frequencies of behavior than lower levels.
(c) Higher levels constrain lower levels through their larger time constraints (Allen and Star, 1982; O'Neill *et al.*, 1986).

Hierarchical levels can either be nested or non-nested. If they are nested, levels contain and constrain and are contained and constrained. Non-nested or "control" hierarchies only constrain or are constrained, as in the examples of trophic levels and river stream systems (Pattee, 1973; Allen and Star, 1982; O'Neill *et al.*, 1986; Salthe, 1985).

Defining a hierarchy of environmental systems inevitably leads to a truncated systems view as one can always think of a level that contains the highest level and a level constituting the lowest level. For a proper truncation of the hierarchy, one must understand the relevance of processes at levels far from the focal level to the focal level holons. This relevance is determined by the degree and way of interactions between holons at different levels. Thus one must know the complexity of the interactions between holons to be able to construct a hierarchically organized set of holons a so-called holarchy.

2.2.3. *Complexity*

In hierarchically organized systems, different holons are linked through the exchange of energy, matter, or information. Generally the main interactions are within the holon, but in some cases an exchange occurs with holons at other levels or sister holons (at the same level). Interactions at one level can occur either through the exchange to a higher level holon, which then constrains the sister holon, or directly to the sister holon (Allen and Star, 1982). For example, the development of the size and the stability of a soil aggregate, in this example being the focal level, generates a change in soil water holding capacity (the next higher level) only if an assemblage of soil aggregates change. The water holding capacity in turn constrains the potentials of the metabolisms controlling the formation of a soil aggregate (the focal level).

Exchanges of information between hierarchical levels is fundamentally asymmetric; holons at higher levels can affect lower level holons but are relatively unresponsive to changes at the lower level because the frequency of the behavior of the lower level holon is much higher then that of the higher level holon (O'Neill *et al.*, 1986). In this way, hierarchically dynamic systems are "nearly decomposable" and only partly interact because different levels have different response times (Simon, 1973).

As a result of this near decomposibilty, one can postulate the existence of emergent properties, which are properties of higher level holons that cannot be deduced from the functioning of their parts. Allen and Star (1982) distinguish emergent properties that are emergent only because of a mismatch of observation set and phenomena and "true" emergent properties that cannot be explained by lower level phenomena. According to Pomeroy *et al.* (1988) many "so called emergent properties" are emergent only because of an insufficient theory of the system's behavior. Whether emergent properties are "true" or not, not all the behavior of a focal level can be explained by properties and processes of lower levels of organization.

Although the higher level is often not directly influenced by the lower level processes, the higher level is constrained by its own spatial structure, which depends on the processes occurring at a lower level (Pattee, 1973). Consequently, energy, matter, and information ascending the hierarchy can also be transferred, but often such a transfer coincides with a considerable time lag. Hence, the exchange of information between hierarchical levels is basically asymmetrical with greater constraints from the higher level on the lower level than the reverse.

A constraint is seen as relative to the defined information exchange and described by the amplitude or significance of the information exchanged and the asymmetry of signal exchange (Allen and Star, 1982). Although the focal level structures and processes are most constrained by the contiguous levels, the total constraint can be decomposed into various levels of residual importance.

Salthe (1985) discriminates three types of constraints:

(a) Higher level constraints define 'what cannot happen' (Allen and Star, 1982).
(b) Focal level constraints are mainly caused by the focal level spatial structure.
(c) Lower level constraints, which are the only truly creative forces that propose possible results.

O'Neill *et al.* (1986) discriminate between structural constraints, which arise from the structural organization in hierarchies, and functional constraints, which originate from a functional control. In practice, hierarchically organized dynamic systems are often controlled by both functional and structural constraints.

2.2.4. *Boundaries and Connectance*

Boundaries between holons, whether functional or structural, are dynamic and transient. They are not absolute separations but must be viewed as relative to the occurring exchanges between either sites of the boundaries (Allen and Star, 1982). As such, both functional and structural boundaries of holons have steep gradients in process rates (Allen and Star, 1982; O'Neill *et al.*, 1986). For example, under noncontaminated conditions the rate of metabolism in the topsoil is functionally controlled by soil moisture and soil temperature. Measuring these two provide essential information on the functional boundaries of different topsoil systems and subsystems.

Functional boundaries do not necessarily coincide with boundaries we perceive as structural and tangible. Sometimes even functional boundaries do not correspond to any tangible object or aggregation of objects at all, as, for example, in the case of the rhizosphere. Defining hierarchical holons and hierarchical levels by considering structural boundaries and tangible objects is considered to be a special case of steep gradients in process rates so that they can functionally be considered steps (O'Neill *et al.*, 1986; Allen and Star, 1982).

The perception of boundaries, both functional and structural, depends on the observer's choice. Boundaries are not right or wrong, true or false. They can be only more or less suitable for a specific purpose.

Across boundaries energy, matter, and information are exchanged. These exchanges, which connect holons, are messages sent by one holon and received by another. The information in the messages can differ for the sender and the receiver, and different receivers may read the signals differently. Thus the content of communicated information between levels is ambiguous, and the focal level perceives only a representation of the other level (Allen and Star, 1982; Salthe, 1985). For the representation of exchanges of energy, matter, and information between levels with different time scales, scaling parameters can be used (Levandowsky and White, 1977). If the reaction rates within subsystems differ greatly compared to their connectance, they will appear as parameters to each other instead of covariables (Levins, 1973). Hence, the connectance between holons is closely related to the time and spatial scales considered and should always be less than the connectance within the holons (O'Neill *et al.*, 1986).

The communication of the effects of processes decreases with the distance between the hierarchical levels over which the communication occurs (Salthe, 1985). The system's connectivity is closely related to the concept of stability. May (1973) and Gardner and Ashby (1970) pointed out that cybernetic systems with many interactions, thus more components with direct interactions, are more likely to be unstable. O'Neill *et al.* (1986) postulate that the system's connectedness rather than the number of components is most important in determining the stability of the system.

2.2.5. *Concepts of Stability of Environmental Systems*

The stability of environmental systems is closely related to temporal and spatial scales. If we want to know if a system operates in a stable way, we need to consider its time constant of characteristic behavior, the length of time a pattern of characteristic behavior exists, the time periods of observation (frequency and total duration), and the spatial characteristics of the observation set. Hence, concepts of stability need to deal with the system's dynamics as well as with the observer's reference.

Culling (1987) distinguishes several types of stability related to the system dynamics. His equifinality s.s., or asymptotic stability, refers to a system that eventually returns to its initial conditions, a point attractor in a phase diagram. A system that returns to a situation close to its initial position is said to have Lyapounov stability or to be locally stable if the allowable magnitude of the disturbance is restricted. This type of stability has a stable limit cycle. Globally stable systems are dynamic systems that do not return to an initial condition but whose behavior is characterized by a certain stable domain with more than one local minimum. The parameter space of dynamic systems can be further classified beyond global stability to the "ultimate" state of chaotic behavior. Environmental systems could exhibit all of these behaviors.

Holling (1973) notes that it is not as important to know how stable an environmental system is within one domain of attraction as it is to know how likely the system is to move from one domain to another and so to persist in a changed configuration. He defines stability as the ability of a system to return to an equilibrium state after a temporary disturbance. The more rapidly a system returns and with the least fluctuations, the more stable it is.

Closely linked to the system's stability is the concept of resilience. Holling (1973) defines resilience as determining "the persistence of relationships within a system." Resilience is a measure of the system's ability to absorb changes of variables and parameters and still persist. An environmental system can fluctuate greatly but still be very resilient. Holling (1973) suggests that the more homogeneous the environment, the more likely it is to only slightly fluctuate (being very stable) but to have a low resilience, being unable to persist in its system configuration after a disturbance.

Stability in hierarchically organized systems is promoted by the organization of direct and symmetric connections in stable subsystems at different levels. The asymmetry of information transfer between different hierarchical levels also contributes to the stability of environmental systems. Furthermore, disturbances at one level can be incorporated in the behavior of the next level (Allen and Star, 1982). The process of incorporation also contributes to a system's stability. Sometimes disturbances at one level even become a prerequisite for the persistence of the system at a higher level, as is the case with fire (Wright, 1974; O'Neill *et al.*, 1986). True perturbations affecting the stability of the system must be rare and transient phenomena, occurring at a scale close enough to that of the focal holon to be connected and to be able to communicate, but far enough away to prevent incorporation (Haigh, 1987).

Identifying the complexity, stability, and resilience of environmental systems is crucial for understanding landscape changes, such as land degradation. An approach to understanding environmental change, therefore, needs to consider the process-pattern duality at different scales. The approach outlined below aims at a synthesis of processes and patterns at different scales, integrating some of the discussed concepts with field practicality.

3. A Combined Top-Bottom and Bottom-up Approach

The landscape system contains multiscaled patterns interacting with scaled processes. For understanding desertification we therefore need to consider both the multiscaled aspects of landscape properties and the close interactions between spatial structures and the environmental key processes controlling desertification. These key processes often cannot be defined by a single discipline. Rather tney are process complexes with different processes dominating desertification at different scales.

As previously discussed, the landscape can be perceived as an hierarchically organized, dynamic, open, environmental system. A multidisciplinary description of the landscape system depends on the range of spatial and temporal scales we choose. Selecting one single spatial and temporal scale is not appropriate because in open environmental systems changes are taking place at many scales (O'Neill, 1988). Therefore, the focus should be on the interactions between processes and patterns at different scales.

Within the context of desertification the selection of a space and timeframe is undefined, spatially between soil aggregate and the region and temporally between months and decennia. Although we often think that we select our scale of observation deliberately to elucidate key phenomena, scales are more often imposed on us by our perceptual, technological, and logistical constraints (Levin, 1992). If we want to address desertification within the framework of open environmental systems, we have to incorporate the following into the analysis:

- Spatial patterns differ at different scales (Turner *et al.*, 1989; Meentemeyer, 1989).
- Changes of spatial patterns differ at different time scales; temporal and spatial scales are linked (O'Neill *et al.*, 1986; O'Neill, 1988).
- Processes controlling changes in spatial patterns differ at different scales; no single mechanism explains observed patterns at all scales (Levin, 1992).
- At different scales different variables are often needed to describe similar processes (di Castri and Hadley, 1988; Jarvis and McNaughton, 1986).

The approach for desertification research outlined here is aimed at incorporating these considerations into a framework that considers both structural and functional landscape properties important for desertification at different scales.

3.1. The Approach

Holistic landscape studies are mainly performed in landscape ecology. They are aimed at classifying the landscape in units that can be recognized as types characterized by only a few well-chosen diagnostic characteristics that are homogeneous for the units (Zonneveld, 1989; Naveh and Lieberman, 1994). Found in all environmental disciplines, reductionistic approaches are used by scientists who pursue the idea that the functioning of an environmental system can be understood only by knowing the functioning of its parts.

The coupled approach described here is reductionistic in the sense that it aims at recognizing the constituents of the dynamic landscape system. The approach is holistic in the sense that it considers more levels, with different kinds of dynamic behavior and different functional characteristics besides the focal level of interest. The approach repeatedly returns to certain topics but each time with a different perspective. In this sense the approach can be seen as a spiral approach in which the dialectic between structure and function is integrated. In this way deductive and inductive research are combined in the various stages.

After defining the research problem and objectives, the coupled approach starts with a holistic landscape analysis, defining the holarchies of interest. Figure 1 outlines the elements of the hierarchical systems analysis.

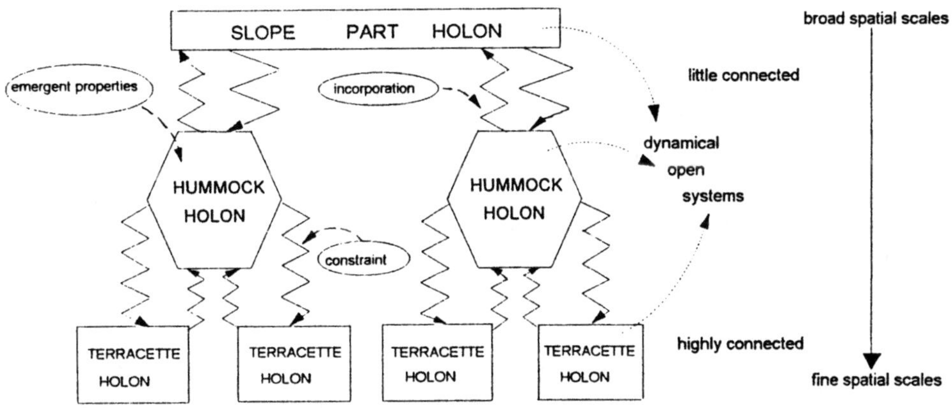

Fig. 1. Overview of holons and their interactions at different hierarchical levels at the research sites.

The landscape analysis is aimed at recognizing spatial patterns at different spatial scales, combining remote sensing with soil and vegetation surveys. Traditionally, holistic landscape studies use the land unit concept (Zonneveld, 1989), which is underlain by scale-invariant soil and vegetation classification systems. These systems are often based on diagnostic rather than functional properties (Zonneveld, 1989; Naveh and Lieberman, 1994), which very much limit the use of this type of landscape analysis within a dynamic systems framework. The soil and vegetation surveys discussed here are more focussed on the functional representation of entities at different spatial scales, although traditional classification systems are also used. The criteria used for separating landscape system units are closely related to the key processes relevant to desertification. The boundaries of selected units coincide with steep gradients in rates of these processes.

Because it can be used only down to the level of physiographical position, the land unit concept cannot provide all the needed detail for hierarchy-oriented desertification research. Moreover, this concept assumes homogeneity for the discriminated unit, and the discriminative criteria arc mostly diagnostic instead of functional. To construct a more complete, preliminary holarchy, we often need detailed surveys that require identifying functional, fine-scale structures, their boundaries, key-processes interacting with these structures, and filter mechanisms between levels.

The different holarchies, constructed for different disciplines, are then preliminarily linked. Holons with similar process rates are ordered at equivalent levels of organization. Although placed at the same hierarchical level, these holons do not necessarily need to be connected, as processes with similar rates do not have to influence each other. To understand the coherence between holons originating from different disciplines, we need to determine coordinating principles and rates of exchange at and between different holons (Müller, 1992). During the top-bottom analysis these principles and rates are identified preliminarily, to be replaced by a more detailed identification at focal levels.

As noted before, desertification research implies selecting a certain range of temporal and spatial scales. At the range of the chosen temporal scales – months to decennia – the spatial connectivity in desertifying landscapes by environmental processes is often limited to certain distances. Therefore, it is justifiable to select a focal level beyond the level of physiographic position. Most processes responsible for pattern generation at broader scales are beyond the scope of desertification research, as, for example, relief formation through tectonics. These processes merely form a constraint.

At the range of the selected focal levels, a bottom-up approach is applied to identify the landscape system's configuration and to measure rates of key processes. Furthermore, the connectance between holons is analyzed in detail with an emphasis on the significance of heterogeneous structures, temporal de-couplings, and representation of holons at other levels. From this analysis, mechanisms of incorporation and emergent properties at various levels can be postulated.

The focal level analysis is supported by dynamic system modeling. In the approach presented here, the modeling is not seen as the focal activity of a hierarchical environmental systems analysis, as proposed by Müller (1992). Rather it is seen as being complementary and especially useful for understanding complex coordinating principles between different holons. But often the models we can construct are more complex then we can study in the field, and their parameterization is often impossible because of methodological, financial, and time limitations (Levin, 1992; Jorgensen, 1994).

Although our understanding of the landscape system has improved significantly when we have applied the coupled approach, we still know little about the evolutionary changes that may occur in the system through desertification. If the study is not constrained by a relatively short duration, a system's stability and resilience can be studied by focusing on the following: (a) gradual change by measuring succession in structures and processes; (b) catastrophic change by measuring near- and beyond-threshold conditions; (c) adaptation by measuring changes in the function of entities. But in practice, although we can identify broader temporal scale changes as from aerial photos or archived data, very often we know too little about the mechanism responsible for the changes that have occurred.

4. Application

The outlined approach was applied during a study in Castilla-LaMancha, central Spain. The area is one of the driest regions in Europe, with average annual precipitation about 400 mm, falling mainly in autumn and spring. Average monthly temperatures range from 28 °C in August to 9 °C in January. The area has been affected by degradation from grazing, land abandonment, and overexploitation of aquifers.

The study was carried out within the framework of the European Field Experiment in a Desertification-threatened Areas (EFEDA). This project focused on soil, vegetation, and atmosphere interactions at different scales (Bolle *et al.*, 1993) with the objectives of identifying (a) key properties characterizing degradation of slopes with seminatural vegetation at different scales and (b) key hydrological processes at different scales. Three sites were selected, forming a degradation sequence: a degraded oak site (encinar), a shrubland site (matorral), and a degraded grassland site (pastizal). Six levels of organization were distinguished: ped-level, terracette-level, hummock-level, slope part-level, slope level, and watershed level (see Table I).

The top-bottom landscape analysis, which is pattern oriented, was aimed at defining structural hierarchies in ecological, geomorphological, pedological, and hydrological properties. Using aerial photo interpretation combined with soil, vegetation, and geomorphological surveys, we selected sites at which we distinguished north- and south-facing slopes. On all slopes, we selected five positions (top, below top, middle, bottom, and valley) at which we determined both diagnostic and functional properties related to minimum and maximum vegetation cover.

The data was analyzed using non-parametric hierarchical cluster analysis (Ward, 1963) and the non-parametric Kruskal Wallis ANOVA (Blalock, 1979). Vegetation diversity was calculated using the Shanon Wiener index.

The bottom-up part of the research focused on one aspect of the ecosystems complexity by studying the dynamics of key hydrological processes at finer scales. Four scales were selected for detailed field measurements: the watershed scale, the slope part scale, the hummock scale, and the terracette scale. Hydrological measurements were performed on the oak site and the shrubland site. At the watershed scale, runoff was continuously monitored at the oak side. At the slope part scale, 20 m wide, open runoff plots were installed. Runoff from natural rainfall was continuously measured using gutters and collection tank.

At the hummock scale, large-scale rainfall simulation experiments were performed using high-intensity rainfall. Runoff was measured continuously using a gutter and collection tank. Infiltration was measured using TDR sensors and tensiometers. Details of the experiments are reported in Bergkamp *et al.* (1996). These experiments were conducted in addition to small-scale rainfall simulation experiments performed at terracette scale. During these experiments time to overland flow using different rainfall intensities was identified by eye observations, and infiltration was measured by excavation and gravimetric soil moisture determination. An overview of the measurement of hydrological processes at different hierarchical levels is given in Table I.

TABLE I

Measurements of runoff and infiltration were performed at different hierarchical levels

Hierarchical level	Holon name	Type	Size m^2	Type of rainfall	Duration event	Rainfall intensity (mm/hour)
5	watershed	drainage basin	500 * 1000	natural	< 10 minutes	<20
4	slope	–	–	–	–	–
3	slope part	open plot	±100 * 20	natural	< 10 minutes	<20
2	hummock	closed plot	2 * 10	simulated	30 minutes	70
1	terracette	closed plot	1 * 1	simulated	< 30 minutes	20, 40, 60, 80

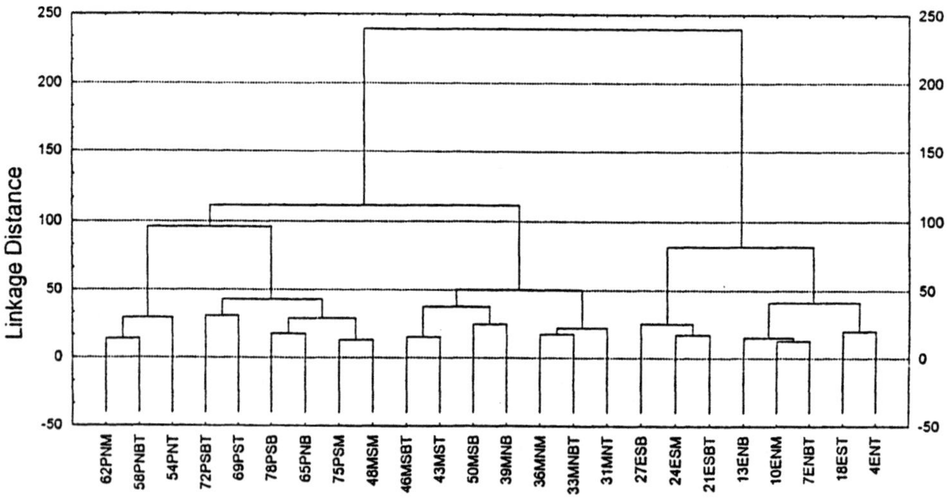

Fig. 2. A hierarchical clustering of sites along a degradation gradient based on vegetation composition. Pastizal (P) (left), Matorral (M) (middle) and Encinar (E) (right) are distinct different groups with respect to vegetation composition.

4.1. Results and Discussion

4.1.1. *Landscape Analysis*

On the basis of vegetation composition and properties related to vegetation patchiness, the locations were ordered in groups at site level (encinar, matorral, and pastizal) as well as at slope level (north, south) (see Figure 2). The ability to distinguish groups from these data shows that these properties can be used to identify diagnostic groups along a degradation gradient. Vegetation diversity revealed differences between sites but was less able to distinguish groups at other levels. As

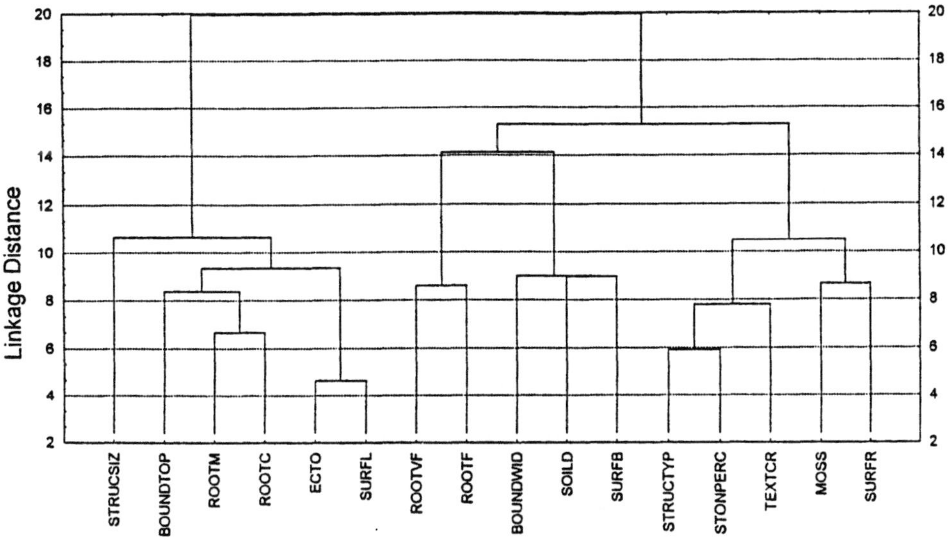

Fig. 3. A hierarchical clustering of soil properties. Three groups are distinguished: left – group of larger roots, litter and soil macro structure and form of boundary of upper soil horizon, middle – group of finer roots, bare soil surface and width of soil horizon boundary, right – group of type of soil structure and properties related to stoniness and moss cover.

such, this study found vegetation diversity less useful in differentiating stages of degradation.

The data can be arranged in clusters for location using properties, and clusters for properties can be made using locations. This type of analysis gives insight into which properties significantly differ at various levels of organization. Generally, clusters of properties show that these properties are more correlated to each other than those in other parts of the cluster diagram. The clustering of the field data in this way helps identify potential functional groups. The preliminary results of the data analysis are presented in Figure 3. The properties shown are all functional properties related to soil biomass, soil texture, and soil profile heterogeneity. Three groups can be distinguished: group 1. cluster of large roots with heterogeneous soil profiles and larger soil structure sizes; group 2. cluster of finer root fractions with boundary width and surface bareness; group 3. cluster of properties mostly related to stones with soil structure type and the occurrence of mosses.

Other results, not shown here, reveal that properties have a significant difference (significance level $p < 0.1$) only at certain levels. As such, a property is useful only for describing differences at a limited set of scales. For example, slope angle did not significantly differ among sites and had little significance among plots at the slope scale. But, of course, slope angle did significantly differ among different slope parts. This example is presented merely to show that static properties such as slope angle differ at different levels but do constrain the functioning of the environment. In this study, static properties that limit the ecosystem possibility

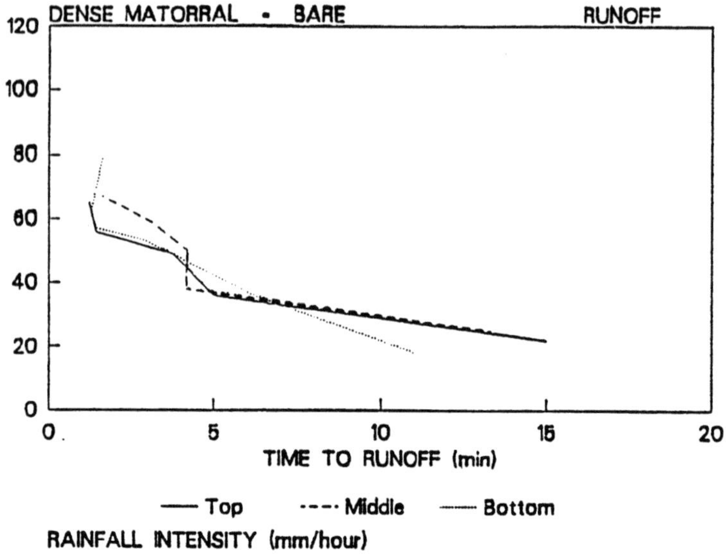

Fig. 4. At terracette scale overland flow is rapidly occurring at bare flat parts of terracettes, shortly after the start of rainfall simulation experiments. Results of different slope parts are shown.

space were found to be regional weather patterns, relationship to groundwater, lithology, slope aspect, slope position, soil profile boundary heterogeneity, and soil stoniness.

These properties can be distinguished from properties that are dynamic with respect to the time scales considered in desertification. The dynamics for desertification originate at finer scales. For different sites different properties were significant, as shown in Table II. Together with results from data analysis, which are not shown here, dynamic properties were identified as soil bulk density (closely related to soil aggregation), root content, soil structure, and forms and sizes of nested patches. These properties are regarded as creative with respect to desertification; they take part in the dynamics that are constrained by the static properties.

4.1.2. *Hydrological Processes at Different Scales*

Under natural rainfall conditions, no runoff was recorded at the slope part-scale at the wide open plots. At terracette scale during the large-scale rainfall simulations no runoff was produced at the hummock scale. Overland flow was observed only close to the gutter. Hortonian overland flow was also rapidly occurring at terracette scale during the fine-scale experiments (see Figure 4). Details of these measurements are reported and discussed in Bergkamp (1996) and Bergkamp *et al.* (1996).

The overland flow at terracette level causes a high connectance between flat and rim parts of the terracettes (see Figure 5). Rapid infiltration at terracette rims, which are stony and sometimes have a higher root content than bare spots, incorporates the overland flow disturbance, which can be generated during extreme events at

TABLE II

Overview of properties which explain differences at different hierarchical levels (significance: $p < 0.1$). Analysis done with Kruskal Wallis ANOVA

Property type	Hummock scale	Minimum and maximum cover		Slope part scale	Slope scale	Site scale
	Encinar	Matorral	Pastizal	Top–foot	North–south	Enc.–Matt.–Past.
Morphology		slope angle		slope angle		
Surface cover	ecto organic mat. % bare % moss cover	ecto organic mat. % bare			ecto organic mat. % bare	
Ecological properties					terracette num. terracette size	Quercus sp. patch size diversity (Shannon Wiener index)
Vegetation composition					Teucrium gnaphalodes Cladonia foliarea Helianthemum cinereu	Quercus sp. Teucrium gnaphalodes Bupleurum suffertico Phlomis lychnitis
Soil properties	% stones bulk density structure type	% stones bulk density structure size % roots (coarse medium, fine)	bulk density % roots (medium	% stones soil depth	% stones % silt, % clay % roots (very fine)	% stones soil depth, % silt structure size boundary topo % root (coarse, medium, fine)

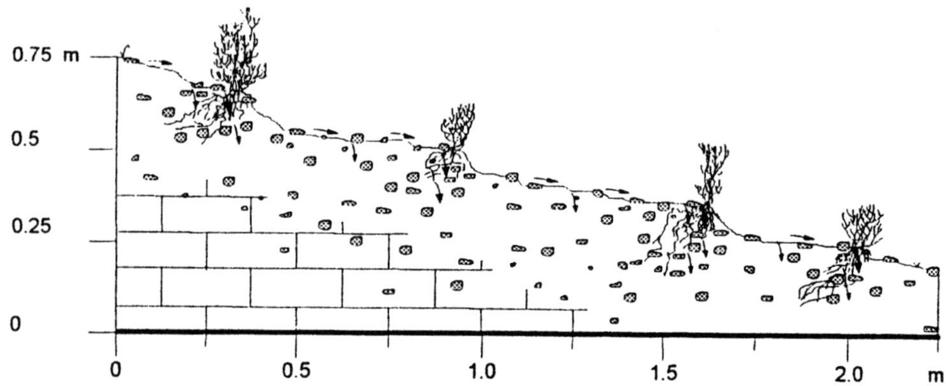

Fig. 5. Overland flow generated at small scale infiltrates near thyme bushes downslope. No exfiltration at downslope site of rims was observed.

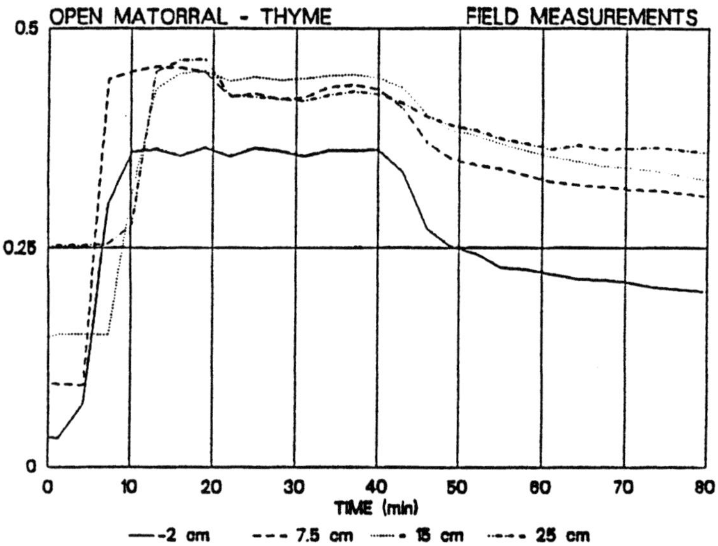

Fig. 6. Soil moisture reaches saturation under thyme bushes rapidly after the start of rainfall simulation experiments. Water infiltrates non-uniformly along macropores and roots. Measurements were performed with TDR equipment.

this level. Only under rare saturated conditions are connections caused by overland flow expected to occur at higher hierarchical levels (Bergkamp, 1996).

The infiltration at the rim parts of the terracettes is extremely heterogeneous. Water infiltrates rapidly along macropores and roots to depths below 25 cm, as shown in Figure 6. Under the flat and often crusted parts of terracettes, however, infiltration was retarded, and the final depth of infiltration was strongly reduced (see Figure 7).

At the hummock scale, water infiltrated rapidly below *Quercus* spp. trees and bushes. Mostly in shallow, stony soils, water was measured to percolate rapidly

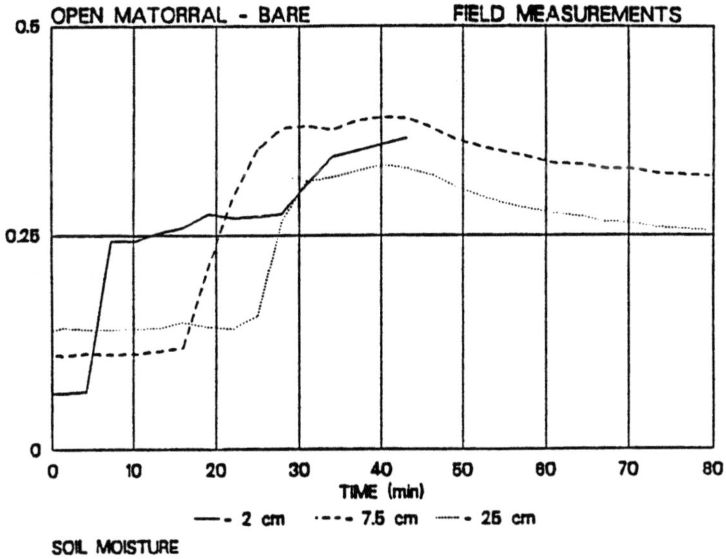

Fig. 7. Infiltration is retarded and more uniform below bare spots of terracettes. Soil moisture changes only slowly in comparison to the changes under thyme bushes during the same experiment.

along roots and cracks into the bedrock. Very little water was stored in the partly hydrophobic litter layers during the rainfall simulation experiments. The boundary between the terracette and hummock holons is sharp in respect to infiltration rates. The properties related to the strongly nonuniform and deep percolation at the hummock scale are seen as emerging at this level. These properties cannot be explained by the interactions at terracette level and as such can be studied only at the hummock scale.

The results show that the heterogeneity of the environment controls runoff and infiltration on semiarid natural slopes. Spatial structures in soil and vegetation at different scales lower system connectivity through which disturbances caused by extreme rainfall events are better incorporated into the functioning of seminatural ecosystems. Destroying the spatial organization at different scales will decrease this ability. On the other hand, the heterogeneity of the lithology, as in this case the fractures in dolomite, affect ecosystem resilience largely by constraining the regeneration of the spatial structures at larger scales. The resilience of ecosystems on dolomite is expected to be much higher than the resilience on, for example, marls because the lithological heterogeneity of the dolomite promotes nonuniform and rapid deep infiltration.

4.1.3. *Synthesis-Indicators for Desertification*

From the above analysis of landscape patterns and finer scale hydrological key processes, indicators for desertification have been selected for assessing landscape changes related to desertification. The indicators should reflect changes in key processes and as such be of the functional type. Choosing diagnostic indicators is therefore less desirable.

As discussed before, the functional properties can be grouped into static and dynamic properties as related to desertification. Static properties constrain land degradation. Dynamic properties reflect the landscape system's dynamics and are related to creative forces in the processes of land degradation. The creative dynamics originate mostly at levels below slope scale. Although changes do occur at broader scales, these changes are perceived as being only changes in factors constraining the creative dynamics at finer scales.

Indicators of these constraining factors are important for recognizing potentials in the landscape. Indicators of dynamics at finer scales serve to identify the status of the actual landscape dynamics and their responses to change. The static properties discussed before are seen in this study as the most important indicators of constraining environmental factors for desertification. These properties are regional weather patterns, relationship to groundwater, lithology, slope aspect, slope position, soil profile boundary heterogeneity, and soil stoniness. Indicators for the creative forces of desertification characterizing the dynamics derived from this study are soil bulk density (closely related to soil aggregation), root content, soil structure, and forms and sizes of nested patches.

These indicators are or are closely related to spatial structures in morphology, soil properties, and vegetation at scales from ped to slope part. The results of the fine-scale hydrological measurements show that especially these structures control the hydrological behavior of the system at different scales. The scaled, spatial structures favor the incorporation of disturbances, originating from strongly varying and fluctuating environmental inputs, and influence the system's connectivity (Bergkamp, 1996). By doing so these structures highly promote the system's stability and its resilience.

4.1.4. *Implications for Desertification Assessment*

As previously discussed, scaled, spatial structures in soils, vegetation, and morphology serve the conservation of water. In the semiarid and arid areas changes in the system's ability to conserve water at different scales are crucial to desertification. Therefore, assessing land degradation should focus on changes in the structures closely related to the key processes of water conservation at different scales. Properties related to these structures differ at the various scales, as well as the environmental processes of water conservation. No single property or process can be used to identify desertification at different scales.

Processes of redistribution of water are governed by the heterogeneity of the environment. The loss of spatial structure at one scale can strengthen the system's functioning at other scales. This is so because, through the loss of structure at finer scales, the incorporation of disturbance at these scales is lost, and the connectance between system elements at larger scales is increased. To improve the assessment of desertification, the focal points in research should be the complexity of the scaled, arid, and semiarid ecosystems and the functional aspects of heterogeneity.

Although larger scales of pattern observation are useful for providing regional overviews and indications of changes in constraining conditions, a true understanding of the changes related to desertification can arise only when we focus on understanding changes in the creative dynamics at finer scales. Therefore, indicators of desertification should be based more on coupled process-structure studies and be specifically linked to spatial and temporal scales. The combined use of scaled, constraining, and dynamic indicators can provide the information needed to identify the coherence between different environmental levels of organization and levels of the socioeconomic and political hierarchies.

Acknowledgements

The research for this paper was carried out as a part of the EFEDA project. Funding was provided by the European Union contract EPOC-CT90–0039 and EV5V-CT93–0282. I kindly acknowledge Anton C. Imeson for his constructive comments throughout the research and Linda Karssies and Wouter van Helden for their help in field work and in processing field data.

References

Allen, T.F.H. and T.W. Hoekstra. 1990. The confusion between scale-defined levels and conventional levels of organization in ecology. *Journal of Vegetation Science* **1**: 5–12.

Allen, T.F.H. and T.B. Star. 1982. *Hierarchy perspectives for ecological complexity*. University of Chicago Press, Chicago.

Bergkamp, G. 1996. The hierarchical organisation of runoff and infiltration at different scales in semi-arid shrublands in central Spain. *Advances in Geo-ecology*. (Submitted).

Bergkamp, G., L.H. Cammeraat and J. Martinez Fernandez. 1995. Water movement and vegetation patterns on shrubland and an abandoned fields in two desertification threatened areas in Spain. *Earth Surface Processes and Landforms*. (In press).

Bertalanffy, L. von. 1950. The theory of open systems in physics and biology. *Science* **111**: 23–29.

Bertalanffy, L. von. 1968. *General system theory*. Braziller, New York,

Blalock, H.M. 1979. *Social statistics*. McGraw Hill, Auckland.

Bolle, H.J. *et al.* 1993. EFEDA: European field experiments in a desertification threatened area, *Annales Geophysicae* **11**: 173–189.

Culling, W.E.H. 1987. Equifinality: Modern approaches to dynamical systems and their potential for geographical thought. *Transactions. Institute of British Geographers. New Series* **12**: 57–72.

Di Castri, F. and M. Hadley. 1988. Enhancing the credibility in ecology: Interacting along and across hierarchical scales. *Geo Journal* **17**: 5–35.

Gardner, M.R. and W. Ross Ashby. 1970. Connectance of large dynamic (cybernetic) systems: Critical values for stability. *Nature* **228**: 784.

Haigh, M.J. 1987. The holon: Hierarchy theory and landscape research. *Catena Supplement* **10**: 181–192.

Holling, C.S. 1973. Resilience and stability of ecological systems. *Annual Review of Ecology and Systematics* **4**: 1–24.

Jarvis, P.G. and K.G. McNaughton. 1986. Stomatal control of transpiration: Scaling up from leaf to region. *Advances in Ecological Research* **15**: 1–49.

Jorgensen, S.E. 1994. Fundamentals of ecologica modelling. In *Developments in environmental modelling*, 2nd edition, Elsevier, Amsterdam.
Koestler, A. 1967. *The ghost in the machine*. Random House, New York.
Koestler, A. 1978. *Janus, a summing up*, Random House, New York.
Levandowsky, M. and B.S. White. 1977. Randomness, time scales, and the evolution of biological communities. *Evolutionary Biology* **10**: 69–161.
Levin, S.A. 1992. The problem of pattern and scale in ecology. *Ecology* **73**(6): 1943–1967.
Levins, R. 1973. The limits of complexity, in H.H. Pattee (Ed.), *Hierarchy theory: The challenge of complex systems*. Braziller, New York, pp. 109–127.
May, R.M. 1973. *Stability and complexity in model ecosystems*. Princeton University Press, Princeton, New Jersey.
Meentemeyer, V. 1989. Geographical perspectives of space, time and scale. *Landscape Ecology* **3**(3): 163–173.
Müller, F. 1992. Hierarchical approaches to ecosystem theory. *Ecological Modelling* **63**: 215–242.
Naveh, Z. and A. Lieberman. 1993. *Landscape Ecology*. Springer-Verlag, New York.
O'Neill, R.V. 1988. Hierarchy theory and global change, in T. Rosswall, R.G. Woodmansee, and P.G. Risser (Eds.), *Scales and global change*, John Wiley & Sons, Chichester, pp. 29–45.
O'Neill, R.V., D.L. DeAngelis, J.B. Waide, and T.F.H. Allen. 1986. *A hierarchical concept of ecosystems*. Princeton University Press, Princeton, New Jersey.
Pattee, H.H. 1973. *Hierarchy theory: The challenge of complex systems*. Braziller, New York.
Phillips, J.D. 1992. The end of equilibrium? *Geomorphology* **5**: 195–201.
Pizzuto, J.E. 1992. The morphology of graded gravel rivers: A network perspective. *Geomorphology* **5**: 457–474.
Pomeroy, L.R., E.C. Hargrove, and J.J. Alberts. 1988. The ecosystem perspective, in L.R. Pomeroy and J.J. Alberts (Eds.), *Concepts of ecosystem ecology*, Springer-Verlag, New York, pp. 1–17.
Prigogine, I. 1945. Etude thermodynamique des phénomènes irréversibles. Thesis, University of Brussels.
Prigogine, I. 1978. Time, structure, and fluctuations. *Science* **201**: 777–785.
Prigogine, I. and I. Stengers. 1985. *Order out of chaos*. Shambala, London.
Renwick, W.H. 1992. Equilibrium, disequilibrium, and nonequilibrium landforms in the landscape. *Geomorphology* **5**: 265–276.
Salthe, S.N. 1985. *Evolving hierarchical systems: Their structure and representation*. Columbia University Press, New York.
Simon, H.A. 1973. The organization of complex systems, in H.H. Pattee (Ed.), *Hierarchy theory: The challenge of complex systems*, Braziller, New York, pp. 1–28.
Strahler, P. 1952. Dynamic basis of geomorphology. *Bulletin, Geological Society of America* **63**: 923–938.
Tansley, A.G. 1935. The use and abuse of vegetational concepts and terms. *Ecology* **16**(3): 284–307.
Trofimov, A.M. and J.D. Phillips. 1992. Theoretical and methodological premises of geomorphological forecasting. *Geomorphology* **5**: 203–211.
Turner, M.G., R.V. O'Neill, R.H. Gardner, and B.T. Milne. 1989. Effects of changing spatial scale on the analysis of landscape pattern. *Landscape Ecology* **3**(3): 151–162.
Ward, J.H. 1963. Hierarchical grouping to optimize an objective function. *Journal of the American Statistical Association* **58**: 236–244.
Weiss, P.A. 1971. The basic concepts of hierarchic systems, in P.A. Weiss (Ed.), *Hierarchically organized systems in theory and practice*, Hafner Publishing Co., New York, pp. 1–43.
Wright, H.E. 1974. Landscape development, forest fires, and wilderness management. *Science* **186**: 487–495.
Zonneveld, I.S. 1989. The land unit: A fundamental concept in landscape ecology, and its applications. *Landscape Ecology* **3**(2): 67–86.

PROPERTIES AND EROSIONAL RESPONSE OF SOILS IN A DEGRADED ECOSYSTEM IN CRETE (GREECE)

C. BOIX*, A. CALVO**, A. C. IMESON*, J. M. SCHOORL*, SORIANO SOTO** and I. R. TIEMESSEN*

* *Fysisch Geografisch Bodemkundig Laboratorium, Landscape and Environmental Research Group, Universiteit van Amsterdam, Nieuwe Prinsengracht 130, 1018 VZ Amsterdam, The Netherlands*

** *Department of Geography, University of Valencia, Avenida Blasco Ibañez 28, 46010 Valencia, Spain*

Abstract. To study desertification processes relating to soil erosion, a climatological and altitudinal gradient from south to north was selected in Crete (Greece) and four locations were selected along the gradient. At the locations precipitation ranged from 1400 mm/year at the highest location to 400 mm/year at the lowest. All locations are affected by the actual land use: intensive grazing, small controlled fires, and abandoned agricultural terraces. Representative soil profiles were described in the field and analyzed in the laboratory, and rainfall simulation experiments in the field measured soil erosion over different soil surfaces and land uses. Data on physical and chemical properties were obtained from the soil profiles and soil hydrology, and erosion data were obtained from the rainfall simulation experiments. Soil aggregation was studied with samples taken from the soil in the rainfall simulation plots and special attention being paid to the aggregate size distribution and the water-stable microaggregation. The interaction between climatological conditions and land use seems to be the main factor controlling soil erosion. This paper describes how the expected erosion along the gradient (from the most humid to the driest site) can be affected and disturbed by specific processes derived from land use.

1. Introduction

Increased soil erosion as a desertification process is expected to result from predicted and alleged climatic change. This climatic change, produced by higher concentrations of CO_2, methane, and other greenhouse gasses in the atmosphere, in some scenarios will increase global temperatures and decrease precipitation (Bultot *et al.*, 1988; Lough *et al.*, 1983). One of the approaches proposed and used by some researchers (Eybergen and Imeson, 1989; Lavee *et al.*, 1991; Imeson *et al.*, 1995; Calvo *et al.*, 1994; Boix *et al.*, 1995) to study the effects of climatic change on soil erosion is to study geomorphological processes along climatological and altitudinal transects or gradients. These transects have similar conditions of lithology, soils, vegetation, slope, and geomorphology, but the amount of precipitation decreases down the gradient.

The effect of present land uses on soil erosion along these gradients, however, has not been totally evaluated. Land use is an important factor causing erosional soil responses different from those expected in response to climate alone.

High values of erosion have been found when conditions become more arid along studied gradients in Israel and Spain (Lavee *et al.*, 1991; Imeson *et al.*, in

Environmental Monitoring and Assessment **37:** 79–92, 1995.

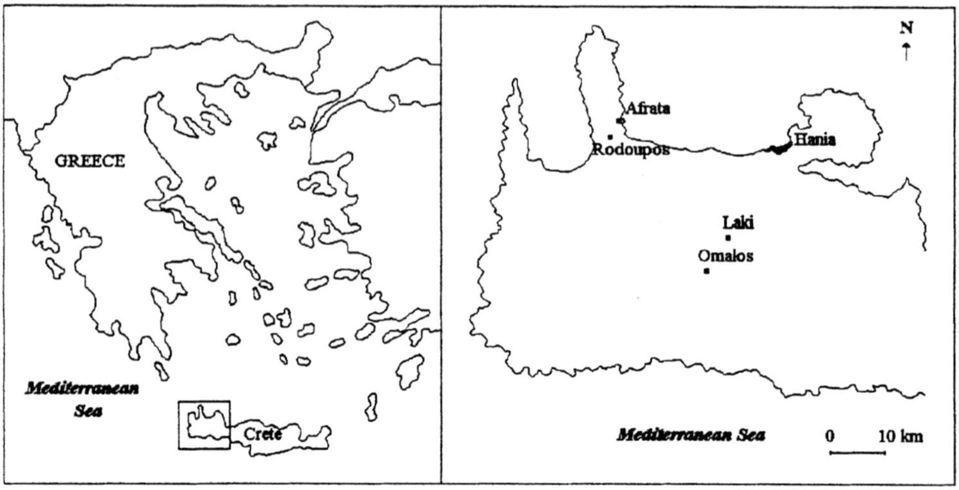

Fig. 1. Map of the study area.

press; Calvo *et al.*, 1994), but the behavior of the soil-vegetation-processes system is heavily influenced by processes occurring at each of the studied sites. And these processes clearly depend on the past and present land uses and their impact on the environment.

To study potential changes caused by climate change on slope processes, physical and chemical soil properties were measured along a climatological gradient in a mountainous zone in Crete (Greece). This research attempted to determine the impact of climate and land use on erosion processes, studying the type of soils and their hydrological response and using some indicators of soil degradation like the water-stable microaggregation (Tiemessen *et al.*, in press).

2. Study Zone

Field sites were selected in northwest Crete near the city of Hania (Figure 1). Four sites were selected at different topographical locations following an altitudinal and climatological gradient from south to north: Omalos, Laki, Rodoupos, and Afrata (see also Table I).

At every site a south-facing slope was chosen for the experiments and sampling. The parent material of all the selected sites is Upper Cretaceous limestone. Land use is an important factor affecting these slopes. The slopes are grazed by goats, and the farmers use small controlled fires to burn parts of a slope to regenerate vegetation for grazing. At the bottom of the slopes are abandoned agricultural terraces.

The density of grazing from the highest to the lowest site is as follows: Omalos seems to be the most intensely grazed site, followed by Laki, Rodoupos, and

TABLE I

Crete sites: general characteristics

Sites	Altitude meters	Aspect °	Slope °	Vegetation	Precipitation mm
Omalos	1100	200	25	*Sarcopoterium spinosum/ Quercus coccifera/ Rhamnus prunifolius/ Asphodelus aestivus/ Colchium/ Thymus capitatus*	1400
Omalos	1100	190	15	*Asphodelus aestivus/ Colchium*	1400
(burnt area)					
Laki	800	180	35	*Sarcopoterium spinosum/ Quercus coccifera/ Asphodelus aestivus/ Colchium/ Salvia triloba*	1000
Rodoupos	400	175	18	*Sarcopoterium spinosum/ Pistacia lentiscus/ Rhamnus prunifolius/ Asphodelus aestivus/ Thymus capitatus*	600
Rodoupos	400	150	30	*Asphodelus aestivus/ Colchium/ Sarcopoterium spinosum/ Thymus capitatus/ Calycotome villosa*	600
(burnt area)					
Afrata	100	165	12	*Pistacia lentiscus/ Sarcopoterium spinosum/ Colchium/ Calycotome villosa/ Lagoecia cuminoides/ Knautia integrifolia/ Albium rubrovitatum/ Ononis reclinata*	400

Afrata. Also, at Omalos, Laki, and Rodoupos abandoned agricultural terraces lie at the bottom of the valley and foot-slope. At Afrata, the lowest and most arid site, the grazing seems not to be so intensive as at the other sites, and there are no abandoned terraces or burnt patches.

3. Methodology

At every slope different units could be recognized by vegetation cover, stoniness, and land use (burnt, grazed). According to these units, five characteristics and representative soil surfaces were chosen: bare patch, vegetated patch, bare and burnt patch, vegetated and burnt patch, and terrace. Rainfall simulation experiments and soil sampling were carried out at these surfaces. To collect data on physical-hydrological properties like infiltration, runoff, sediment concentration, and soil moisture, rainfall was simulated with a portable sprinkler rainfall simulator (rain intensity of 55 mm/h during 45 minutes) in closed plots of 0.24/m^2 (Calvo *et al.*, 1988).

During the rainfall simulation experiments researchers directly observed the time to ponding (time in which water is organized in little ponds around the plot but is not yet moving) and time to start surface runoff (the water on the soil surface is moving to the lower part of the plot). Researchers also measured the runoff output every 30 seconds or every minute and sampled runoff every 15 minutes. From these data the runoff coefficient was calculated as a relation between the rainfall input and the runoff output.

Also measured were such parameters as sediment concentration, sediment yield, and erosion. Before and after each simulation a soil moisture sample was taken from the upper 3 cm of the soil surface. At the end of the experiment the plot was removed, and the soil profile was opened to describe and examine the wetting front.

Soil samples of the surface layer (0–5 cm) were taken at the representative surfaces mentioned above. From these samples aggregate size distributions were determined by mechanical sieving in the fractions 10–5, 5–2, 2–1, and 1–0 mm. In addition, the amount of water-stable microaggregates and clay and silt proportions for the different soil surfaces were determined in the fraction < 0.105 mm. The water-stable microaggregation was determined with a Quantachrome Microscan particle analyzer (Edwards and Bremner, 1964; Imeson and Verstraten, 1989).

Seven soil profiles were studied at the slopes. Chemical and textural analyses were carried out to determine the main characteristics of the soils. Soils were described in the field (FAO, 1977; FAO-Unesco, 1988) and analyzed in the laboratory for the more significant parameters to relate soil properties with operating processes.

Laboratory analyses were performed for most of these samples with the following determinations: texture (Ministerio de Agricultura, 1986), pH, salinity (Richards, 1954), calcium carbonate content (Jackson, 1958), and organic matter (Walkley and Black, 1934).

4. Results

4.1. Soil Study: Physical, Chemical, and Morphological Characteristics

All the soils are developed on limestone material, and the soil profiles are considered representative for the study area. At some sites abandoned terraces are situated

TABLE II

Soil profile descriptions at the different sites

Profile	Hor	Depth	Color dry	Texture	Structure	$CaCO_3$ content	Pores	Stone fragments	Roots	Class soil
OM	A1	0–2	7,5YR4/4	Loam Clay	Mid. Subangular	Non Calcareous	Frequent Fine	Abundant Subangular	Frequent	RGe
	A2	2–15	10YR5/4	Loam Clay	Subangular	Non Calcareous	Frequent Fine	Abundant Subangular	Frequent	
	C	15–45	7,5YR5/6	Loam Clay	Subangular	Non Calcareous	Frequent Mid	Abundant Subangular	Few	
	2C	+45	7,5YR5/6	Loam Clay	Subangular	Non Calcareous	Few Fine	Abundant Subangular	Few	
LA	A1/AE	0–2	10YR5/4	Loam Clay Silt	Subangular	Non Calcareous	Few Fine	Abundant Subangular	Very Fine Few	RGe
	A2	2–20	10YR5/4	Loam Clay Silt	Subangular	Non Calcareous	Frequent Fine Mid	Abundant Subangular	Frequent Fine Mid	
	B/C	+20	7,5YR4/4	Loam Clay Silt	Subangular	Non Calcareous	Few Fine	Frequent Subangular	Few Very Fine	
AF	AE	1.5–2	5YR5/6	Clay Silt	Subangular	Non Calcareous	Few Fine Few Fine	Abundant Subangular	Common Very Fine	
	AB	2–5	2,5YR3/6	Loam Clay Silt	Subangular	Non Calcareous	Few Fine Few Fine	Abundant Subangular	Few Fine	
	Bt1	5–12	2,5YR4/6	Clay	Subangular	Non Calcareous	Very Few Fine	Abundant Subangular	Very Few Fine	LVx
	Bt2	+12	2,5YR4/6	Clay Silt	Subangular	Non Calcareous	Very Few Fine	Abundant Subangular	Very Few Fine	

Table II. Continued

Profile	Hor	Depth	Color dry	Texture	Structure	$CaCO_3$ content	Pores	Stone fragments	Roots	Class soil
RO1	A1	0–5	5YR5/6	Clay Silt	Subangular	Non Calcareous	Very Few Fine	Abundant Subangular	Common Fine Mid	LVx
	AB	5–10	5YR4/6	Clay Silt	Subangular	Non Calcareous	Few Very Fine	Abundant Subangular	Few Fine	
	Bt1	10–30	2,5YR4/4	Clay	Subangular	Non Calcareous	Few Very Fine	Abundant Subangular	Few Fine	
	B/C	+30	2,5YR4/6	Clay Silt	Subangular	Non Calcareous	Few Fine	Abundant Subangular	Very Few Fine	
RO2	AB	0–2	5YR4/4	Clay Silt	Subangular	Non Calcareous	Abundant Fine	Abundant Subangular	Very Few Fine	LVx
	Bt2	5–10	5YR4/6	Loam Clay	Subangular	Non Calcareous	Few Fine	Few Subangular	Very Few Fine	
RO3	A1	0–2	7,5YR5/4	Clay Silt	Granular Mid.	Non Calcareous	Abundant Fine	Abundant Subangular	Common Mid	LVx
	A2	2–20	5YR5/4	Loam Clay	Subangular	Non Calcareous	Few Fine Few Fine	Abundant Subangular	Few Very Fine	
	Bt	+20	5YR5/6	Loam Clay Silt	Subangular	Non Calcareous	Very Few Fine	Abundant Subangular	Few Very Fine	
RO4	A1	0.4–2	5YR5/6	Clay Silt	Subangular	Non Calcareous	Abundant Fine	Abundant Subangular	Common Fine	LVx
	Bt	2–15	5YR5/4	Clay	Subangular	Non Calcareous	Few Fine	Abundant Subangular	Few Very Fine	

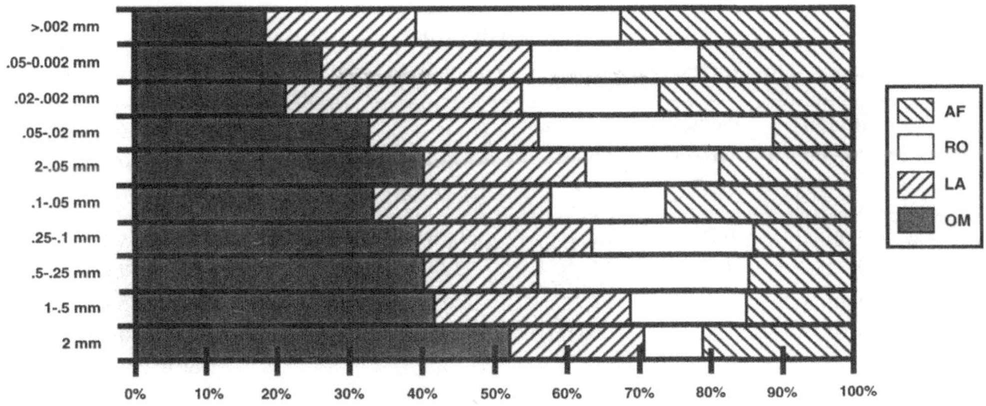

Fig. 2. Average values of soil texture for the four studied slopes (OM:Omalos, LA: Laki, RO: Rodoupos, AF: Afrata).

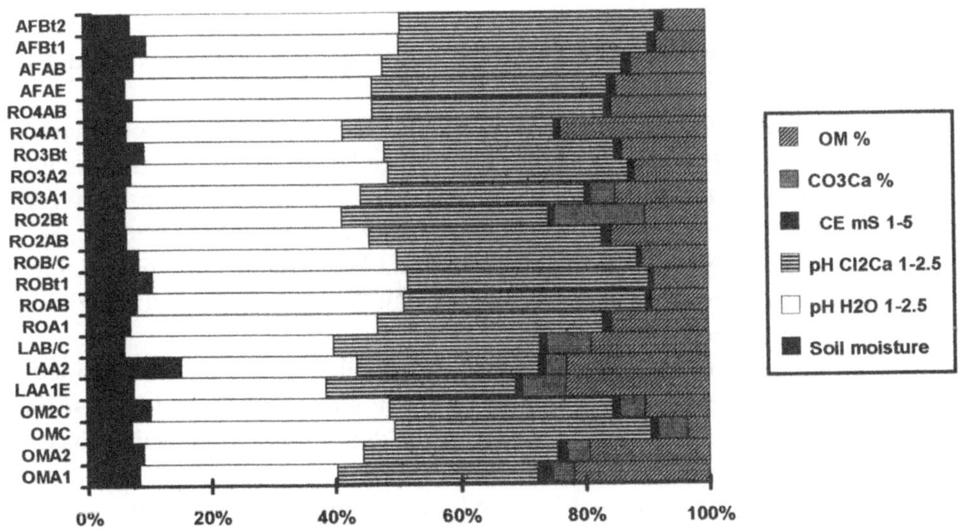

Fig. 3. Values of chemical properties and soil moisture content for the different soil horizons (OM: Omalos, LA: Laki, RO: Rodoupos, AF: Afrata).

downslope, and at the upper part of the slope are rock outcrops, resulting in many superficial rock fragments and more stoniness. The seven profiles and all horizons are described in Table II.

The surface layers of all the soils have a granular structure, varying from moderate to fine, changing into subangular blocks in more profound layers. The block size decreases with depth. The most developed structures can be found in the

TABLE III
Average values for the four sites (OM: Omalos, LA: Laki, RO: Rodoupos, AF: Afrata)

	Runoff coefficient	Sediment concentr. (gr/l)	Sediment yield (gr)	Erosion (gr/m^2/h)	Bulk Density (gr/cm^3)	Porosity (%)	Soil moisture content (%)
OM	0.66	9.26	61.41	356.19	0.94	64.45	15.43
LA	0.68	1.95	15.74	97.59	0.90	65.70	17.69
RO	0.69	1.72	12.53	70.99	0.95	64.35	22.41
AF	0.37	1.40	6.00	33.40	0.91	65.80	10.04

soils of Rodoupos due to a high clay content. In general, clay but mainly silt make up the dominant fractions (Figure 2).

Soil texture at Laki and Omalos are comparable. At Afrata and Rodoupos the texture is more clayey in layers where clay illuviation has occurred. The average values of the clay content vary respectively from 25.64% and 25.26% at Laki and Omalos to 34.33% and 33.00% at Afrata and Rodoupos. Maximum values are obtained in the Bt-horizon of the Afrata soil profile (63.74%).

The values of some soil chemical properties are expressed proportionally in Figure 3. The organic matter content of the surface layers of the soils is quite high, corresponding to forest soils, especially for Laki and Omalos, with an average organic matter content of 5.16% at Laki and 3.04% at Omalos (Figure 3).

In the whole study area the carbonate content is insignificant although a light recarbonization can be detected in the superficial layers, possibly due to a carbonate supply by overland flow.

The mean maximum values of electrical conductivity (EC) are obtained at Omalos with an average of 0.32 mS and a maximum of 0.57 mS in the surface layer. Afrata and Laki have similar values. At Rodoupos the lowest values have an average of 0.21 mS.

The mean pH varies between 7.32 and 7.84, with the higher values at Afrata and Rodoupos. The maximum value has been found in the A2-horizon of the soil profile 3 (RO3) in Rodoupos.

According to the FAO (1988) the soils can be classified as chromic Luvisols at Afrata and Rodoupos and eutric Regosols at Laki and Omalos.

4.2. Hydrological Response and Soil Aggregation

Different values for runoff, sediment, and erosion were found among the sites. At slope scale the highest value of erosion was found at Omalos, the highest and most humid site. The lowest erosion value was found at Afrata, the least humid site. Average values for each slope are given in Table III. Omalos shows the highest

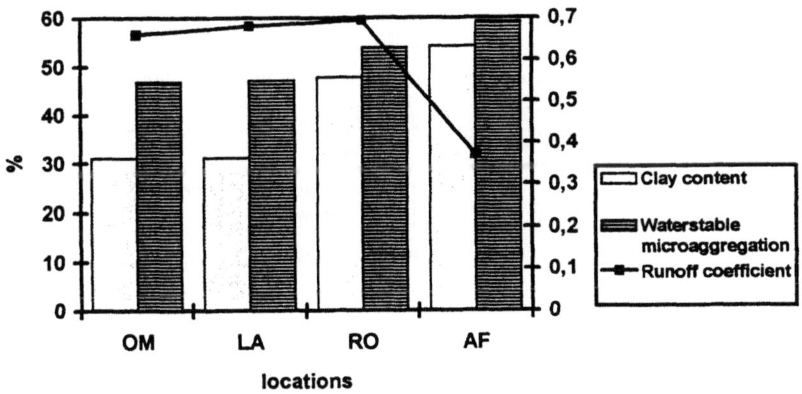

Fig. 4. Averages of clay, water-stable microaggregation and runoff coefficients at the four slopes (OM: Omalos, LA: Laki, RO: Rodoupos, AF: Afrata).

erosion rate as well as the highest sediment concentration and sediment yield and a very high bulk density. In the same way, Omalos shows the lowest water-stable microaggregation and the lowest clay content very close to Rodoupos (Figure 4). Although Omalos shows the highest erosion rate, the highest runoff coefficient is found at Rodoupos, an intermediate site. During the field work the soil at Rodoupos was very wet, showing the highest soil moisture content and the highest bulk density. Afrata, the lowest and most arid site, presents the lowest values of erosion and runoff coefficient. At Afrata the clay content is the highest as well as the water-stable microaggregation and the average diameter of the microaggregates. During the field experiments Afrata also showed the lowest soil moisture content, so the time to ponding and producing runoff was longer than at the other sites. The values of erosion, sediment concentration, and sediment yield decrease with altitude in the gradient. Omalos showed the highest erosion, followed by Laki, Rodoupos, and Afrata.

Besides variability between the average values of Omalos, Laki, Rodoupos, and Afrata, the values also vary among locations on one slope, depending on conditions (bare, vegetated, burnt patches, terraces).

Another approach to analyzing the data is to study the values for each specific condition on the slope. All experiments were classed into the following five groups according to soil surface and land use characteristics: bare patch, vegetated patch, bare and burnt patch, vegetated and burnt patch, and abandoned terrace (Table IV). Only at the intermediate sites (Laki and Rodoupos) were experiments conducted and samples taken for all five groups. Afrata, the lowest site, lacked surfaces of abandoned terraces and burnt patches. At Omalos, the highest location, patches had been so recently burned that no revegetated burnt patches were found. When more than one experiment was performed in surfaces with the same characteristics at the same slope, the average values were calculated to classify all the experiments in the five groups explained above and in the methodology (Table IV).

TABLE IV

Soil moisture condition before the experiments and rainfall simulation results under different conditions on the slopes. (OM. Omalos, LA: Laki, RO: Rodoupos, AF: Afrata; b: bare patch, v: vegetated patch, bb: bare and burnt patch, vb: vegetated and burnt patch, t: abandoned terrace)

Plots	Soil moisture (%)	Wetting front (cm)	Time to ponding (minutes)	Time to runoff surface (minutes)	Runoff coefficient	Sediment concen. (gr/l)	Sediment yield (gr)	Erosion (gr/m^2/h)
OM b	4.49	10	2	2.35	0.76	22.98	155.75	945.55
OM v	17.59	14.75	2.73	3.2	0.4	6.07	36.92	203.66
OM bb	20.66	19.5	0.37	0.45	0.79	6.05	39.94	203.96
OM t	18.96	10	1.22	2.33	0.69	1.94	13.04	71.57
LA b	19.67	16.5	2.23	3.86	0.64	5.09	28.91	174.61
LA v	24.86	13	3.22	1.91	0.52	0.43	1.22	7.5
LA bb	21.47	19	1.56	2.1	0.8	2.04	15.09	87.37
LA vb	11.35	12	0.57	0.57	0.78	1.13	8.04	46.85
LA t	11.08	19.5	1.56	2.1	0.66	1.08	25.46	171.6
RO b	32.75	20	1.69	1.3	0.85	3.21	26.11	146.24
RO v	27.43	19	0.52	1.04	0.58	1.54	9.21	51.86
RO bb	11.73	4	2.05	4.5	0.77	0.67	5.1	27.11
RO vb	21.2	4	2.33	3.22	0.6	2.37	14.11	76
RO t	18.92	10.25	1.53	1.65	0.67	0.81	8.11	53.73
AF b	9.84	10.25	2.3	4.01	0.52	1.75	9.42	52.16
AF v	10.25	9.5	4.73	5.87	0.22	1.04	2.58	14.64

The runoff coefficients are quite high for all the sites but are highest for the bare and burnt plots. The bare plots also have the highest erosion rates and the highest values of bulk density for all the sites. Abandoned terraces also present high values of runoff and erosion. But considering that the steepness at the terraces is very low, these values are always lower than in bare patches on the slope and higher than in vegetated patches.

At the higher sites, Omalos and Laki, the burnt patches and the abandoned terraces quickly reached the time to ponding, and the surface runoff started much sooner than in the nonburnt areas. At Rodoupos the time to ponding was reached quicker by the nonburnt than the burnt patches. The soil moisture content during the experiments, however, was lower in the burnt area than in the nonburnt area, increasing the infiltration capacity of the burnt soils. In Afrata, the lowest site, the ponding and the surface runoff started sooner in bare patches than in vegetated patches.

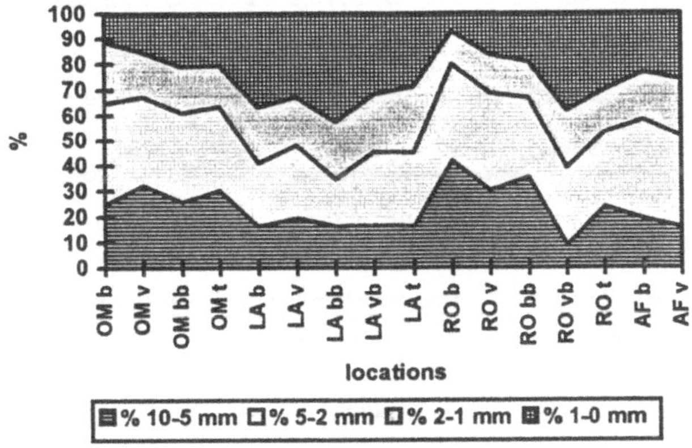

Fig. 5. Aggregate size distribution on the slopes under specific conditions. (OM. Omalos, LA: Laki, RO: Rodoupos, AF: Afrata; b: bare patch, v: vegetated patch, bb: bare and burnt patch, vb: vegetated and burnt patch, t: abandoned terrace).

With respect to soil aggregation, the lowest and more arid site (Afrata) presents the highest water-stable microaggregation (fraction < 0.105 mm) and the highest clay content in the soil. In general, the fraction 5–2 mm seems to be related to the fraction 1–0 mm in the aggregate size distribution, so an increase or decrease in the fraction 5–2 mm decreases or increases the fraction 1–0 mm (Figure 5). This fact is especially clear for Laki, where a very high value of aggregates in the fraction 1–0 mm is found together with a low value in the fraction 5–2 mm. An explanation for this occurrence lies possibly in the cohesion forces of the 5–2 mm aggregates, which seem to be weaker with a lower clay content. When these aggregates break down, they are incorporated into the 1–0 mm fraction.

Some authors (Edwards and Bremner, 1967; Imeson and Verstraten, 1989) have demonstrated a positive influence of clay and organic matter content on the amount of water-stable microaggregates. In this case the organic matter was found not to influence the water-stable microaggregates, but a positive relation was found between clay content and water-stable microaggregation. An inverse relation seems to exist between water-stable microaggregation and runoff production. In this way, Afrata, the most arid site, with a high water-stable microaggregation and a high clay content, shows the lowest runoff coefficient.

Under specific soil surface conditions the soil aggregation is showing some differences. The burnt patches, especially the burnt and bare patches, have the lowest values of water-stable microaggregation, whereas vegetated patches and terraces have higher values. Vegetated patches with higher values of water-stable microaggregation have lower runoff coefficients and lower erosion rates than do bare patches.

Figure 6 shows the clay content, the total amount of water-stable microaggregates, and the average diameter of the microaggregates under specific conditions on the studied slopes. These values seem to follow a parallel trend. Vegetated areas

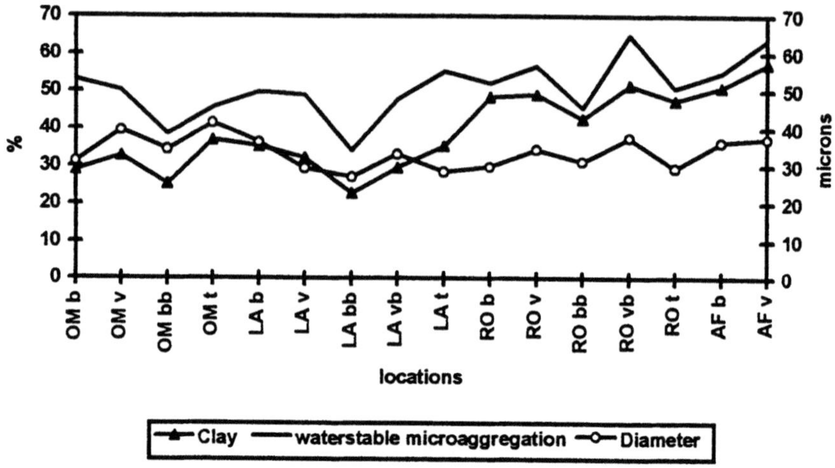

Fig. 6. Clay content, water-stable microaggregation, and diameter of the microaggregates under specific conditions on the slopes. (OM. Omalos, LA: Laki, RO: Rodoupos, AF: Afrata; b: bare patch, v: vegetated patch, bb: bare and burnt patch, vb: vegetated and burnt patch, t: abandoned terrace).

show higher water-stable microaggregation, higher clay content, and higher diameter of microaggregates. For Omalos the abandoned terrace also has high values in these parameters. The combination of a burnt area with a new young vegetation (Rodoupos) seems to guarantee a good soil microaggregation.

5. Discussion

In examining the results of this study, one must consider the limitation presented by the scale of the field experiments. Nevertheless, the results are valid in establishing comparisons between the soil surfaces and slopes along the selected climatological and altitudinal gradient. The erosion rates along this climatological gradient in Crete are the greatest at the highest and most humid location and the least at the lowest and driest location.

The most important influence on soil erosion is land use. Omalos, the highest site, is suffering intensive grazing by goats, which are attracted by the higher biomass production resulting from good soil moisture conditions. As a result of grazing, the soil is trampled and shows a high value of bulk density. Aggregates in large fractions are incorporated in the soil surface, being easily carried by runoff and producing very high values of sediment concentration.

Afrata, the lowest site, was intensively grazed in the past. Because of more arid conditions at present, however, the biomass production is much lower than at the other sites, and grazing is less intensive. In Afrata we found no signs of the small fires lit by farmers to regenerate vegetation.

The water-stable microaggregation has been taken as an indicator of soil degradation in this study. A correlation between clay content and water-stable microaggregation shows that the sites with higher values of clay content also have higher values of water-stable microaggregation. At slope scale higher erosion rates have been found at the sites with lower clay content in the soil and lower water-stable microaggregation. In this way the less grazed slope in Afrata, with less soil moisture content and the highest values of clay and water-stable microaggregation, has the lowest runoff coefficient and the lowest erosion rate (Table III).

At plot scale, bare patches and bare burnt patches show higher runoff coefficients, higher erosion rates, and lower stable microaggregation than vegetated patches. Vegetated patches always showed the lowest runoff coefficients and the best water-stable microaggregation among the soil conditions studied at each slope. The abandoned terraces at the medium sites (Laki and Rodoupos) have erosion rates higher than the vegetated patches, but these erosion rates are lower than expected, probably because of quite high clay contents and a very low steepness.

The chromic Luvisols found at Rodoupos and Afrata, the most arid but less intensively grazed sites with high clay content and high values of water-stable microaggregation, produce lower erosion rates and show quite a high infiltration capacity. The eutric Regosols at the more humid Omalos and Laki sites, with a low clay content and low water-stable microaggregation, are much more erodible, presenting a low infiltration capacity, especially in trampled and burnt soil patches.

Apart from these results, the soil moisture content at the moment of the rain has an important influence on values of infiltration and runoff coefficients. So despite its low erosion values and sediment concentration, Rodoupos presents very high runoff coefficients due to the high soil moisture content at the moment of the rainfall experiments in the field (Table III).

6. Conclusion

Climatic conditions and land use practices are interacting in Crete to favor soil erosion in humid ecosystems, where such high values of soil erosion were not expected. Land use in Crete is an important cause of soil degradation. Favorable climatic conditions for biomass production cause indirectly more soil degradation due to the damage of grazing. The overgrazing in Crete results in high erosion rates in humid ecosystems with potentially unfavorable conditions for soil erosion. The case of Crete is demonstrating how land management practices are directly responsible for soil erosion processes. Controlling grazing in the humid ecosystems of the Cretan mountains would avoid the high soil losses found by our study.

Acknowledgements

This work was financially supported by the Commission of the European Communities in the Climatology and Natural Hazards Program, EV5V-CT91-0023 ERMES project. The first author thanks the European Community for the concession of a Human Capital and Mobility Fellowship (EV5V-CT-94–5228), which allowed her to continue her research.

References

Bultot, F., A. Coppens, G.L. Dupriez, D. Gellen, and F. Meulenberghs. 1988. Repercussions of a CO_2 doubling on the water cycle and on the water balance – A case study for Belgium. *Journal of Hydrology* **99**: 319–347.

Boix, C., A. Calvo, A.C. Imeson and M.D. Soriano. 1995. Relationships between physical soil properties and runoff in a mountain zone with a climatological gradient in Spain. In *Desertification in a European context: Physical and socioeconomic aspects*, Commission of the European Communities, Brussels, pp. 399–409.

Calvo, A., J.B. Gisbert, E. Palau and M. Romero. 1988. Un simulador de lluvia portatil de fácil construcción, in M. Sala and F. Gallart (Ed.), *Métodos y Técnicas para la edición en el campo de procesos geomorfológicos*, Monografía no. 1, Sociedad Española de Geomorfología, Barcelona, pp. 6–16.

Calvo, A., M.D. Soriano Soto, C. Boix Fayos and I. Tiemessen. 1994. Suelos y procesos geomórficos en ungradiente climático altitudinal (Alicante). In J. Arnaez Vadillo, J.M. Garcia Ruiz and A. Gomez Villar (Eds.), *Geomorfologia en España*, tomo II, Sociedad Española de Geomorfologia, Logroño, pp. 29–40.

Edwards, A.P. and J.M. Bremner. 1964. Use of sonic vibration for separation of soil particles. *Canadian Journal of Soil Science* **44**: 366.

Edwards, A.P. and J.M. Bremner. 1967. Microaggregates in soils. *Journal of Soil Science* **18**: 65–73.

Eybergen, A. and A.C. Imeson. 1989. Geomorphological processes and climatic change. *Catena* **16**: 306–319.

Food and Agriculture Organization of the United Nations (FAO). 1977. *Guía para la descripción de perfiles de suelos*, Rome.

Food and Agriculture Organization of the United Nations (FAO)-Unesco. 1988. *Soil map of the world.* 1:5,000,000, I, revised legend, Rome.

Imeson, A.C., H. Lavee, H., A. Calvo and A. Cerdá. In press. The erosional response of calcareous soils along a climatological gradient in Alicante. *Geomorphology.*

Imeson, A.C. and J.M. Verstraten. 1989. The microaggregation and erodibility of some semi-arid and Mediterranean soils. *Catena Supplement* **14**: 11–24.

Jackson, M.L. 1958. *Soil chemical analysis*, Prentice Hall, Englewood Cliffs, New Jersey and London.

Lavee, H., A.C. Imeson, S. Pariente and Y. Benyamini. 1991. The response of soils to simulated rainfall along a climatological gradient in an arid and semi-arid region. *Catena* **19**: 19–37.

Lough, J.M., T.M.L. Wigley and J.P. Palutikof. 1983. Climate and climate impact scenarios for Europe in a warmer world. *Journal of Climate and Applied Meteorology* **22**: 1673–1684.

Ministerio de Agricultura. 1986. *Métodos oficiales de análisis de suelos*, tomo III, Madrid, pp. 178–188.

Richards, L.A. (editor). 1954. *Diagnosis and improvement of saline and alkali soils.* Agriculture Handbook 60. U.S. Department of Agriculture, Agriculture Research Service, Government Printing Office, Washington.

Tiemessen, I.R., C. Boix Fayos, A.C. Imeson, M.D Soriano Soto and A. Calvo Cases. In press. Monthly variations in microaggregation in relation to climate and aspect at an altitudinal transect, in A. Cerdá and D.J. Oostwould Widjeness (Eds.), *Dynamical geomorphology of slopes and riverbeds in alpine catchments: The dilemma of climate change*, Amsterdam.

Walkley, A. and I.A. Black. 1934. An examination of the Degtjarett method for determining soil organic matter and proposed modification of the chromic acid titration method. *Soil Science* **37**: 29–38.

IDENTIFICATION AND OBSERVATION OF DESERTIFICATION PROCESSES WITH THE AID OF MEASUREMENTS FROM SPACE: RESULTS FROM THE EUROPEAN FIELD EXPERIMENT IN DESERTIFICATION-THREATENED AREAS (EFEDA)

HANS-JÜRGEN BOLLE
Freie Universität Berlin, Institut für Meteorologie, Carl-Heinrich-Becker-Weg 6–10, D-12165 Berlin; present address: Stücklenstrasse 18c, D-81247 München

Abstract. The ECHIVAL* Field Experiment in Desertification-threatened Areas (EFEDA) addresses the question of desertification from the viewpoint of changing interactions between the land surface and the atmosphere under varying climatic conditions. The basic tool to improve our understanding of these processes are Soil-Vegetation-Atmosphere-Transfer (SVAT) and climate models. In testing techniques for deriving the needed input data from observations from space, EFEDA requires high-precision data sets that can be used to aggregate desertification-related land-surface characteristics into the scale up to the grid width of global climate models (10^4–10^5 km^2). In this context schemes have been developed to infer from satellite measurements fluxes at the surface. To validate the information inferred from observations in space, ground measurements have been performed on 2500 km^2 of Castilla-La Mancha, Spain, during the drying periods of the summers of 1991 and 1994. Ultimately EFEDA aims to determine cumulative fluxes over longer periods to allow discrimination between natural variability and trends over large areas, such as the land around the Mediterranean. To recalibrate and adjust the algorithms used to infer information about the land surfaces from satellite measurements, "anchor stations" are proposed for critical areas to provide collateral information and continuous quality control of inferred information.

1. Introduction

Desertification is understood to result from processes that, because of climate change or human activities, reduce the land's ability to support biomass production. Starting from points and proceeding in patches, desertification develops in phases and accelerates during droughts. Soil and vegetation, however, may regenerate during wetter periods. As an essentially nonlinear process and not as a final state, desertification cannot be assessed by occasional inspection of the landscape and spotty measurements such as the progress of dune front movement over the years. Highly accurate, long-term observations covering large areas are needed to decide if a desertification process is intensifying, spreading, covering more and more ground, or moving with precipitation patterns, covering roughly a constant fraction of the vulnerable landscape.

The process of desertification cannot be understood without considering the interaction among soil, vegetation, and the atmosphere. This interaction in turn

* ECHIVAL = European International Project on Climatic and Hydrological Interactions between Vegetation, Atmosphere and Land surfaces.

Environmental Monitoring and Assessment **37:** 93–101, 1995.

cannot be described without models, which continuously need updated input data at intervals varying from weeks to seasons. The ECHIVAL Field Experiment in Desertification-threatened Areas (EFEDA) established under the Environmental Programme of the Commission of the European Communities (DG XII) directs research into regional processes, specifically those related to the hydrological cycle but seen in the context of larger scale changes of weather patterns (climate) and with socioeconomic boundary conditions (Bolle *et al.*, 1993). Its group on REmote Sensing and RAdiometric Properties of the Surface (RESRAPS) is responsible for using observations from space to assess land surface characteristics needed for conducting diagnostic studies and for operating these models.

This can be considered a three-prong approach:

- (1) A *sequential* or *static approach* that requires inspecting the state of the land surface in sequential time steps (months, years, decades) with the aim of empirically describing changes of land surface characteristics.
- (2) A *process-oriented dynamic approach* that studies desertification as a process enabled and initiated by climatic conditions and initiated or accelerated by human and animal activities.
- (3) A *budget approach* that considers the change of mass budgets (water, soil, minerals, biomass) over specific time intervals as a measure of desertification.

2. Applying Satellite Data in the Sequential Static Research Approach to Desertification Research

To detect changes (e.g. long-lasting conversion of land surfaces after fires or gradual changes by erosion or overgrazing), simple methods have been proven to be adequate. Such measures include intelligent false-color presentation, classification, reflectance analysis, and vegetation index data series. More sophisticated methods may be applied in small selected areas if one wants to measure the changing composition of land cover. The detailed component analysis requires knowledge of the "pure" signals of the single components (soil, vegetation types, shadows). These "pure" signals are generally not known and often can be determined only by field measurements and change in time.

Vegetation Index data series are often proposed as a basis for recognizing changes. But such data series based upon one quantity alone may lead to misinterpretations. By computing the NDVI, as an example, one index is formed out of two independent information sets. Thus one information set, in this case reflectance, gets lost. This loss of information makes discriminating between certain soils and sparse vegetation difficult if not impossible. Furthermore, the NDVI, though extremely useful in other respects as we shall see, is not yet a quantity to be used in models. A "Change Indicating Data Base" must consequently contain more and, as far as possible, all available information.

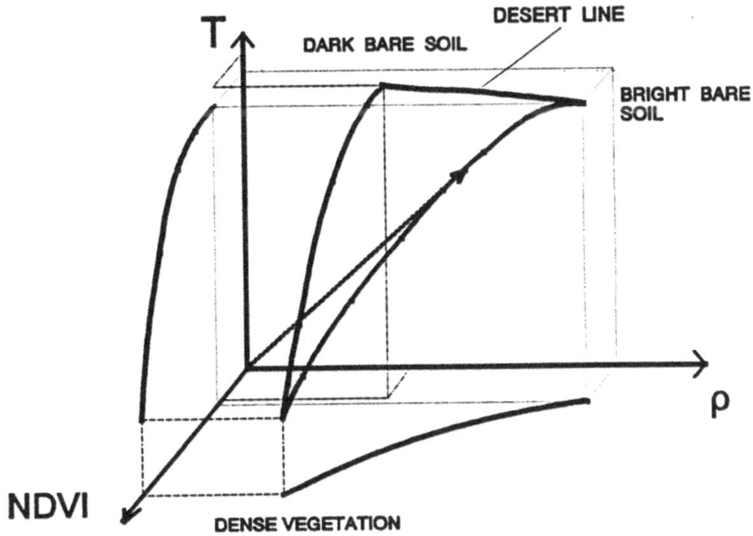

Fig. 1. Three-dimensional example of a "land surface state vector".

The goal therefore is to develop a multi-component vector scheme that can characterize the state of the surface and can be expanded as new measuring techniques are developed. In such a scheme the state of the surface will be characterized by a vector in a multi-dimensional space with time-dependent components that can be derived from periodical satellite observations:

$$\{\rho_{\mathrm{VIS}}(t_i), \rho_{\mathrm{NIR}}(t_i), ..., \mathrm{NDVI}(t_i), ..., T_s(t_i), ...\} .$$

Here ρ_{VIS} and ρ_{NIR} are the reflectances in the visible and the near infrared, respectively, normalized to a specific observation angle θ and a specific solar aspect angle $\zeta_{\odot}$, which in turn are related to daytime t_i. T_s is the surface temperature inferred from measurements in the thermal infrared. Other spectral components can be added as they become available. One of the components to be added in due time is the microwave brightness temperature $T_{MW}(t_j)$. This component may refer to a different time, $t_j = t_i + \Delta_j t$, if the sensor is mounted on a separate satellite.

Land-surface changes in a time series of this vector will show up by characteristic variations of trends (or jumps as in the case of fire). Figure 1 shows a three-dimensional representation of this vector, using only the components ρ_{VIS}, NDVI, and T_s. In the case of desertification the vector will move from high vegetation indices, low reflectance values in the visible, and low temperatures towards low vegetation indices, eventually higher reflectance in the visible (depending on the soil type), and high temperatures. This trend may not be noticeable in a small assemblage of data because a seasonal variability is superimposed. The process is not straight forward.

Because of geographical differences the "vector," which describes the land-surface changes, has to be related to the state of surfaces by observations in test areas (anchor stations) and field experiments.

3. Applying Satellite Data in the Process-Oriented Dynamic Approach

The interaction of the land surface and the atmosphere is determined by the exchange of energy, momentum, and matter, specifically water vapor. Thus the surface temperature measured from space cannot simply be interpreted as a measure e.g. of climate if the processes that determine surface temperature are not fully considered. In conjunction with SVAT models, changes in these transfers give insight into processes that occur at the surface and finally determine the state of the surface. To study the desertification process with the aid of measurements made from space, one must follow with high precision changes of land-surface properties that are governing this interaction. Needed are multiannual measurements that among other things allow the study of variability and phase shifts of vegetation periods. The main question is to what degree and with what accuracy these quantities can be inferred from remotely sensed radiative signals.

To answer these questions, complete multidisciplinary data sets have been compiled in EFEDA test areas on the ground. The experimental setup allowed the comparison of land-surface quantities extracted from measurements made in space with field measurements. The setup also allowed the aggregation of ground data over areas that are compatible with lower resolution and aggregated satellite data. Thus in addition, the results of scaling satellite data could be tested with ground data.

The most critical quantity is surface evaporation and vegetation cover because they determine the availability of water for biomass production. To estimate evaporation with the aid of remote sensing measurements, different approaches have been taken, but generally one must compute the complete energy budget of the surface.

To determine the components of the energy budget (compare Figure 2), one has to start with the net radiative flux density M_R^* (W m^{-2}) at the surface with its four components: downwelling solar radiation flux $M_{SW}^{\downarrow}$, upwelling shortwave radiation flux $M_{SW}^{\uparrow} = \rho M_{SW}^{\downarrow}$, with ρ = surface albedo, downwelling longwave radiation $M_{LW}^{\downarrow}$, and upwelling longwave radiation $M_{LW}^{\uparrow} = \sigma T_{eff}{}^4 = \sigma \varepsilon T_0{}^4$, where σ is the Stefan-Boltzmann constant, ε the emissivity, T_{eff} the radiometric, and T_0 the thermodynamic surface temperature. Only in spectral regions without absorption could the net solar flux be directly determined from measurements at the top of the atmosphere because net radiation fluxes at the top and bottom of the atmosphere must equal each other in this case. But measuring the outgoing TOA flux is very

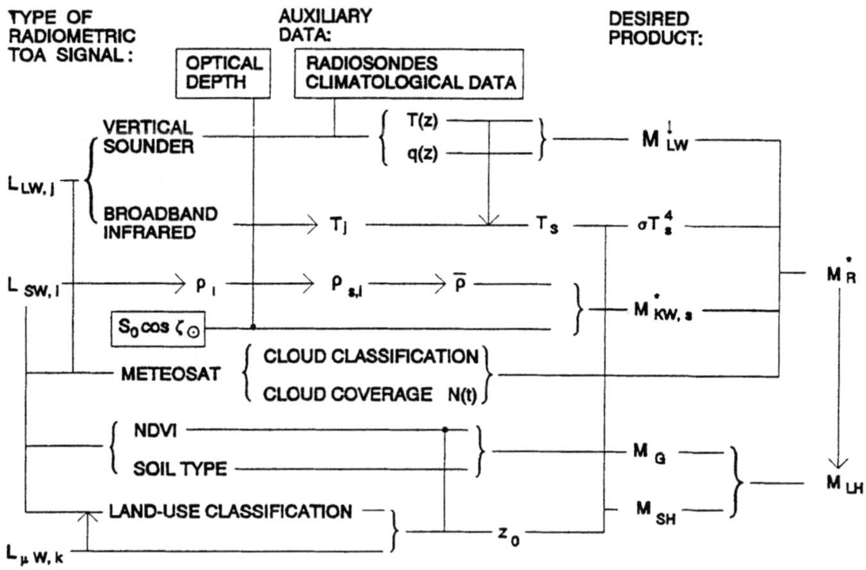

Fig. 2. Schematic diagram of satellite and collateral data flow in determining the surface energy budget.

difficult with directional observations because of the anisotropic angular behavior of the TOA flux. Only specially designed instruments like the ERBE (Earth Radiation Budget Experiment) mission can determine time averages of this quantity with sufficient accuracy. It seems therefore more attractive to infer the surface albedo, which in itself is an indicator for changes.

Though in principle the components of the net radiant flux may all be computed from remote sensing data for the overpass time of a meteorological satellite, in practice the deduction of these quantities requires solving problems arising from insufficient absolute calibration of the spaceborne instruments, the degradation of their detectivities, the atmosphere's interference with the signal, and the dependence of the signal on the measuring geometry.

Several algorithms are known to compute instantaneous net radiation fluxes at the surface from radiances measured in space. To obtain diurnal or even monthly net fluxes from these data, we need a combination with more frequent information that we can gain only from geostationary satellites. This data gathering involves a scaling problem because of the different spatial resolution. If the net radiation flux is known at a certain time for a cloudless sky and maybe even for two times – day and night – then its diurnal variation may be estimated from the known behavior of the fluxes. But then we have to consider the modification due to cloudiness. Diurnal trends of cloudiness and cloud types can be estimated only by evaluating data from geostationary satellites, which observe the scene every half hour. But these measurements have a resolution of only about 8 km, which is quite a different scale from the scanning radiometer data.

The heat flux into the ground can be estimated from the diurnal surface temperature amplitude if the soil's thermal properties are known from collateral data. Integrated over 24 hours, the flux is generally small but may not be neglected in early summer, when the soil is warming, and in autumn, when the flux's direction is reversed and the soil is cooling. Integrated over 1 year, this flux must be 0.

For computing the sensible heat flux into the atmosphere, we basically must know the temperature difference between the surface and the air at a reference level as well as the surface resistance. Since under no circumstance can the vertical temperature difference be measured by remote sensing methods, the solution of this problem requires parameterization of this quantity in which net radiation flux and vegetation cover play major roles.

The latent heat flux is determined in the scheme of Figure 2 as residuum of the energy budget equation. But latent heat flux can also be independently treated, requiring a more sophisticated approach.

4. Satellite Data and the Budget Approach

Desertification can alternatively be described as an input-output system representing the soil layer. The inputs consist of radiative energy, precipitation, carbon dioxide, and nitrogen from the atmosphere as well as water pumped from deep reservoirs. The outputs consist of evaporation of water into the atmosphere and water runoff through streams and aquifers into the ocean (or deep subsurface reservoirs) accompanied by a loss of soil, carbon, and minerals. The equilibrium of this system may be disturbed by the following:

(i) Climatic boundary conditions, including changed net radiation (due to a change in cloudiness).
(ii) Expansion of the unsaturated zone.
(iii) Reduced rainfall.
(iv) Increased runoff.

The above phenomena result in a loss of water and minerals for biomass production and a loss of energy for photosynthesis. The area integrated effect may thus be studied best by assessing the carbon, mineral, and water budgets through improved measuring of precipitation, evaporation, water transport, carbon, and minerals in streams, and biomass at the land surface.

The present methodology for determining budget quantities from space is still in its infancy. With advanced techniques that are gradually developed with the advent of a new generation of Earth-observing satellites, this approach may become an important element of global change research in the near future. Sediment plumes at the mouths of rivers can already be observed from space although quantitative relationships between color changes and sediment deposition still need to be established. The estimate of continental biomass changes through time series of suitable vegetation indices is being implemented as part of the global carbon cycle research. Studies are ongoing to estimate cumulative precipitation from cloud observations in the visible and infrared parts of the spectrum. In preparation are observations with microwave sensors, which will greatly improve this technique. In addition, the estimate of atmospheric water vapor is in progress. Also of interest in this connection is the import and export of water vapor, which require wind data.

The budget approach calls for pilot studies in which essential ingredients are models, ground and aeronomic measurements supported by remote sensing. The technology to improve our knowledge of the water budgets of large areas presently is developed within the GEWEX project, which is part of CGIP.

5. Satellite Data Products, Their Validation and Scaling

This section distinguishes between primary and secondary data. After corrections have been applied, primary data can more or less be directly inferred from observations in space. Secondary data products, on the other hand, are computed from the primary data and collateral information. Primary information consists of quantities such as reflectance, vegetation index, photosynthetic active radiation, or surface temperatures. Secondary information contains quantities that are indirectly deduced from satellite observations and collateral knowledge. For the application discussed here, secondary data are needed for computing parameters that govern the fluxes and the fluxes themselves.

To determine the accuracy of remote sensing data, data were collected at a network of ground stations in Castilla-La Mancha, Spain, during the drying period of the summers of 1991 and 1994. These data were supported by measurements from aircraft, and this research was supplemented by smaller scale activities in Italy and Crete in 1994. Maps of surface spectral reflectance, albedo, emissivity, and temperature have been derived from LANDSAT, NOAA-AVHRR, and METEOSAT data. High spatial resolution primary data have been directly compared with radiometric measurements at the ground. By upscaling the high-resolution data to the scale of the lower resolution data, the lower resolution data could also be validated.

The changes during the drying period have been followed by sequences of reflectance and NDVI satellite data, which reveal on the one hand the effect of drying the soil surface and on the other hand the increasing effect of irrigation. The surface emissivity was derived from multispectral day and night measurements from space combined with measurements at the surface. Thus we could produce maps of thermodynamic surface temperatures, which are the basis for heat flux calculations (Nerry *et al.*, 1988; Nerry, 1993; van de Griend, 1993).

Methods have been developed to infer fluxes or flux-related quantities from the remote sensing data using a minimum of collateral information from the field experiment. The sensitive heat flux, for example, is a complex function of the net radiation, soil moisture, and the structure of the atmospheric boundary layer. Through the use of satellite data, these complex functional relationships have to be simplified to an equation that contains the air density, ρ, the specific heat at constant pressure, c_p, the air temperature at a reference height, T_a, the surface temperature, T_s, and the surface resistance, r_s:

$$M_{SH} = \rho c_p \frac{T_s - T_a}{r_s}$$

One approach (Bastiaanssen *et al.*, 1994; Bastiaanssen and Roebeling, 1993) identifies the air temperature with the surface temperature over wet (e.g. irrigated) land. The surface resistance has to be parameterized using the vegetation index and experimental collateral data.

Energy transfer processes can be analyzed only if physical-biological parameters are observed at an appropriate frequency. The highest temporal resolution is achievable from geostationary satellites, presently at horizontal resolutions between 5 and 10 km. Interpreting land-surface signals for areas of 50 km^2 is difficult for Mediterranean landscapes, but measurements from the geostationary satellite METEOSAT provide such information as daily changes of land-surface temperatures and cloud cover from which the surface radiation budget can be derived (see e.g. Dedieu *et al.*, 1987). Land-surface parameters, such as surface temperature and spectral reflectance, are also measured at specific times of the day from NOAA satellites with a resolution of about 1.4 km^2. Thus only radiation budget information is available for different hours of the day. To obtain the daily evaporation, the energy budget equation is solved for the time of a NOAA-AVHRR overflight, and the evaporative fraction $M_{LH}/(M''_R - M_G)$, calculated from ground measurements, is made constant for the whole day.

6. Conclusions

The EFEDA effort of the Commission of the European Communities provided a broad basis for advancing the evaluation and application of process studies of remote sensing data, and the EFEDA modelling group has begun using remote sensing data in mesoscale models. The unique data base compiled during field experiments allowed inferred information to be validated against ground measurements. Problems and errors could thus be detected and algorithms developed to correct for them. Substantial progress could be made to derive secondary data products like fluxes and to validate them against flux measurements at the surface and from aircraft. Cumulative data is being produced, but more research is needed in the future. In addition, anchor stations are needed in critical areas to provide collateral information and continuous quality control of the inferred data.

Acknowledgements

The EFEDA project is granted within the Environmental Programme of the Commission of European Communities DG XII under the program management of Dr. Fantechi and Dr. Balabanis. The team leaders of the RESRAPS group and scientists who made major contributions to this research are W.G.M. Bastiaanssen, Winand Staring Center; N.L Wageningen and F. Nerry, University Louis Pasteur of Strasbourg; F.A. van de Griend, University Amsterdam; J. Meliá Miralles, E. López-Baeza, and J. Moreno, University of Valencia; and P. Rossini, Telespazio, Rome.

References

Bastiaanssen, W.G.M., D.H. Hoekman and R.A. Roebeling. 1994. *A methodology for the assessment of surface resistance and soil water storage variability at mesoscale based on remote sensing measurements*, IAHS Special Publication No. 2. International Association of Hydrological Sciences Press, Wallingford.

Bastiaanssen, W.G.M. and R.A. Roebeling. 1993. Analysis of land surface exchange processes in two agricultural regions in Spain using Thematic Mapper Simulator data, in H.J. Bolle, R.A. Feddes, and J.D. Kalma (Eds.), *Exchange processes at the land surface for a range of space and time scales*, IAHS Publication No. 212. International Association of Hydrological Sciences Press, Wallingford, pp. 407–416.

Bolle, H.J. *et al.* 1993. EFEDA: European field experiments in a desertification threatened area. *Annales Geophysicae* **11**: 173–189.

Dedieu, G., P.Y. Deschamps and Y.H. Kerr. 1987. Satellite estimation of solar irradiance at the surface of the Earth and of surface albedo using a physical model applied to METEOSAT data. *Journal of Climate and Applied Meteorology* **26**: 79–87.

Nerry, F. 1993. Longwave infrared measurements. In *EFEDA: Final report*. EFEDA-Project Office, Berlin, pp. 312–319.

Nerry, F., J. Label and M.P. Stoll. 1988. Emissivity signatures in the thermal IR band for remote sensing: Calibration procedure and method of measurement. *Applied Optics* **27**: 758–764.

van de Griend, A.A. 1993. On the relationship between thermal emissivity and the normalized difference vegetation index for natural surfaces. *International Journal of Remote Sensing* **14**(6): 119–1131.2

WHERE, WHY, AND TO WHAT EXTENT HAVE RANGELANDS IN THE KAROO, SOUTH AFRICA, DESERTIFIED

W. R. J. DEAN, S. J. MILTON and M. A. DU PLESSIS
FitzPatrick Institute, University of Cape Town, Rondebosch 7700, South Africa

Abstract. We used simple questionnaire surveys to obtain rough categorical information on the status of semiarid and arid rangelands in the Karoo, South Africa. We sent questionnaires to extension officers of the Department of Agricultural Development, asking them to rank the condition of the rangeland in their extension districts (usually about 500 farms) on a three- or five-point scale. The questionnaire also sought information on the possible reasons for changes in the condition of the rangelands; vegetation types most at risk on mismanaged rangelands; and changes in the status of plants, birds, and mammals. Extension officers rated 35% of the semiarid rangelands in southwestern South Africa (the Karoo and southern Kalahari Desert) as being in poor condition. There was some congruency between the perceived condition of the agricultural extension district and the percentage that the stocking rate had decreased. Rangeland condition in more arid areas tends to be poor. Stocking rates have decreased disproportionately more in arid areas than in the more mesic areas of southwestern South Africa. Grasses were rated as the most threatened group of plants, and succulents the least threatened group. Overgrazing in the past was considered to be the most important cause of present-day rangeland condition, followed by present overstocking of rangelands. Antelope were considered more threatened by rangeland management than mammalian carnivores. Avian raptors and bustards, gamebirds, and tortoises were all perceived to be threatened by rangeland mismanagement.

1. Introduction

Selective grazing by livestock is probably the major threat facing rangelands in semiarid and arid southwestern South Africa (the Karoo *sensu lato*) (Roux and Vorster, 1983). The current carrying capacity of the Karoo is estimated to be exceeded by 40%, or 3 million small stock units (Department of Agriculture, 1986). Grazing trials in the Karoo have shown that grazing can be a major determinant of vegetation composition, depending on intensity and season (Roux and Vorster, 1983), and state-and-transition models have shown that karroid vegetation has shifted to new, stable, almost intractable states by inappropriate management of rangelands (Milton *et al.*, 1994; Milton and Hoffman, 1994). Recent governmental concern about the state of the rangelands in South Africa (not solely in the Karoo) has led to a national grazing strategy for South Africa (Bruwer and Aucamp, 1991), but we still need to investigate the nature and geographical extent of changes in the biota and secondary productivity of the rangelands in the Karoo.

In this paper we report on the results of a survey using questionnaires to obtain categorical information on the state of the rangelands in southwestern South Africa. We present the results of the survey and briefly review the desertification debate with reference to the Karoo.

Environmental Monitoring and Assessment **37:** 103–110, 1995.

2. Study Area

The southwestern arid zone of Africa has traditionally been divided into three major areas: the Namib Desert, the southern Kalahari Desert, and the Karoo (Werger, 1985). The Karoo *sensu lato* consists of two biomes (Rutherford and Westfall, 1986): the southern semiarid and western arid succulent shrubland Karoo (Succulent Karoo) and the grassy central, upper, and "false Karoo" (Nama-Karoo). The area is divided into magisterial (administrative) districts, which are also used by the Department of Agriculture as their extension areas. Agricultural extension officers work within these districts, advising farmers and pastoralists about livestock and proper farming methods. The districts thus provide convenient units for collecting data on the state of rangelands, nature and extent of rangeland degradation, and changes in the relative abundance of plants and native animals.

3. Perceptions of Agricultural Extension Officers

We used a simple questionnaire survey to obtain rough categorical information on the status of semiarid and arid rangelands in the Karoo and southern Kalahari. Questionnaires were sent to extension officers of the Department of Agriculture, asking them to rank the condition of the rangeland in their extension districts on a three- or five-point scale. The questionnaire also obtained information on the possible reasons for changes in the condition of the rangelands; vegetation types most at risk on mismanaged rangelands; and changes in the status of plants, birds, and mammals.

4. The State of Rangelands in the Karoo

4.1. Extent of Rangelands in Poor Condition

Agricultural extension officers rated 35% of the semiarid rangelands in the Karoo and southern Kalahari as being in poor condition. The percentage of the district rated by the extension officers as being in poor condition has been plotted against the percentage decrease in stocking rate for that district given by Dean and Macdonald (1994) (Figure 1). In general, with the exception of two points where the stocking rate has not changed, the perceived condition of the agricultural extension district was congruent with the percentage that the stocking rate had decreased. But extension officers tended to rate rangelands as being in rather better condition than the stocking rates data suggest. Of interest is a recent study (Düvel and Scholtz, 1992) that shows that pastoralists in semiarid parts of South Africa generally view their rangelands in a more favorable light than do local extension officers. So the potential for overstocking is high.

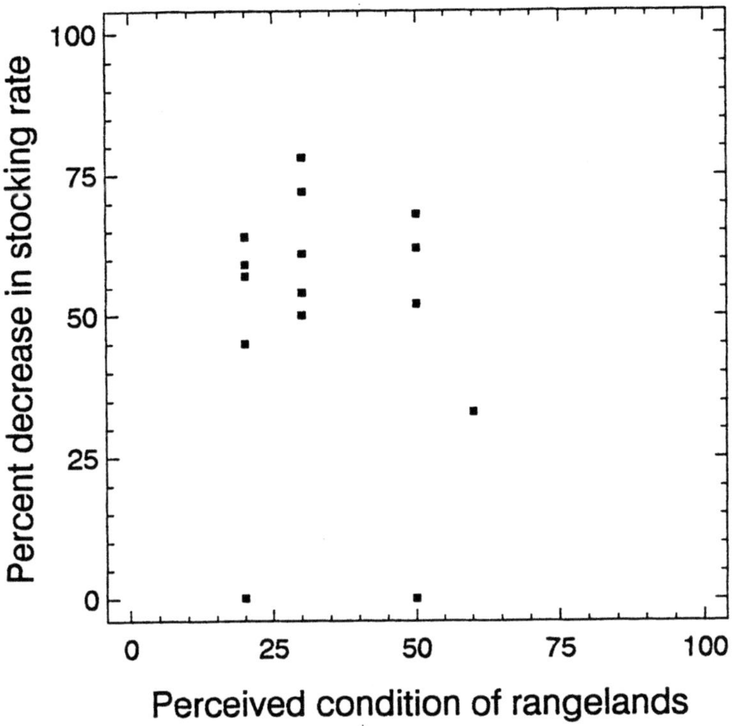

Fig. 1. Changes in stocking rates plotted against the perceived condition of farms in the Karoo.

TABLE I

Average scores given to plant groups by agricultural extension officers, ranked in decreasing order from most to least threatened. The "n" is the number of extension officers who responded to the question

Plant group	n	Average score
Grasses	29	7.8
Annual plants	23	4.9
Succulent plants	23	3.9

4.2. Plant and Animal Groups Threatened by Rangeland Management

The extension officers were asked to score on a scale of 1 to 10 the plant groups that they considered most threatened by rangeland management in their agricultural district. The results are given in Table I. Predictably, grasses were rated as being

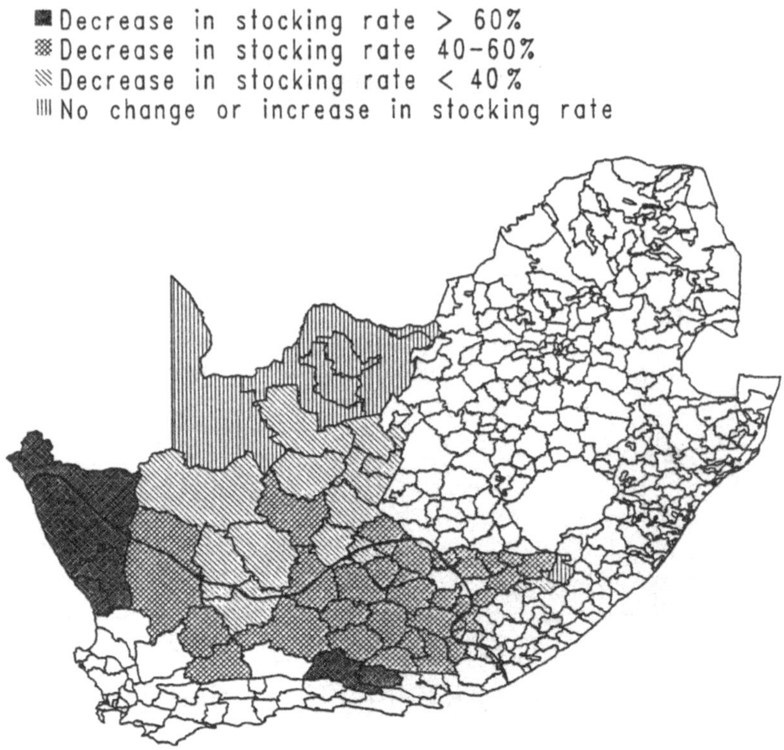

Fig. 2. Position of frontier in 1806 (from Christopher, 1982) shown by bold line, and areas where stocking rates have decreased in the Karoo (data from Dean and Macdonald, 1994).

the most threatened group of plants, but it is surprising that succulent plants did not score higher, given their vulnerability and sensitivity to livestock trampling.

Extension officers were also asked to rank on a scale of 1 to 6 the causes of poor rangeland condition in their districts. Overgrazing in the past was considered to be the most important cause of present rangeland condition, followed by present overstocking of rangelands (Table II). Drought and selective grazing were also perceived to be important in causing poor condition of rangelands. Overstocking in the past, particularly of stock unsuitable for Karoo shrublands, may have contributed substantially to the present condition of rangelands. Figure 2 shows the condition of rangelands relative to the frontier in 1800 and the parts of the Western Cape Province that were first settled.

Agricultural extension officers were also given a prescribed list of animal groups and asked to score on a scale of 1 to 10 what they considered to be the groups most threatened by rangeland management in their district (Table III). The generally low scores show a diversity of opinion among the extension officers, but in general antelope were considered more threatened by rangeland management than

TABLE II

Average scores given to possible causes of poor rangeland condition by agricultural extension officers, ranked in decreasing order of perceived importance. The "n" is the number of extension officers who responded to the question

Cause of poor condition	n	Average score
Overgrazing in the past	30	5.4
Present overstocking	30	4.1
Drought	30	3.7
Selective grazing	30	3.0
Unsuitable livestock	30	1.6
Burning	30	0.5
Present understocking	30	0.3

TABLE III

Average scores given to groups of animals by agricultural extension officers, ranked in decreasing order from most to least threatened. The "n" is the number of extension officers who responded to the question

Animal group	n	Average score
Antelope	27	4.7
Carnivores	26	4.6
Raptors	27	4.3
Bustards	25	4.0
Termitivores	27	3.9
Gamebirds	25	3.5
Tortoises	24	3.1
Primates	26	2.9
Snakes	19	2.0
Insects	21	2.1
Songbirds	26	2.0
Lizards	19	1.6
Frogs	19	0.7

mammalian carnivores and avian raptors. Bustards, gamebirds, and tortoises were all perceived to be threatened by rangeland management.

4.3. THE DESERTIFICATION DEBATE

The nature, extent, rates, and mechanisms of postcolonial vegetation change in the Karoo have been frequently debated (Acocks, 1953; Jarman and Bosch, 1973; Hoffman and Cowling, 1990a; Dean and Macdonald, 1994; Dean *et al.*, 1995). The precolonial eastern Karoo was believed to be dominated by perennial grasses such as *Sporobolus fimbriatus* and *Themeda triandra* (Acocks, 1953; Roux and Theron, 1987), and it has been generally assumed that overstocking in the recent past has largely been responsible for the dwarf shrubland evident today. It has been suggested that the vegetation of the Karoo changed somewhat before European colonization as a function of phases of climatic shifts and the impacts of both *Khoi San* hunter-gatherers and *Khoi Khoi* herders (Sampson, 1986).

This issue is likely to remain incompletely resolved for some time (Dean *et al.*, 1995). The idea of a grass-dominated Karoo forms an important part of the current semiarid rangeland management philosophy and suggests that the Karoo should support more perennial grasses than it currently does.

The nature of vegetation change in the Karoo has attracted much interest. Models focusing on the dynamics, rates, and broad patterns of vegetation change in the Karoo (Acocks, 1953; Jarman and Bosch, 1973; Hoffman and Cowling, 1990b) are briefly discussed below.

The perception that the Karoo has expanded in a wave northeastward along a broad front into the more productive grasslands (Acocks, 1953) has considerably influenced pastoral management policy in South Africa.

A second model suggests that vegetation changed in the Karoo as the result of the coalescence through time of degraded patches over extensive regions. When rangeland degradation is examined more closely, the patchiness of the process appears evident. Rangeland condition appears to differ between and within farms. Vegetation patchiness within farms is associated with vegetation clearing and plowing (Dean and Milton, 1995), with localized overgrazing (Fuls, 1992) and areas where animals congregate, including water points (Roux and Opperman, 1986; Bosch and Gauch, 1991), pans (Milton *et al.*, 1992), and holding camps and feedlots.

A third model suggests that the edge of the Karoo is dynamic and that Karoo vegetation shifts between dominance by shrubs and dominance by grasses but that the overall direction of vegetation change is towards decreased productivity.

In general, these models and views are largely untested (Dean *et al.*, 1995). But recent studies on the use of stable carbon isotope ratios to determine C_3 and C_4 grass fractions in soil organic matter (Stock *et al.*, 1993; Bond *et al.*, 1994) have shown that grasses were once more abundant in the Karoo than they are now.

The notion that the future for the biota of Karoo rangelands is bleak is supported by perceptions of agricultural officers, published reports on the Karoo (Department of Agriculture, 1985a, 1985b, 1986), the progressive decrease in stocking rates in the Karoo (Dean and Macdonald, 1994), and studies showing recent changes in the cover and composition of the vegetation of the Karoo (reviewed by Dean *et al.*, 1995).

Although Karoo shrublands are strongly resistant, they are no longer likely to be at equilibrium. For many years livestock production in the Karoo has apparently exceeded the carrying capacity of the natural environment (Roux and Vorster, 1983; Roux and Opperman, 1986; Dean and Macdonald, 1994), causing the vegetation in the Karoo to shift through transitions to new states. At the root of these shifts is a perception by the early European settlers that the rangelands were more productive than they actually were and that drought years were exceptional rather than the norm (Talbot, 1961). There is no doubt that the secondary productivity of Karoo rangelands has recently changed (Downing, 1978; Roux and Opperman, 1986; Dean and Macdonald, 1994). Whether these changes are irreversible remains an open question.

Acknowledgements

This paper is a contribution to the Desertification Programme of the FitzPatrick Institute, University of Cape Town. The Programme is funded by the Foundation for Research Development, the Department of Environment Affairs, and the Southern African Nature Foundation. We thank the Directorate of Resource Conservation (Department of Agriculture) for data on grazing capacity in the Karoo. Attendance at the International Symposium and Workshop on Desertification in Developed Countries was funded by the Bureau of Land Management, U. S. Department of the Interior. We are grateful to Beaumont McClure and Walt Whitford for facilitating our attendance at the symposium.

References

Acocks, J.P.H. 1953. Veld types of South Africa. *Memoirs of the Botanical Survey of South Africa* **28**: 1–192.

Bond, W.J., W.D. Stock and M.T. Hoffman. 1994. Has the Karoo spread? A test for desertification using carbon isotopes from soils. *South African Journal of Science* **90**: 391–397.

Bosch, O.J.H. and H.G. Gauch. 1991. The use of degradation gradients for the assessment and ecological interpretation of range condition. *Journal of the Grassland Society of Southern Africa* **8**: 138–146.

Bruwer, J.J. and A.J. Aucamp. 1991. The National Grazing Strategy of the RSA – Objectives, achievements and future challenges. Paper presented at "Meeting Rangeland Challenges of Southern Africa in the 1990s," Pretoria, 6–10 May 1991.

Christopher, A.J. 1982. Towards a definition of the nineteenth century South African frontier. *South African Geographical Journal* **64**: 97–113.

Dean, W.R.J. and I.A.W. Macdonald. 1994. Historical changes in stocking rates of domestic livestock as a measure of semi-arid and arid rangeland degradation in the Cape Province, South Africa. *Journal of Arid Environments* **26**: 281–298.

Dean, W.R.J. and S.J. Milton. 1995. Plant and invertebrate assemblages on old fields in the arid southern Karoo, South Africa. *African Journal of Ecology* **33**: 1–13.

Dean, W.R.J., M.T. Hoffman, M.E. Meadows and S.J. Milton. 1995. Desertification in the semi-arid Karoo, South Africa: Review and reassessment. *Journal of Arid Environments* **30**: 247–264.

Department of Agriculture. 1985a. *Landbou-ontwikkelingsprogram.* Noordwestesubstreek. Elsenburg: Department of Agriculture and Water Supply, Winter Rainfall Region.

Department of Agriculture. 1985b. *Landbou-ontwikkelingsprogram.* Klein Karoosubstreek. Elsenburg: Department of Agriculture and Water Supply, Winter Rainfall Region.

Department of Agriculture. 1986. *Landbou-ontwikkelingsprogram.* Karoostreek. Middelburg: Department of Agriculture and Water Supply.

Downing, B.H. 1978. Environmental consequences of agricultural expansion in South Africa since 1850. *South African Journal of Science* **74**: 420–422.

Düvel, G.H. and H.P.J. Scholtz. 1992. The incompatibility of controlled selective grazing systems with farmers' needs. *Journal of the Grassland Society of Southern Africa* **9**: 24–29.

Fuls, E.R. 1992. Ecosystem modification created by patch-overgrazing in semi-arid grassland. *Journal of Arid Environments* **23**: 59–69.

Hoffman, M.T. and R.M. Cowling. 1990a. Vegetation change in the semi-arid Karoo over the last 200 years: An expanding Karoo – fact or fiction? *South African Journal of Science* **86**: 286–294.

Hoffman, M.T. and R.M. Cowling. 1990b. Desertification in the lower Sundays River Valley, South Africa. *Journal of Arid Environments* **19**: 105–117.

Jarman, N.G. and O. Bosch. 1973. The identification and mapping of extensive secondary invasive and degraded ecological types test site D, in O.G. Malan (Ed.), *To assess the value of satellite imagery in resource evaluation on a national scale*, C.SR., Pretoria, pp. 77–80.

Milton, S.J. and M.T. Hoffman. 1994. The application of state-and-transition models to rangeland research and management in arid succulent and semi-arid grassy Karoo, South Africa. *African Journal of Range and Forage Science* **11**: 18–26.

Milton, S.J., W.R.J. Dean, M.A. du Plessis and W.R. Siegfried. 1994. A conceptual model of arid rangeland degradation: The escalating cost of declining productivity. *BioScience* **44**: 70–76.

Milton, S.J., W.R.J. Dean and C.P. Marincowitz. 1992. Preferential utilization of pans by springbok (*Antidorcae marsupialis*). *Journal of the Grassland Society of Southern Africa* **9**: 114–118.

Roux, P.W. and D.P.J. Opperman. 1986. Soil erosion, in R.M. Cowling, P.W. Roux and A.J.H. Pieterse (Eds.), *The Karoo biome: A preliminary synthesis, part 1 – Physical environment, South African National Scientific Programmes Report* **124**: 92–111.

Roux, P.W. and G.K. Theron. 1987. Vegetation change in the Karoo biome, in R.M. Cowling and P.W. Roux (Eds.), *The Karoo biome: A preliminary synthesis, part 2 – Vegetation and history, South African National Scientific Programmes Report* **142**: 70–95.

Roux, P.W. and M. Vorster. 1983. Vegetation change in the Karoo. *Proceedings of the Grassland Society of Southern Africa* **18**: 25–29.

Rutherford, M.C. and R.H. Westfall. 1986. Biomes of southern Africa – An objective categorization. *Memoirs of the Botanical Survey of South Africa* **54**: 1–98.

Sampson, C.G. 1986. Veld damage in the Karoo caused by its pre-Trekboer inhabitants: Preliminary observations in the Seacow Valley. *The Naturalist* **30**: 37–42.

Stock, W.D., W.J. Bond and D. le Roux. 1993. Isotope evidence from soil carbon to reconstruct vegetation history in the south-western Cape Province. *South African Journal of Science* **89**: 153–154.

Talbot, W.J. 1961. Land utilization in the arid regions of southern Africa. Part I: South Africa, in L.D. Stamp (Ed.), *A history of land use in arid regions*, *Arid Zone Research*, vol. 17. Unesco, Paris, pp. 299–338.

Werger, M.J.A. 1985. The Karoo and southern Kalahari, in M. Evenari, I. Noy-Meir and D.W. Goodall (eds.), *Hot deserts and arid shrublands*, vol. 2, *Ecosystems of the World*, 12 A & B, Elsevier, Amsterdam, pp 283–359.

DESERTIFICATION CONTROL: A FRAMEWORK FOR ACTION

H. E. DREGNE
International Center for Arid and Semiarid Land Studies, Texas Tech University, Lubbock, Texas, USA

Abstract. Desertification is a little-understood term that aggregates several land degradation processes occurring in the arid regions of the world. The major processes are vegetation degradation, water erosion, wind erosion, salinization, and soil compaction. Water erosion is the principal threat to environmental stability in both arid and humid climatic zones. Land degradation is generally reversible unless damage is very severe or soils are shallow. Practices to control land degradation are widely available but are not put to use for many reasons. Absence of a food crisis in developed countries is one of the important reasons. A related reason is the perception that no real problem exists. A third reason is that degradation control is not cost effective, except for controlling salinity and compaction. It is time to change emphasis from reducing on-site damage to reducing off-site damage.

1. Introduction

Desertification is the result of a complicated interplay of social, economic, and technical factors. Desertification control is a fairly straightforward procedure, technically, but one that commonly is difficult to carry out. The difficulties sometimes arise because of the variability of climate and other physical factors or because of a lack of political will. But the most important factors are economic and social. My framework for desertification control is based primarily on the ease with which control can be accomplished and the economic benefits of repairing land degradation damage.

Before pursuing the matter of desertification control, we need to understand what is meant by desertification. As used here, desertification is land degradation in the drylands due largely to human activities. Degradation refers to a reduction in the biological potential. The principal desertification processes are vegetation degradation, water erosion, wind erosion, salinization, and soil compaction (Table I). My global estimate of desertification according to land use is that approximately three-quarters of the world's rangelands are at least moderately degraded, as are about half of the rainfed cropland and around one-quarter of the irrigated land (Dregne and Chou, 1992). Only about 0.01 percent of the drylands are estimated to be very severely (irreparably) desertified.

2. Potential for Land Reclamation

Desertification control can mean restoring degraded land to its original productivity or only stopping further degradation. The difference depends upon the objectives

Environmental Monitoring and Assessment **37:** 111–122, 1995.

TABLE I

Criteria for desertification class

Desertification class potential	Plant cover	Water erosion	Wind erosion	Salinization	Soil compaction	Productivity % of	
						Rangeland	Cropland
Slight 90–100	Excellent range condition	None to little	None to little	$EC_e \times 10^3 < 4 dS/m$	None to little	75	100
Moderate 75–90	Good range condition	Moderate sheet and rill erosion, few shallow gullies	Soil deposits in furrows and along roads, uneven road surface	$EC_e \times 10^3 < 4–8 dS/m$	Reduced water penetration thin plow plan	50	75
Severe 50–75		Severe sheet and rill erosion, gullies common	Sand blasting of plants, soil accumulation in fencerows and furrows, blowouts and hummocks common	$EC_e \times 10^3$ 8–15 dS/m	Restricted water and root penetration, thick plow pan	25	50
Very severe 0–50	Poor range condition	Soils very shallow over parent material, many deep gullies	Calcrete exposed, dune formation, many blowouts and hummocks soils	Thick salt crust, nearly impermeable	Thick and hard plow pan	0	25

TABLE II
Potential for economically reversible land restoration[a]

Desertification process	Desertification class			
	Slight	Moderate	Severe	Very severe
Vegetation degradation, rangeland	High	High to low[b]	Low to very low[b]	Very low
Water erosion	High	Medium	Very low	Very low
Wind erosion	High	Medium	Low	Very low
Salinization	High	High	Medium	Low
Soil compaction	High	High	Medium	Low

[a] Restoring land to approximately its original productivity.
[b] Depends heavily on climate.

of the reclamation project and on the degradation processes (Table II). Salinized and compacted soils can be restored to something like their original condition in nearly every case. Vegetation degradation can be fully compensated, in principle, unless the soil also has been degraded. In practice, however, vegetation restoration of rangelands where the average annual precipitation is less than about 150 to 200 mm may be so slow as to be economically impractical.

The main reason is climate variability. If shrub increase or invasion is extensive, restoration of a perennial grass cover may well be impossible in a human lifetime unless an expensive, long-lasting improvement project is undertaken, even if the annual precipitation amounts to as much as 350 to 400 mm.

Wind erosion damage to soil is reversible if the soil is uniform and deep and if the soil surface can be smoothed to eliminate ridges and hummocks. The damage becomes economically irreversible on rangelands when blowouts (hummocks interspersed with depressions scoured by wind) have formed. Long-term wind damage to cropland usually consists of sand deposits from locally eroded fields and scouring that exposes unproductive subsoil layers. Both forms of damage may occur in the same field.

Water erosion is the most damaging land degradation process. Soil loss, whether due to sheet, rill, or gully erosion, is a degrading process whenever the rate of loss exceeds the rate of new soil formation. Since soil formation is mostly a process that forms new soil at rates of a centimeter in several decades or centuries, even low erosion rates of a few tons per hectare per year are soil degrading. Deep soils free of root-inhibiting layers, such as the aeolian deposits on the Loess Plateau of China, may remain highly productive after many years of severe sheet and rill erosion. Gullies cause immediate and easily observable reductions in productivity. Rills (very shallow gullies) also are readily visible, but sheet erosion does not leave an obvious trace of its occurrence. Sheet erosion may go on for years, attracting attention only when soil is deposited on roads or the tops of hills become lighter in color. By the time the knobs of hills change color, permanent losses of soil productivity will have occurred. All that erosion control can do in that case is to stop further productivity losses. While the eroded soil cannot be returned to its exact original condition, the soil can often be restored to its original productivity level, given good management.

3. Land Reclamation Priorities

There is no likelihood in the foreseeable future that governments, donor agencies, or land managers are going to stop and reverse desertification on the 3.5 billion ha of degraded drylands. The cost simply is too high. Furthermore, exaggerated claims about the extent of food shortages that land degradation will cause have dulled responses to a very real threat. Land degradation is an insidious process that can easily be ignored over the short term.

The United States would probably not have an effective and broadly supported soil conservation program if Hugh Hammond Bennett had not been a superb publicist as well as a good scientist. He was helped tremendously by the 1930s depression and the Dust Bowl black blizzards. A story that may or may not be correct says that dust fallout in Washington, D.C. from a west Texas storm persuaded Congress to establish the Soil Conservation Service in 1935. A thick layer of dust in offices and homes and on automobiles made a greater impression than endless statistics on soil and crop damage in the Great Plains.

3.1. Setting Priorities

Table II, indirectly, is a priority list for land reclamation if restoring productivity is the paramount concern. Highest priority would be given to salinity control and reducing soil compaction on croplands. These are the easiest measures to take, and the yield response is fast. Preventing further degradation by erosion, overgrazing, and woodcutting is best undertaken when the damage is moderate.

A priority list would be different if off-site damage were the primary consideration. Water and wind erosion are the only degradation processes that generate significant off-site damages. Salinization of irrigated land can sometimes degrade downstream water quality, but the effect is not globally extensive.

3.2. Off-Site and On-Site Effects

Off-site water erosion impacts are related to increased runoff from eroded land and the deposition of eroded particles at lower elevations (Clark, 1985). Increased runoff leads to greater flooding downstream as well as to sediment deposition on adjoining fields and on highways, urban areas, and floodplains and in rivers, lakes, and reservoirs. Runoff from farm fields carries nutrients in solution and on

TABLE III

Percent of degraded agricultural land capable of providing net economic benefit from reclamation*

Land use	Percent
Irrigated land	98
Rainfed cropland	70
Rangeland	30

*At least moderately desertified.

soil particles. The nutrients can pollute water supplies. By comparison, the on-site detrimental effect of water erosion is principally to reduce soil productivity.

Off-site wind erosion impacts are many and varied (Huszar and Piper, 1986). They include damage to (1) human and livestock health from air pollution; (2) crops, machinery, and exposed surfaces from sand blasting; (3) highways, railroads, water bodies, fields, and landscaped areas from the deposition of saltating particles and surface creep; and (4) air and automobile traffic from suspended particles that severely reduce visibility. Detrimental on-site effects mainly consist of sand blasting and burial of crops, unevening of field surfaces, reduction in soil fertility, and, sometimes a reduction in long-term soil productivity (Davis and Condra, 1989).

3.3. Reclamation Benefits

My own estimate of the developmental priority for desertified land is shown in Table III, but a strong element of opinion enters the calculation. My evaluation of published economic studies and comments by colleagues in other countries leads me to believe that the 98 percent figure for irrigated land is reasonably accurate. This figure is high because returns from irrigated drylands can be high if good management is practiced. Failure of large irrigation projects is legendary, but not because the potential land productivity is overrated. That is demonstrated by the high productivity of small plots of land irrigated by shadufs or individual pumps. Management is the key.

Rainfed cropland reclamation is less certain of being profitable. There is a consensus that soil conservation projects usually are not profitable in the short term of 5 to 10 years insofar as onsite benefits are concerned. These projects may, however, have a positive benefit/cost ratio over the long term.

Dryland projects that combine soil conservation with yield-improving water conservation are the exception. Such projects can increase yields immediately. Reducing the off-site impacts of erosion, particularly flooding and soil deposition on highways, these projects are probably profitable in certain places. But they benefit a broad base of taxpayers instead of identified beneficiaries.

The economic benefits of rangeland improvement become more speculative as the climate becomes drier. There is no question about that. There has been mounting pressure on the world's better grasslands for centuries. During relatively wet periods, grasslands are plowed and converted to rainfed croplands. Sometimes these lands are abandoned when the inevitable droughts recur, but these lands are usually not abandoned or are cropped again when another wet period comes along. The consequence of this global land use change is that rangeland area is reduced in the favorable regions and expanded in the less favorable regions, making degradation control even more speculative.

4. Land Degradation Control Problems

A distressing aspect of action programs to control land degradation is the record of the U.S. soil conservation agencies. A 1983 analysis of the effectiveness of the U.S. conservation program concluded that after 50 years and an expenditure of $18 billion, soil erosion remained "a formidable problem" (General Accounting Office, 1983).

On the other hand, virtually instant success followed the introduction of the 1985 Conservation Reserve Program (CRP). Within 4 years of beginning the CRP, erosion was estimated to have declined by about one-third. The decline was the result of returning highly erodible cropland to native vegetation and the requirement for having conservation plans on farms receiving federal support payments (Weber, 1990).

The 1950s Soil Bank program also retired cropland from cultivation. It, too, had an immediate impact. Unfortunately, former grasslands were plowed up again after the 10-year program ended (Wight *et al.*, 1983). That will also occur after 1995 if financing for land retirement under the CRP ends as scheduled. The cross-compliance requirement, if continued, will partially offset the end of land retirement.

The CRP experience demonstrated clearly that if a nation has the political will to control erosion and has the required financial resources, it can stop erosion immediately. In this case, government payments to farmers to take cropland out of cultivation were attractive. Only the most highly erodible land was targeted for conversion to grass and forest land. The targeted land was the source of a disproportionately large amount of the nation's erosion. It also was some of the least productive land, and farmers were amenable to taking it out of cultivation at a fairly low cost. Reducing erosion on the more productive land where obvious erosion is less apparent may be more difficult.

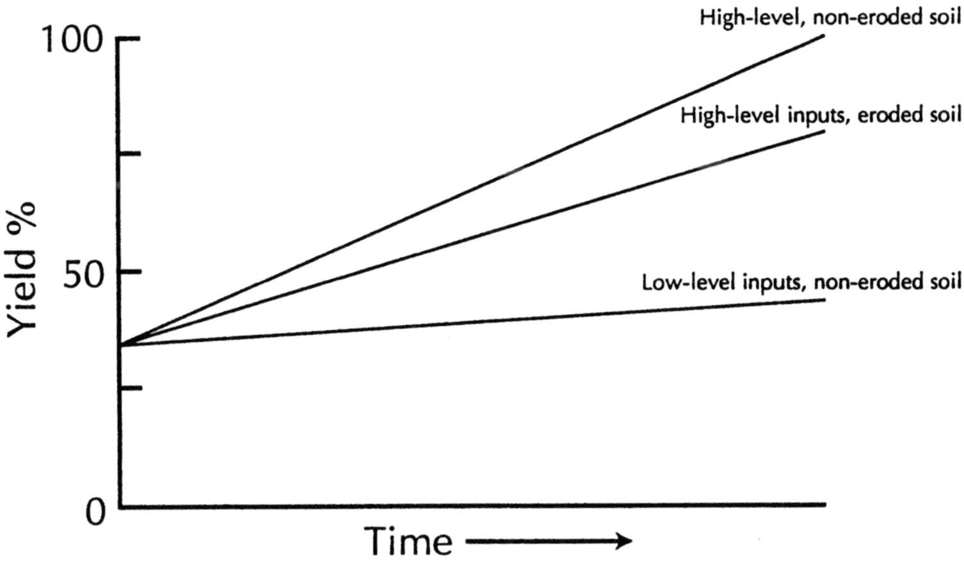

Fig. 1. Hypothetical relation of crop yields to erosion and inputs.

Erosion is only one of the land degradation processes. Fortunately, well-known practices can also control the other processes, although these practices are not as striking in their impact as changing cultivated land into grassland.

Given that desertification control practices are known, why is land degradation such an apparently intractable problem? Many reasons have been proposed. Poverty is the one that is cited over and over again by nontechnical people, in keeping with the contention that desertification is a socioeconomic problem, not a technical problem. Other socioeconomic constraints may be price controls, tenancy arrangements, belief that there is no problem or that the problem is not important now, lack of a conservation ethic, prestige associated with ownership of large herds of livestock, suspicion of governments, greater investment potential of livestock, concern for day-to-day family survival, unavailability of loans to finance conservation practices, reduction in income during first years of reclamation, negative short-term benefit/cost ratios of conservation practices, perceived excessive risk of failure to maintain family well-being, government disincentives for conservation, and lack of management skills.

There are valid technical reasons for the unwillingness of land operators to initiate land degradation control. One is the absence of proven techniques that increase productivity while improving the land. Another is the lack of advisors familiar with the recommended practices. A third is the insufficiency of proof that yields are declining, particularly where technology brings yield increases even as degradation is destroying the resource base (Figure 1).

Poverty may be the reason for mismanagement of some lands, but it certainly is not the sole reason. If it were, there would be little or no land degradation in

Australia, Canada, and the United States. Those three relatively wealthy countries have conducted national surveys clearly demonstrating that land degradation is continuing despite the money governments have spent, the availability of proven practices, and a competent advisory service.

A study of farmer perception of erosion on their farms in the Palouse region of the U.S. Pacific Northwest gave dismaying results (Osterman and Hicks, 1988). Farmers reported that water erosion was not significant even though it was readily observable during interviews. Farmers have no incentive to halt erosion that they do not believe is occurring.

Government policies sometimes encourage erosion. A prime example is the U.S. Department of Agriculture's urging of farmers to put every available hectare of land into food crop production in the mid-1970s. The incentive for that drastic change in farm policy was the shortage of wheat in the Soviet Union and the subsequent high world prices (Sachs, 1994). Conservation lost its attraction when the opportunity to reap economic benefits by disregarding erosion hazards came along.

5. Framework for Action

Recent years have seen an outpouring of concern about environmental degradation, including land degradation. An International Convention to Combat Desertification was adopted in 1994. Its objective is to combat desertification and mitigate drought impacts. Africa is the primary focus of the Convention, but all countries experiencing desertification or drought are to participate. As one reads the preparatory documents, however, it is obvious that developing countries want the Convention to help their economic development more than to combat degradation. Whether the Convention should attach economic development to desertification is a contentious issue between developed and developing countries.

5.1. Land Intensification

My proposed framework for combatting desertification is oriented toward enhancement of the agricultural productivity of a country's best land. The proposed action applies to soil erosion and rangeland degradation, not to salinization and soil compaction. Salinization and compaction are relatively easy and economic to control. My premise is that the preferred, if not only, way to reduce pressure on degrading marginal lands is to reduce the need for products of those lands.

Known techniques can be employed to intensify crop and livestock production on lands that are more favorably endowed with potentially highly productive land. A well-conceived program of land intensification will not usually be an overnight success. The proposed production methods must be tested on farm fields, if that has not already been done, and adapted to specific site needs. In addition, a support infrastructure will be required to make available technical assistance, roads, fertilizers, and other contributors to development. An effective organization is called for to strengthen agricultural research and extension.

Agricultural policies, also, must contribute to a betterment of production, not hinder it. Simply allowing the profit motive to operate can rapidly increase crop yields. Price controls are not compatible with productivity increases. The International Food Policy Research Institute (IFPRI) has come to believe that intensification is the route to follow (Pinstrup-Andersen and Pandya-Lorch, 1994).

Along with intensification of crop production on better lands goes the need for governments to set priorities on spending resources for both increasing production (Mellor *et al.*, 1987) and controlling land degradation. As noted previously, few governments are in a position to make a broad-scale attack on land degradation. Such an attack would be too expensive, unlikely to be successful in the short term, and excessively demanding on planners and managers. It would also require knowledge that is presently unavailable.

Setting priorities at both the political and technical level is not easy. Yet priorities must be set if resources are not to be frittered away on an uncoordinated attack that would be wasteful. High on the list of essential actions is selection of the lands that would receive the first attention. One reason for choosing better lands initially is to provide the early success that would generate continued support. After the relatively easy success has been achieved, the more difficult environments can be attacked.

Economic development should accompany land reclamation so that the poverty factor in the occurrence of land degradation can be minimized. Agriculture, in all its aspects (production, marketing, research, extension, agribusiness, transportation, credit, etc.), must be the driving force for development in the near future, at least for most developing countries. High levels of crop production are fully compatible with sustainable agriculture if good management is practiced. Lands with marginal climates, soils, and slopes are the ones for which it is difficult to devise a sustainable and economically profitable cropping or pastoral system.

In western Europe and the southeastern United States crop production has been shifted to the most stable landscapes where the land degradation threat is minimal. The advent of big machinery and the ability to drain and crop low-lying level lands has brought about abandonment of cultivated land on highly erodible sloping lands. The badly eroded cotton soils in the Piedmont region of the U.S. South that helped Hugh Hammond Bennett generate support for establishing the Soil Erosion Service in 1933 are now covered with trees (Trimble, 1974). Hidden by the trees are enormous gullies and severely eroded soils.

Intensive cultivation has shifted to the low-lying soils that were formerly too wet and too heavy to farm. A similar shift occurred in western Europe, allowing sloping land to be used for pastures and woods, with a consequent drastic reduction in erosion. Productivity is so high now in western Europe that the European

Community is trying to find alternative uses for abandoned croplands (Carruthers, 1986). This concern is with too much production, not too little.

5.2. Government Actions

To facilitate sustainable agricultural development governments can take many actions that will reduce pressures on marginal lands and improve degraded land. Actions will differ in industrialized societies and the poor less-developed countries.

Wealthy industrialized countries have the luxury of being able to finance conservation of natural resources for future generations. It is wise for them to continue to do so, although with a different focus. The new focus should be on reducing the off-site damage that pollutes water and air resources, threatens lives and property by flooding, reduces biological diversity, harms terrestrial and ocean fisheries, hastens eutrophication of water bodies, silts harbors and estuaries, and causes a host of other problems. An effective program requires a priority determination of where the off-site damages are worst and which degraded lands are responsible for those damages. Practically all offsite damage due to land degradation results from water and wind erosion. Human-induced salinization can also significantly affect downstream water quality, but such occurrences probably are rather inextensive.

Focussing on off-site damages can generate greater public support for on-site soil conservation because off-site effects are more easily seen to be important. A sand storm that closes airports and triggers allergenic reactions in people is easier to identify with than sand-filled furrows in a farmer's field. Wealthy countries also can concern themselves with important issues of biological diversity, endangered species, resource pollution, and developing a conservation ethic. There is little worry that continued land degradation threatens their food supply in the immediate future.

Poor developing countries should take a radically different approach. Governments should stop making a fuss about desertification except as it provides them a rallying cry for what they really need: economic development. Although increasing agricultural production to raise purchasing power is the key factor for rural-dominated societies, economic development may benefit more from building a good all-weather road system than from hiring more extension agents. Perhaps the most difficult change to achieve is to persuade political leaders that rural well-being is essential for a strong economy. The more usual objective of politicians in developing countries is to cater to urban populations because city people can start riots and overthrow governments.

Emphasizing intensification of agricultural production on the better lands makes it easier to control degradation and attain the goal of sustainable development. Desertification – as a subject of major concern – can be shunted aside for the present. One compelling reason for doing so is the lack of knowledge at higher levels of government about the extent and severity of land degradation in their countries. African countries, in particular, decry desertification even though they may have little understanding of the relative importance of different desertification processes, the timeframe and costs of reversing degradation, the difference between drought and desertification, and the magnitude of the damage desertification has done. Governments do understand food shortages and believe they know how to increase agricultural production.

Land degradation control in poor developing countries probably can best be left to nongovernmental organizations (NGOs). NGOs are action oriented, think small, involve local people, usually learn from their mistakes, and operate on low budgets.

6. Conclusions

Land degradation is a global problem whose precise dimensions are unknown and for which there is no easy solution. The principles of desertification control are well known. Just as the causes and effects of land degradation in the drylands are site specific, so are the local control practices.

Despite the availability of control methods, land degradation continues almost unabated. The two major reasons land degradation control is not practiced appear to be a perception that the problem is not a serious one and a belief that control is not economic. The latter reason is a good one because it is generally true. The perception may or may not be correct.

My contention is that the best way to improve degraded marginal land is to reduce the pressure to exploit those lands. An obvious need is to control population growth, which can have a long-term effect. For the immediate future, land pressures can be reduced only by increasing the productivity of the best land. Practically every country has some land with a good potential for crop and livestock production. Capitalizing on that situation by intensifying production there will improve the economic condition of the agriculturalists and accelerate national development. Such action will do nothing directly to reclaim degraded marginal lands but will allow that land to be used less intensively. Less intensive land use, in turn, will help the land heal.

Poor developing countries would do well to forget about desertification and focus attention on economic development. Desertification control can be left to NGOs, which are more attuned to local needs and priorities.

Developed countries are advised to emphasize off-site effects of land degradation, principally erosion, and generate support for on-site control. Off-site impacts are easier for the urban dweller to appreciate than a future threat of food shortages.

References

Carruthers, S.P. 1986. *Alternative enterprises for agriculture in the UK*, Centre for Agricultural Strategy, Reading, United Kingdom, pp. 12–16.

Clark, Edwin H. II. 1985. The off-site costs of soil erosion. *Journal of Soil and Water Conservation* **40**(1): 19–22.

Davis, Bob and Gary Condra. 1989. The on-site costs of wind erosion on farms in New Mexico. *Journal of Soil and Water Conservation* **44**(4): 339–343.

Dregne, H.E. and Nan-Ting Chou. 1992. Global desertification dimensions and cost, in H.E. Dregne (Ed.), *Degradation and restoration of arid lands*, Texas Tech University Press, Lubbock, pp. 249–282.

General Accounting Office. 1983. *Agriculture's soil conservation programs miss full potential in the fight against soil erosion*. General Accounting Office, Washington.

Huszar, Paul C. and Steven L. Piper. 1986. Estimating the off-site costs of wind erosion in New Mexico. *Journal of Soil and Water Conservation* **41**(6): 414–417.

Mellor, John W., Christopher L. Delgado and Malcolm J. Blackie. 1987. Priorities for accelerating food production growth in sub-Saharan Africa, in John W. Mellor, Christopher L. Delgado and Malcolm J. Blackie (Eds.), *Accelerating food production in sub-Saharan Africa*, Johns Hopkins University Press, Baltimore, pp. 353–375.

Osterman, Douglas A. and Theresa L. Hicks. 1988. Highly erodible land: Farmer perceptions versus actual measurements. *Journal of Soil and Water Conservation* **43**(2): 177–182.

Pinstrup-Andersen, Per and Rajul Pandya-Lorch. 1994. *Alleviating poverty, intensifying agriculture, and effectively managing natural resources*. International Food Policy Research Institute, Washington, pp. 2–3.

Sachs, Aaron. 1994. Dust to dust. *World Watch* **7**(1): 32–35.

Trimble, Stanley W. 1974. *Man-induced soil erosion on the southern Piedmont*. Soil and Water Conservation Society, Ankeny, Iowa, pp. 1–6.

Weber, Peter. 1990. U.S. farmers cut soil erosion by one-third. *World Watch* **3**(4): 5–6.

Wight, J. Ross, C. Kerry Gee and R.J. Karchner. 1983. Integrated rangeland and cropland management, in H.E. Dregne and W.O. Willis (Eds.), *Dryland agriculture*, American Society of Agronomy, Madison, Wisconsin, pp. 435–460.

POLICY ROOTS OF LAND DEGRADATION IN THE ARID REGION OF THE UNITED STATES: AN OVERVIEW

KARL HESS, JR.* and JERRY L. HOLECHEK**

* *Cato Institute, 1000 Massachusetts Avenue, NW, Washington, D.C. 20001–5403, USA;*
** *Department of Animal and Range Sciences Box 30003, New Mexico State University, Las Cruces, NM 88003, USA*

Abstract. The roots of land degradation in the arid region of the United States are set firmly in the era of open range conditions that led to the depletion of rangeland resources in the late 19th and early 20th centuries. Upon closure of the open range and establishment of the U.S. Forest Service and the Bureau of Land Management, public lands in the arid region were protected from the classic tragedy of the commons. Private lands, in turn, fell under the stewardship guidance of such organizations as the Soil Conservation Service, the Agricultural Stabilization and Conservation Service, and the Cooperative Extension Service. Despite these institutional protections, arid land degradation continues at an ecologically and socially unacceptable rate. The most likely explanation for this degradation resides in the institutions and public policies that shape and control livestock grazing. Flaws and disincentives endemic to the grazing permit system have encouraged destructive grazing. Federal subsidies to public- and private-land ranchers have encouraged rangeland stocking rates in excess of carrying capacity on economically and ecologically marginal lands and consistently favored unsound land management practices over superior ones. Policy-generated distortions of market forces have lead to unanticipated and environmentally undesirable land use outcomes. Land management, technical support, and educational institutions have also inadvertently contributed to land degradation in the arid region. Reversing the trend of land degradation in an advanced country like the United States demands that we understand the role of public policy in land use outcomes and that we structure future public policies in accordance with the lessons of recent history.

1. Introduction

Land degradation is ubiquitous in the history of the arid region of the western United States. During the late Pleistocene hunters crossed the Bering land bridge from Asia to Alaska and moved en masse south along an ice-free corridor into the heart of what paleontologists describe as the North American garden of Eden. In blitzkrieg fashion, the new immigrants hunted mammoths, mastodons, saber-toothed tigers, native horses, ground sloths, camels, giant beavers, dire wolves, and four-pronged antelopes to extinction. Ten millennia later, the Anasazis of the American Southwest farmed fertile bottomlands into wastelands and left as a legacy of their failed agriculture abandoned cliff dwellings and ghost-like canyon settlements. Today, places like Chaco Canyon, New Mexico, stand as monuments to the environmental misdeeds of early Americans.

Widespread depletion of natural resources in the arid region of the United States, however, did not begin in earnest until the arrival of European settlers. Beginning in the 17th century with Spanish settlement and then escalating a hundred fold

Environmental Monitoring and Assessment **37:** 123–141, 1995.

with post-Civil War Anglo-American migration, the forests and grasslands west of the 100th meridian succumbed to the axes, plows, and livestock of loggers, homesteaders, and ranchers. American pioneers were, in the words of Alexis de Tocqueville (1835), chronicler of *Democracy in America*, "insensible to the wonders of inanimate nature and they may be said not to perceive the mighty forests that surround them till they fall beneath the hatchet."

Nowhere was de Tocqueville's observation truer in a metaphorical sense than on the arid grasslands of the American West. From 1872 until the final closing of America's public lands to settlement in 1934, millions of cattle and sheep overgrazed hundreds of millions of hectares of open rangelands, devastated riparian zones along thousands of kilometers of streams, and eliminated wildlife and wildlife habitat on a scale not seen since the late Pleistocene. Arid soils were eroded of their fertility and water-holding power; plant communities were simplified or simply transformed into deserts; and a landscape that once supported tens of millions of free-roaming bison and antelope soon supported less than half that number in equivalent biomass of cattle and sheep. Since 1934 federal laws and regulations, public policies, public and private institutions, and an emerging public sentiment in favor of environmental protection have funneled millions of workhours and billions of dollars into stopping land degradation – into halting the environmental destruction that commenced when the ancestors of native Americans first crossed the Bering land bridge. And indeed, over a period of 60 years, the most conspicuous signs of land degradation have faded from the western landscape.

The half of the arid West that was once ruled by the harsh logic of the tragedy of the commons is now under the supervision of the U.S. Forest Service, the Bureau of Land Management (BLM), the National Park Service, and other federal and state land-managing agencies. On the other half – the privately owned half of the arid West – the Natural Resources Conservation Service (formerly the Soil Conservation Service); Consolidated Farm Services Agency; the Bureau of Reclamation; state land grant universities; and reams of laws, regulations, and policies shape and direct in ways visible and invisible the land use patterns that currently dominate the region.

Paradoxically, the highly centralized and well-intentioned legal and institutional framework that now governs land use in the arid region has not stopped the forward march of resource depletion. The bulk of grazing lands in the arid region remain in relatively degraded condition with millions of hectares reported deteriorating at this moment (U.S. General Accounting Office, 1988a, b). The best available data shows that about two-thirds of Forest Service and BLM grazing lands are in less than good ecological condition. A total of 14% of the two agencies' lands – almost 10 million ha – are estimated to be declining in ecological condition, and more than 11 million ha are classified in poor ecological condition.

Riparian areas are also degraded. BLM figures show that more than half of the area along the agency's 8,480 km of perennial streams in Colorado are in poor condition, and more than 80% of the area along its 18,987 km of streams in Idaho are degraded. Equivalent range and riparian estimates are not available for private lands in the arid region, but anecdotal data supports the observation that depletion and degradation are regionwide problems.

A number of causes can be advanced to explain the ecological stagnation, and in some cases decline, of this arid region. Among the most popular possibilities are a dearth of biological information, inadequate technology, cultural impediments to a land ethic, market failures, economic pressures for exploitation, and a residual tragedy of the commons – a function of inadequately financed and empowered regulatory agencies.

But none of these explanations are compelling. Biological information and technology are advancing exponentially and are available at relatively low costs to *willing* land users. Impediments to an earth-friendly land ethic are collapsing in the wake of a tide of national environmental sentiment. Market failure, at least as a causal factor in land degradation, is insignificant in the arid region. Economic markets are not signaling ranchers to produce more cattle and sheep. If anything, markets are delivering an unambiguous message to reduce livestock production in the West (Holechek and Hawkes, 1993; Holechek *et al.*, 1994).

Finally, public lands are in theory fully protected from the environmental ravages of the tragedy of the commons by virtue of federal control over livestock numbers, distribution, and seasons of grazing use. Further additions to agency power, personnel, and budgets are unlikely to yield benefits commensurate with costs. Federal authority, employees, and funding in the arid region have increased dramatically over the past 50 years without any clear evidence of environmental amelioration beyond that attained with the closing of the open range.

It is alarming that advanced science, advanced management, and advanced cultural attitudes have failed to halt the spread of land degradation in one of the most advanced nations on earth. This failure suggests that the institutions and policies forged in the 20th century to cure rampant land abuse in the arid region are not working – or even worse, working to the detriment of the land. This failure challenges the conventional wisdom that science and technology, when hitched to aggressive governmental policy, can and will give adequate safeguards to the integrity and well-being of the natural world. And this failure makes the compelling case that the roots of land degradation at the threshold to the 21st century are firmly entrenched in a century of ecologically dysfunctional public policies.

2. Historical Origins of Land Degradation

Settlement of the arid region in the United States – *the lands west of the 100th meridian* – followed the political course set by the Homestead Act of 1862 and then concluded by the expanded homestead acts of 1904 (Kincaid Act), 1909 (Enlarged Homestead Act) and 1916 (Stock-Raising Homestead Act) (Holechek *et al.*, 1989). Central to these acts, and to the massive depletion of western rangelands that followed their passage, were the size limitations placed on homestead units. Neither the 65 ha allowed under the 1862 Act nor the 259 ha permitted under the 1916 Act were adequate to support sustainable livestock operations in the arid region (Holechek *et al.*, 1989). Moreover, many of these arid lands were incapable of being homesteaded because of the absence of water. Although these findings were publicized by John Wesley Powell (1878) in his *Report on the Lands of the Arid Region of the United States*, Congress held tenaciously to a vision of an agrarian West of small yeoman farmers. As a result, public policy set the preconditions for the tragedy of the western commons.

Livestock operators who settled first in the arid West were unable to legally acquire sufficient lands to support economically sustainable ranching units (Holechek *et al.*, 1989). They were forced to rely on unclaimed, open rangelands to provide yearlong forage for their herds. To protect their rangelands – to enclose them from the commons – stockmen resorted to fencing, control of waters, and informal policing of customary grazing areas by associations of cattle growers. But with the escalation of homesteading and the arrival of large migrating herds of sheep in the 1880s, stockmen faced a crisis of rangeland control.

To ensure open access to western rangelands for itinerant sheep herders and newly arrived homesteaders, the Federal Government mobilized its legislative, executive, and judicial powers against recalcitrant cattlemen (Gates, 1968). It outlawed the enclosure of the open range by either fencing or control of scarce waters, disarmed ranching associations of their authority to restrict entry onto the public domain, and instructed military troops to enforce open range conditions. These actions quickly received the judicial blessing of the Supreme Court. In its 1890 decision on *Buford v. Houtz*, the court declared the public lands of the United States "free to the people who seek to use them where they are left open and unenclosed...." By the unanimous decree of all three branches of government, the official policy of the United States was to sustain and perpetuate the tragedy of the commons on the lands of the arid region (Gates, 1968).

3. Institutional Response to Land Degradation

The immediate cause of land degradation in the arid region up until 1934 is clear. Unable to control access to the open range, ranchers, sheepmen, and homesteaders raced to overgraze arid grasslands. "The only protection a stockman has," wrote a rancher, "is to keep his range eaten to the ground, and the only assurance that he will be able to secure the forage crop any one year is to graze it off before someone else does." As described by Albert Potter, first director of the Forest Service's Grazing Section, "flocks passed each other on the trails, one rushing in to secure what the other had just abandoned as worthless, feed was deliberately wasted to prevent its utilization by others, the ranges were occupied before the snow had left them" (O'Toole, 1988).

Range wars swept the arid region. Grazier fought grazier, depleting in the struggle the vegetation and soils of the arid region. In its 1936 report on *The Western Range*, the Forest Service depicted a grim legacy of nearly a half century of overgrazing (U.S. Department of Agriculture, 1936). "There is perhaps," the agency observed, "no darker chapter nor greater tragedy in the history of land occupancy and use in the United States than the story of the western range." In total, the Forest Service estimated that 75% to 90% of all public domain and private and state lands had been severely depleted and were in declining ecological condition.

Federal response to land degradation came slowly and in incremental steps. Open range conditions on many forest lands in the arid region were successfully combated with the creation of forest reserves in 1891 and the subsequent transfer of those reserves to the newly created U.S. Forest Service in 1905. Thirty years later, the Taylor Grazing Act put an end to open range conditions on all remaining public domain lands. In 1946 the Bureau of Land Management was established through a merger of the Grazing Service and the General Land Office. Finally, a quarter century of environmental statutes, beginning with the Multiple-Use Sustained-Yield Act of 1960, rounded out the federal response to land degradation in the arid region.

The laws, institutions, and policies inaugurated in the 20th century to combat overgrazing in the arid region have succeeded to the extent they have been able to contain the tragedy of the western commons. Yet, as evidence indicates, they are not stopping depletion and deterioration of rangeland soils, vegetation, wildlife habitat, and wetlands. Stocking in the arid region continues, in aggregate, to exceed carrying capacity despite federal mandates to the contrary. Land management practices remain abysmal, even with unprecedented levels of financial and educational investment in land conservation and restoration. Overgrazing, poor management, and land degradation survive and prosper in the arid region at levels that have never been ecologically acceptable and that today are no longer socially tolerable.

4. Institutional Framework for Land Degradation

To understand the paradox of modern land degradation in the arid region – *to appreciate its institutional roots* – it is crucial to understand the structure and function of the grazing permit system. Grazing permits are the principal tools by which land management agencies oversee and regulate private grazing on federal lands (Hess, 1992; O'Toole and Hess, 1994). In addition, grazing permits are the legal devices that join the public and private lands of the arid region into productive units called grazing allotments.

Under the regulatory direction and guidance of grazing permits, agencies and livestock producers have been able to forge order out of the chaos of the open range. Grazing permits have effectively enclosed the open range, granting security of tenure to ranchers and requisite authority to land managing agencies to mitigate resource depletion.

Yet, the grazing permit system has also had unexpected consequences. Besides forming the foundation of grazing management in the West, the grazing permit system has set the preconditions for sustaining and perpetuating contemporary land degradation in the arid region. In essence, this system has institutionalized overgrazing on rangelands, and it has done so in three specific ways:

- Formalized Overstocking. The grazing permit system has formalized overstocking on arid lands by over-obligating grazing privileges to resident stockmen. The permits originally issued by both the Forest Service and BLM authorized livestock numbers that exceeded the sustainable carrying capacity of native vegetation. As a result, land agencies have struggled over the past 75 years to bring permitted numbers of cattle and sheep in line with available forage. In part, they have succeeded: stocking on public lands has been nearly halved. That success, however, has been diminished by the fact that calf crops cattle weights coming off of arid lands have nearly doubled since 1935. Moreover, since the definition of carrying capacity has been expanded in recent years to accommodate all elements of an ecosystem, the effective overstocking of public ranges may well be greater now than it was following the closing of the open range.
- Perverse Incentives. The grazing permit system has provided strong incentives to stockmen to resist – *most often by political means* – ecologically justified stocking reductions. This is because grazing permits, or more accurately the authorized number of livestock (*grazing preference*) attached to them, are the only tangible assets ranchers have in public lands (Holechek and Hess, 1994a). Ranchers have no legal right in any physical attribute of federal ranges. Yet, because grazing permits are marketable, because low grazing fees are capitalized into their value, and because grazing permits enhance the worth of private base properties, ranchers see and treat these permits as quasi-private property. As a result, the primary incentive of public land ranchers is to protect their property interest – to steward and conserve the one thing they own, not the land, but the grazing preference attached to their permit. Since the stewarding and conserving of authorized numbers is best facilitated by lobbying, the *political* carrying capacity of grazing allotments almost universally takes precedence over their *biological* carrying capacities.
- Institutionalized Market Failure. The grazing permit system has institutionalized and constrained the choices available to public-land ranchers. Riparian areas, for example, have marketable value as habitat for recreation and fisheries. Upland plant communities have marketable value as habitat for big game. Yet, grazing permits constrain the marketable uses of riparian and upland communities to water and forage for domestic stock. Ranchers cannot dedicate rangeland water and forage to competing money-making activities such as recreation, hunting, and fishing. What ranchers can do, and what

the grazing permit system persuades them to do is to allocate riparian and upland resources for the exclusive use of domestic stock. Public law *compels* stockmen to see and treat recreation, hunting, and fishing as competing – *not complementary* – activities to their arid land livestock operations. Market failure, in the most profound sense, is the outcome. Multiple use regulations aside, the incentives endemic to the grazing permit system encourage ranchers overall to use water and forage only for activities that have paying benefits.

Destructive overstocking of western rangelands is just one of the unintended consequences of the grazing permit system. Land degradation in the arid region is also associated with other features of the grazing permit system, specifically the historic use of this system as a tool of social engineering. Federal policies regarding allotment size, community allotments, and grazing nonuse are cases in point:

- Small Allotments. The vast majority of public land grazing allotments are small and support fewer than 200 head of livestock. Small grazing allotments reflect the Federal Government's historic commitment to small ranching operations, a social equity position compatible with the agrarian thrust of its western land policy since 1862. Not surprisingly, the grazing programs of both the Forest Service and BLM have been and continue to be driven by the dual objectives of conservation and social reform.

  The 1986 environmental impact statement for the Santa Fe National Forest Plan, for example, set new standards and guidelines for resource conservation. It also acknowledged that "current range management policy recognizes the importance of the smaller permit in supporting the socio-cultural environment of northern New Mexico." More recently, Secretary of the Interior Bruce Babbitt announced a new conservation reform initiative for public grazing lands. In doing so, he argued passionately that "It's good public policy to make sure the small guys stay on the land."

  Sentiments in favor of small ranching operations may be politically astute, but they are not necessarily ecologically sound. Small grazing allotments are rarely economically sustainable – at least not without subsidy supports, whether from government or from a rancher's second job. Average, annual net income per permittee from federal land ranching is less than $5,000 (Holechek, 1992; Holechek and Hess, 1994a). In addition, studies from New Mexico suggest that federal land operations smaller than 200 head are marginally profitable at best and may entail, at worst, net losses as great as $60 per unit of livestock (Torell and Word, 1991). Assuming these studies are valid throughout the West, more than half of all public land ranches may be economically nonsustainable.

  Pressed by debt, high costs, and low revenues, ranchers on small allotments are frequently unable to invest in range improvements. Moreover, their small allotments often lack sufficient forage reserves to weather extensive drought periods. These operations cannot afford to graze conservatively in normal times or to reduce stocking in abnormal times. The marginal nature of their operations demands that every available unit of authorized forage be fully used every year. Cuts in permitted stocking, whether due to lack of rainfall, shifts in multiple-use planning or agency monitoring, are potentially disastrous for their marginal operations. For these reasons, small allotments are frequently in the poorest ecological condition. Land degradation, even under the strictest supervision of land managing agencies, increases in probability as the economic needs of the rancher clashes with the ecological needs of the land.

- Community Allotments. Many grazing permits are allocated to combinations of two or more ranchers to form community grazing allotments. Such allotments are the remnants of the open range where the livestock of entire communities of ranchers once overgrazed public lands in unison. As a practical matter, community allotments offer federal agencies a vehicle to implement the social and managerial agendas of public land policy: they provide a structural option to economically strapped small allotments and an equitable means to protect the public land position of small ranchers.

  In theory, community allotments avoid the environmental fate and abusive treatment associated with common use lands by vesting control over grazing to agency-sponsored grazing associations and oversight committees. These institutional protections, however, do not fully shield community allotments from the tragedy of the commons. In some community allotments areas of high forage value are overgrazed by the livestock of members concerned that if they don't exhaust the available grass someone else will. Elsewhere, members of community allotments are reluctant to invest in range stewardship, fearful that their associates will take a "free ride" and not make commensurate contributions.

  Free-rider concerns extend to ranchers who might graze their livestock conservatively on individual allotments but are unlikely to do so on community allotments, where stocking decisions are relegated to the group. These ranchers worry that any individual move to conserve forage will be countered by an increase in stocking by other members of the community allotment. The tragedy of the commons they face is far more subtle than the one that preceded them on the open range, but it is nonetheless a contributing factor to ongoing land degradation in the arid region.

- Nonuse Policy. The terms of Forest Service and BLM grazing permits require their holders to make substantial use of range forage, especially on ranges deemed in good condition, or face possible loss of grazing privileges. In recent years both agencies have warned permittees to make full use of their allotted forage or face reallocation of excess forage to outside livestock operators. This policy has led to aggregate levels of grazing on arid lands higher than might otherwise have occurred given prevailing market conditions and conservation attitudes of some public land ranchers.

> Although authorized stocking levels are designed in theory to promote sustainable use, incentives that discourage more conservative grazing practices encourage, in turn, the institutional trend toward land degradation. In particular, nonuse requirements in both federal law and regulation act as barriers to alternative land uses that could either accelerate restoration of degraded ranges or better protect arid lands with unique environmental values. For example, an administrative law judge ruled in April 1993 that The Nature Conservancy had failed to comply with the Taylor Grazing Act and BLM's nonuse requirements by not fully stocking two grazing allotments in southeast Arizona. In a decision reaffirming the government's social commitment to sustaining the monopoly position of graziers on arid lands, the court ordered BLM to reassign The Nature Conservancy's grazing permits to ranchers who would fully stock the allotments. Not incidently, the allotments were mostly in fair to poor ecological condition.

The institutional framework of land degradation in the arid region, however, extends beyond the regulatory parameters of the grazing permit system. Factors such as limited grazing tenure (10-year maximum permits) and prohibitions on private ownership of range improvements on public lands tend to sever the self-interest of stockmen from the ecological interests of the land.

Multiple use policy, in particular, has erected institutional barriers to sounder land stewardship practices. In theory, multiple use plans allocate land resources to their highest and best use. Recreation and wildlife are increasingly viewed by society as the highest and best use. However, because multiple use policy also mandates livestock grazing and because conditions tied to the grazing permit system prevent ranchers from benefiting from "highest and best uses," resource allocations invariably turn on political rather than economic or ecological considerations. Multiple use rules insulate ranchers from the true costs of grazing – the opportunity costs generated by forgone recreation and wildlife. In effect, these rules shield stockmen from competing land uses; they accommodate grazing levels that public opinion and market forces would not otherwise tolerate.

5. The Policies of Land Degradation Subsidies

Land degradation in the arid region is influenced by an array of public policies. Among those policies, federal subsidies to livestock producers – *on both private and public lands* – are most notable. Their cumulative effect is to alter land management outcomes in ways that are both predictable and, more importantly, unexpected. For example, subsidies to livestock producers tend to keep the most marginal – *and often most environmentally fragile* – lands in production. Because profit margins are lowest on these lands, subsidies often make the difference between grazing and no grazing or light grazing versus heavy grazing. Subsidies raise incomes and lower costs sufficiently to justify activities, and levels of activities, that would not otherwise be cost effective or environmentally sound.

Subsidies sustain what is nonsustainable. In the arid region this translates into elevated stocking rates – *stocking rates that are neither economically nor ecologically stable and lasting*. Such stocking rates invariably lead to environmental outcomes that conflict with reasonable range management goals. Land degradation becomes more, not less, probable. Subsidies also generate an environmental "snowballing" effect where ranchers become less likely to invest in and adopt land use technologies that are most appropriate to the land. Subsidies remove the economic incentives for stockmen to seek better land management tools; they *make it easier* for stockmen to oppose crucial management innovations such as the reintroduction of fire into rangeland ecosystems.

By the same logic, subsidies make ecologically inappropriate technologies economically sound. They transform failed tools into acceptable tools by distorting cost/benefit ratios – by making unsustainable management strategies and technologies profitable ventures. At the very least, subsidies reward ranchers for bad management; at the very worst, they put stockmen who manage well at a competitive disadvantage. Examples of private- and public-land subsidies include the following:

- Intensive Technology Subsidies: Range Reclamation. Following the end of World War II, and escalating through the 1950s and 1960s, Congress provided substantial funding to the Forest Service and BLM for massive range rehabilitation programs. The Vale Rangeland Rehabilitation Program in southeastern Oregon is a case in point. Prior to the start of the program in 1963, BLM's Vale District was severely degraded, a result of overobligation of grazing privileges and local allotment history (Heady, 1988). Livestock numbers exceeded estimated carrying capacity by 40%. Allotments were both small and communal. Stockmen invested little in their lands because of ranch size and the free-rider problem of community allotments. And ranchers were uncompromising in their protection of grazing preference.

  When faced with the prospect of BLM-enforced livestock reductions, ranchers joined with agency employees and local politicians to lobby for a range reclamation program. Congress responded by appropriating $18 million for the Vale Project between 1963 and 1985. When the program began, there were 332 permittees; when it ended in 1985 only 184 permittees remained, each receiving an average of $97,000 in range improvement benefits. Put into 1992 dollars, the project cost $56 million total – or $304,348 per permittee (Holechek and Hess, 1994a). In all, 304,000 ha were involved in the Vale project: 48,000 ha seeded, 69,000 ha sprayed, and 187,000 ha of native range (Heady, 1988). Cost per hectare treated in 1992 dollars was $30.

  The Vale project succeeded in raising estimated carrying capacity on program lands from 285,000 animal unit months (AUMs) in 1963 to 485,000 AUMs in 1986 (with actual stocking peaking at 400,000 AUMs) (Heady, 1988).

However, long-term benefits of the program are unclear. Already, productivity of sprayed areas has returned to pre-1963 levels, riparian areas are stagnating in fair to poor condition, and wildlife are on balance faring no better than they did in 1963. There is growing concern that the 200,000 AUMs created between 1963 and 1986 will not be sustained over the next 30 years and that *land degradation will resume its historic trend.*

The option of destocking Vale rangelands and restructuring allotments was never considered even though overobligated preference could have been purchased at market value for about 16% of the final cost of the program (Holechek and Hess, 1994a). Instead, the government chose to spend $280 in 1992 dollars for every AUM created – *AUMs that are likely to prove ephemeral in the long run.* Land degradation in Vale was not halted; it was simply postponed. In the interim, bad grazing practices and inappropriate grazing policies and institutions have been rewarded and given a 30-year reprieve. No lessons have been learned from the past, making likely the repetition of the past in the future. Marginal lands are being overcome by brush as overstocking inches forward yearly. Complacency borne from lavish federal appropriations is discouraging individual innovation. And sustainable ranching in southeastern Oregon is fading as a probable scenario.

- Intensive Technology Subsidies: Herbicidal and Mechanical Brush Control. Unlike the massive range reclamation programs of the past, herbicidal and mechanical brush control programs remain an ongoing option for both private and public land ranchers in the arid region. In most cases, brush control is practiced on historically overgrazed ranges – ranges where overstocking has eliminated grass as a fuel for fire and as a competitor to brush invasion and expansion. On public lands the costs of brush control are paid for by the land managing agencies or shared by the agencies and permittees. On private lands these costs are normally divided between the Natural Resources Conservation Service (NRCS) and participating ranchers, with the government assuming as much as 65% of the cost.

  As a general rule, intensive brush control on public and private lands of the arid region are not cost-effective to ranchers when they alone must pay the full economic price (Holechek and Hess, 1994b). Costs on a per-hectare basis for herbicidal and mechanical treatments average $5–8/hectare and $10–20/hectare respectively.

  Moreover, the income flow of benefits from control treatments is generally short lived. On less productive sagebrush ranges and on most mesquite and shinnery oak ranges in the arid region, benefits rarely exceed costs long enough to make investments profitable. Often, returns never cover costs, and in many cases ranchers would do much better by investing their brush control money in Treasury bonds.

  In addition, intensive brush control treatments have hidden costs that are often ignored in assessments of profitability and desirability: environmental costs like loss of soil stability in shinnery oak sites and loss of wildlife habitat on sagebrush ranges. Without cost-sharing subsidies, ranchers would not and could not afford such widespread and costly rangeland treatments.

The effects of government brush control subsidies are numerous. They discourage ranchers from seeking more environmentally friendly solutions to brush encroachment. They reward bad management, giving poor ranch managers a competitive edge over those whose practices are better. They encourage overstocking by masking the true costs of resource depletion. And they sustain livestock operations at levels that are otherwise nonsustainable. In all, brush control subsidies make land degradation a bearable cost of arid land ranching.

- Emergency Feed Subsidies. The Agricultural Stabilization and Conservation Service distributes emergency feed relief during periods of drought to both private and public land ranchers in the arid region of the United States. Nationwide, the program spends more than $0.5 billion in some years in direct cash assistance and millions more in direct distribution of corn feed. In the western states, ranchers who participate in the program receive annual average cash payments ranging from $4,000 in New Mexico to $11,000 in Oregon, plus varying amounts of free corn feed. Interestingly, many public land ranchers receive two to three times as much in direct cash relief from the government as they pay in federal grazing fees.

In theory, the feed program is intended to supplement rangeland forage during unusual climatic periods of extreme drought. Yet, the historic record of feed relief does not match the frequency of drought in the arid region. From Oregon to New Mexico, ranchers have received drought assistance every 6 to 7 years out of a 10-year cycle.

Interestingly, ranchers in the Vale Project are among the most highly paid participants of the emergency feed program in the arid West. In 1994 they received an average relief payment of $25,000 each to sustain stocking levels estimated at 125–150% above the long-term carrying capacity of their total ranching operations. This program encourages ranchers to stock their ranchlands for years of above-average precipitation rather than for average years. This argues persuasively that the feed program is little more than federal subsidy for overstocked and overgrazed rangelands.

New Mexico is a case in point. Emergency feed relief represents 10 to 15% of the net income of private and public land ranchers in that state. In addition, economic studies by the authors found that aggregate stocking rates in New Mexico are 15 to 25% higher than they would be in the absence of feed subsidies. In other words, government assistance encourages and allows levels of stocking that are economically nonviable and ecologically destructive to the grass resource. The result is a spiralling effect where overgrazing creates demand for feed subsidies and where feed subsidies expand otherwise nonsustainable numbers of livestock. The net effect is to make land degradation an ongoing program of public policy.

- Range Betterment Funds. Of grazing fees collected on Forest Service and BLM lands, 50% are returned to ranchers through their regional, state, and local grazing districts for spending on rangeland improvements. In 1992, for example, BLM collected $22 million in grazing fees, $11 million of which were invested in developing waters, building fences, and implementing other grazing management practices. Although the sum is relatively minor, it does play a role in the economic decisionmaking of ranchers.
  Combined with a grazing fee alleged to be below market value, this subsidy contributes to making higher stocking rates economical. This subsidy is particularly significant when public land ranchers earn only $3 in profits for every $4 BLM spends (1992 figures) on managing rangelands. By assuming a major share of range management costs, federal agencies artificially tip the scale of sustainability in favor of marginal operations that are not economically viable and in many cases are ecologically nonsustainable – *at least in the absence of generous rangeland operating subsidies.* Together, range betterment and associated subsidies – which in aggregate pay public land ranchers more than three times their payments in federal grazing fees – make land degradation an unwanted, unanticipated, but highly probable outcome.
- Experimental Stewardship Programs (ESPs). In 1978 the Federal Government launched a program of experimental stewardship to encourage and foster better land management practices. Test areas were selected in Tonopah (Nevada), Modoc-Washoe (California), Challis (Idaho), and East Pioneer (Montana). At Modoc-Washoe, Challis, and East Pioneer, for example, $3 million was spent over 10 years on range improvements for a total of 85 ranchers – or $35,000 per participant (Hess, 1992). The subsidies, however, had unintended outcomes. Rancher investments in rangeland management fell appreciably, and rangeland conditions did not rebound as anticipated (U.S. General Accounting Office, 1988).
  Like other subsidy programs, ESPs shift responsibility for resource use away from users and disperse the costs of resource depletion among the larger population of taxpayers. The effective result is to reward stockmen for poor management practices and to do so without changing the public policies and practices that encouraged and fostered land degradation in the first instance. For that reason the conditions for resource depletion remain unaltered in the arid region and the prospects for land improvement remain dim.

5.1. Market Intrusions

Market intrusions encompass an array of subsidy-free public policies that distort land use outcomes through intervention into and distortion of market processes. Market-intrusionary policies are endemic to agricultural policy in the United States. Their history is well documented in numerous publications. Although less prominent in the arid land livestock industry, intrusionary policies are significant.

In general, the environmental impacts of these policies are similar: they promote and sustain land degradation in the arid region. The following are some examples of some market intrusionary policies:

- Selected Conservation and Natural Resource Laws. Many laws governing land and resource use in the arid region have unanticipated consequences. In 1985, for example, The Farm Security Act was passed by Congress with stiff "sodbuster" provisions to reduce cultivation of highly erodible lands. Close to the same time, farmers in east-central New Mexico, seeking new irrigation sources, challenged the water rights of local stockmen. The New Mexico State Engineer ruled that livestock use was not a beneficial use in the area in question and that rancher water rights would be vacated within a 5-year period if they were not put to some other approved use.

  Adding complexity to the situation, The Farm Security Act created special quotas for Valencia peanuts, a farm crop grown mainly in eastern New Mexico. As a result of these converging streams of events, ranchers put their rangelands to beneficial use by selling them to peanut growers, who in turn responded to the lucrative Valencia peanut quotas by plowing up what constituted some of the most erodible lands in the arid region. Through the unexpected chemistry of a federal law aimed at stopping sodbusting and a state water law postulated on beneficial use, land degradation made a major advance in the high plains of New Mexico.
- County Property Taxes. Privately owned ranching lands in the arid region are favored with low property taxes in comparison to per-acre assessments on nonagricultural lands. Although the property tax break is capitalized into a ranch's market value, the tax break can and does influence land use outcomes in an environmentally adverse fashion. The reason is straightforward. Ranchers who wish to take grazing lands out of livestock production face increases in per-acre assessments as high as 1000% or more. This increased assessment is not a hardship for land owners who dispose of their property for commercial development (a type of transaction, incidently, not impeded by high property taxes); it is, however, a hardship and a disincentive for ranchers wishing to maintain open rural space for nonagricultural purposes.

  For example, in southwestern New Mexico an association of ranchers have agreed to set aside private lands as "grass banks." These banks are designed to provide supplementary forage during periods of drought or in conjunction with range management practices on nonbank lands that require temporary rest from grazing use. Because these lands will not be used on a regular basis for agricultural production, they may be taxed at a higher rate – *a factor discouraging some ranchers from participating in the innovative land conservation plan.* The same disincentive applies to landowners who may want to set aside former grazing lands for nature preserves but who lack the requisite nonprofit status to shield themselves from the resulting economic liability of higher local taxes. Designed to promote and foster an economically healthy agricultural sector, property tax breaks discourage market-driven environmental options that might arrest land degradation, create new land uses, and put arid rangelands on the road to ecological recovery.

- Free Access to Public Lands. Livestock grazing is not always the best and highest use for many private lands in the arid region. Moreover, livestock grazing is in many cases a cause of land degradation. This situation presents a dilemma to the residents of the arid region. Building a sustainable West demands that land uses be found and developed that can support the region's people while at the same time ensuring the long-term ecological integrity of plant communities, wildlife populations, and free-flowing waters. A major obstacle to this goal, however, is free access to public lands.

  Free access to public lands puts private lands at a competitive disadvantage in the production and sale of nonconsumptive goods and services such as camping, backcountry hiking, wildlife viewing, and wildland photography. Because these activities have no direct cost to users of public lands, owners of private lands who wish to develop them and sell them to the public face a monumental marketing problem: they cannot sell goods and services that a neighbor, in this case the Federal Government, is giving away free – or at least below cost. Federal underpricing of resources skews decisionmaking on private lands, tending to perpetuate land uses that are neither economically optimal nor ecologically appropriate for an arid landscape. Underpricing stands as a barrier to a sustainable West, forcing the arid region to cling to traditional activities and to pursue them at levels – often under the influence of subsidies and other market intrusions – that historically have culminated in land degradation.

5.2. Management and Educational Institutions

A broad array of management and educational institutions have evolved over the past 100 years in response to land degradation in the arid region. The relative effectiveness of these institutions in combatting resource depletion is generally acknowledged and broadly accepted among professionals in the field of rangeland management. Consensus exists that land management agencies such as the Forest Service and BLM, technical assistance organizations such as the Natural Resources Conservation Service and Cooperative Extension Service, and university-based natural resource schools have played a key role in reversing destructive land use practices on the grazing lands of the arid region. Although these organizations have clearly contributed to the reversal of mass land and resource depletion in the West, their roles have not been consistently beneficial. Both in the past and present, management, technical, and educational institutions have inadvertently encouraged, promoted, and helped perpetuate land use practices that are incompatible with sustainable human industry on western rangelands. Among these practices are the following:

- Federal Land Management Agencies. Federal land managing agencies have supported and implemented the massive subsidy programs that have distorted land use outcomes in the arid land ranching West. Although those programs have not on balance benefitted land conditions, they have paid political dividends in the form of expanded operating budgets to both the Forest Service and BLM. Institutional self-interest has also motivated land managing agencies to engage in practices inconsistent with conservation and stewardship goals. The support of the U.S. Forest Service for environmentally questionable below-cost timber sales was predicated in part on the budget maximizing incentives offered by the Knutson-Vandenberg Act of 1930 (O'Toole, 1988). BLM, in turn, used the apparent success of its early allotment management plans (AMPs) to justify an expanded range management budget even though the Department of the Interior's Office of Policy Analysis attributed the successes to the fact that the AMPs "were implemented on the better rangelands grazed by more energetic and progressive ranchers." Finally, both agencies have raised the specter of forage shortages time and again to justify expanded appropriations for range management programs – *programs which, ironically, have exacerbated rather than resolved the problem of diminishing grazing resources.*
- Technical Support Organizations. Agencies such as the Natural Resources Conservation Service (NRCS) and the Consolidated Farm Services Agency (CFSA) Conservation Service (ASCS) are the conduits for private land subsidies in the arid region – subsidies that have promoted overstocking and sustained through federal relief environmentally damaging land practices.

  In addition, the NRCS, CFSA, and Cooperative Extension Service of the Department of Agriculture have provided arid land ranchers with free or nearly free information. Although good information is vital to land conservation, its gratuitous provision distorts land use outcomes. Normally, obtaining information is a cost of doing business, a cost that most businesses must consider when making decisions on what and how much to produce. Subsidized information to ranchers – *even if provided for the best of environmental reasons* – tends to shift the cost-revenue balance in favor of less conservative rangeland stocking.

  Lastly, subsidized information in the public sector discourages innovations in and proliferation of information from the private sector. Fewer sources of information, in turn, diminish the overall menu of ideas available to stockmen to cope with and mitigate historic trends of land degradation in the arid region.
- Academic Institutions. Natural resource and range management schools in leading western universities share institutional responsibility for perpetuating the policies of land degradation in the arid region. Their curriculums, historically geared toward intensive technologies that expand domestic grazing use on arid lands, presume (1) that the highest and best use of rangelands, even in advanced countries such as the United States, is for food production and (2)

that rangeland forage is essential to providing adequate supplies of red meat in the United States.

Economic trends, however, argue against such assumptions. First, the profitability of arid land ranching has declined in recent years. Rates of return on ranch investments are now lower than the historic levels of 2 to 3% (Holechek, 1992; Workman and Evans, 1993). In addition, land appreciation that traditionally supplemented low returns to ranch investments is now negative in many areas of the arid West (Holechek and Hawkes, 1993). Driven by consumer demand, market forces are making it clear that the highest and best use of many western ranges is no longer red meat production. Second, supplies of red meat are abundant and at low prices and will remain so irrespective of arid land contributions (Holechek *et al.*, 1994). Like food stuffs in other sectors of the agricultural economy, a shortage of red meat is not a foreseeable problem. Market forces, if left unfettered, would likely retire the most marginal – and most environmentally sensitive – grazing lands in the arid region, would encourage technologies and management strategies favoring long-term economic survival, and would provide powerful incentives for many graziers to explore and develop alternative land uses that are both economically and ecologically sustainable. The curriculums of natural resource and range management schools have had the opposite effect of unfettered markets, however. With few exceptions, the framers of those curriculums have not considered or challenged the institutional and policy roots of land degradation. Academic researchers have not examined in scientific light the ecological implications of the grazing permit system, the institutional flaws that promote poor stewardship, the subsidy systems that award bad management practices, and the intrusions into market processes that drive arid land users to undesirable land use outcomes. Instead, researchers have become institutional bulwarks and apologists for flawed institutions, unworkable public policies, and badly motivated ranchers. They facilitate land degradation in the arid region by an academic infrastructure that has built its reputation around accommodating the growth and proliferation of a nonsustainable livestock industry in the American West. They defeat with their intransigence the noble conservation goals that their institutions symbolize.

6. Ecosystem Thinking and a Sustainable West

The role of institutions and public policies in the structure and function of rangeland ecosystems is barely understood. But existing evidence points irrefutably to a strong ecological link between the nature of institutions and policies and the state of the land resource. In an advanced country such as the United States that link is compelling: the roots of land degradation in the arid region are securely entrenched in more than a century of environmentally destructive public policies.

This observation has broad implications for global arid land management, particularly in advanced nations where the lessons of the arid region of the United States best apply. The conservation and protection of arid ecosystems cannot proceed without an awareness and understanding of the policies and institutions that impinge on land use. Moreover, the conservation and protection of arid ecosystems cannot succeed without an appropriate policy and institutional environment – one that *expands* rather than *restricts* flows of information, one that *assigns* rather than *shields* the costs and benefits of land use to land users, one that *acknowledges* rather than *ignores* the vital role of incentives in shaping the decisionmaking process, and one that *recognizes* rather than *obfuscates* the severe limitations of engineering human and natural ecosystems that are inherently complex and stubbornly uncertain. In the final analysis, the prevention of land degradation in both developing and advanced countries will depend on the willingness of nations to abide more by the verdicts of market processes than by the well-intentioned designs of planners and politicians. For discussions of how public land policy might be changed to correct the flaws we have previously discussed, see Anderson and Leal (1991), Hess (1992), Holechek (1993), Holechek and Hess (1994a), and O'Toole and Hess (1994).

References

Anderson, T.L. and D.L. Leal. 1991. *Free market environmentalism.* Westview Press, Boulder, Colorado.

de Toqueville, Alexis. 1835. *Democracy in America.* Saunders and Otley, London.

Gates, P.W. 1968. *History of public land law development.* Zenger Publishing Co., Washington.

Heady, H.F. (editor). 1988. The Vale rangeland rehabilitation program: An evaluation. *Forest Service Resource Bulletin PNW-RB-157.* U.S. Department of Agriculture, Forest Service, Pacific Northwest Forest and Range Experiment Station, Portland, Oregon.

Hess, K., Jr. 1992. *Visions upon the land: Man and nature on the western range.* Island Press, Covelo, California.

Holechek, J.L. 1992. Financial aspects of cattle production in the Chihuahuan Desert. *Rangelands* **14**: 145–149.

Holechek, J.L. 1993. Policy changes on federal rangelands: A perspective. *Journal of Soil and Water Conservation* **48**: 166–174.

Holechek, J.L. and J. Hawkes. 1993. Desert and prairie ranching profitability. *Rangelands* **15**: 104–109.

Holechek, J.L., J. Hawkes and T. Darden. 1994. Macroeconomics and cattle ranching. *Rangelands* **16**: 118–123.

Holechek, J.L. and K. Hess. 1994a. Free market policy for public land grazing. *Rangelands* **16**: 63–67.

Holechek, J.L. and K. Hess. 1994b. Brush control considerations: A financial perspective. *Rangelands* **16**: 193–196.

Holechek, J.L., R.D. Pieper and C.H. Herbel. 1989. *Range management principles and practices.* Prentice-Hall, Englewood Cliffs, New Jersey.

O'Toole, R. 1988. *Reforming the Forest Service.* Island Press, Covelo, California.

O'Toole, R. and K. Hess. 1994. Reforming the western range. *Different Drummer* **1**(2): 1–62.

Powell, J.W. 1878. *Report on the lands of the arid region of the United States with a more detailed account of the lands of Utah.* Government Printing Office, Washington.

Torell, L.A. and W. Word. 1991. *Range livestock cost and return estimates for New Mexico*, 1990. New Mexico Agricultural Experiment Station Research Report 659. Las Cruces, New Mexico.

United States Department of Agriculture (USDA), Forest Service. 1936. *The Western Range*. U.S. Congress 74th, Second Session, Senate Doc. 199. Washington, D.C.

United States General Accounting Office. 1988a. *Public rangelands: Some riparian areas restored but widespread improvement will be slow*. Report B-230548. Washington, D.C.

United States General Accounting Office. 1988b. *Rangeland management: More emphasis needed on declining and overstocked grazing allotments*. Report B-204997. Washington, D.C.

Workman, J.P. and S.G. Evans. 1993. Utah ranches – An economic snapshot. *Rangelands* **15**: 253–255.

USE OF EARTH OBSERVATION SATELLITE DATA FOR LAND DEGRADATION MAPPING AND MONITORING IN MEDITERRANEAN ECOSYSTEMS: TOWARDS A SATELLITE-OBSERVATORY

J. HILL, S. SOMMER, W. MEHL and J. MEGIER
Institute for Remote Sensing Applications, Environmental Mapping and Modelling Unit (EMAP), Joint Research Centre – Ispra Establishment, Commission of the European Communities, I-21020 Ispra (Va), Italy

Abstract. The degradation of the permanent seminatural vegetation and the resulting acceleration of soil degradation and erosion processes constitute major elements of land degradation in the Mediterranean basin. Given the European Commission's interest in mapping and controlling desertification phenomena in the Mediterranean member states, the "Environmental Mapping and Modelling Unit" (EMAP) of the Joint Research Centre is investigating possibilities for using operational earth observation satellites for mapping and repeated monitoring of vegetation and soil characteristics. In the context of previous experiments, approaches have been developed that can already be routinely applied to large regions. This paper discusses the problems of standardized retrieval of remotely sensed primary parameters (such as reflectance), concepts relating to the thematic interpretation of reflectance data, and the definition of satellite-derived degradation indices. In addition, this paper presents requirements for designing an operational satellite observatory for monitoring Mediterranean land degradation.

1. Introduction

Land degradation processes – which imply a reduction of the potential productivity of the land (e.g. soil degradation and accelerated erosion, reduction of the quantity and diversity of natural vegetation) – are widely spread in the Mediterranean basin. In continuing a long history of human pressure upon land resources, the main environmental impact results from interactions between climatic characteristics and ecologically unbalanced human interventions, which, in the sense of recent definitions of the United Nations Environment Programme*, are often summarized as "desertification processes." An overview of the ecological, physical, social, economic, and cultural issues that are collectively contributing to the increasing risk of further degradation of Mediterranean lands has recently been presented by Perez-Trejo (1994).

* UNEP(1991) has redefined desertification in the sense of considering it as "land degradation in arid, semiarid and dry-semihumid areas resulting mainly from adverse human impact," where the term "land" includes land and local water resources, the land surface, and its natural vegetation.

Environmental Monitoring and Assessment **37:** 143–158, 1995.

2. Main Sources and Processes of Mediterranean Ecosystem Degradation

The criteria that have been associated with desertification processes are in some ways inappropriate descriptions of Mediterranean landscapes but biased towards African examples. Perez-Trejo (1994) therefore concludes that we need a reconceptualization of desertification – one more appropriate for the European situation, in which the roles of urban-industrial expansion, tourism, and agriculture in relation to the allocation of water resources are seen as significant contributors to the problem. Inadequate land use practices (e.g. excessive grazing, fuelwood collection, uncontrolled fires) further contribute to the acceleration of degradation processes, which result mainly from complex interactions of plant growth and erosion processes. It is now widely agreed that accelerated water erosion is one of the most important sources of soil degradation, which, together with the destruction of vegetation cover and structure, contributes to the potential increase of desertification in Mediterranean Europe. Soil degradation and vegetation destruction are often connected, since the degree of soil degradation in many ways reflects the state of vegetation that covers and conditions it (Perez-Trejo, 1994). We conclude that the most important *physical* indicators of land degradation in the Mediterranean basin are therefore related to the destruction of soil and vegetation resources.

3. Extension and Dynamics of Degradation Processes

The degradation processes exhibit an enormous spatial variability within the Mediterranean basin. Degraded areas are found in the direct vicinity of apparently stable ecosystems, and both degrading and recovering systems may occur under a large variety of climatic and physiographic conditions. In other words, to understand whether and where desertification in the European Mediterranean is mainly driven by climatic conditions or adverse human impacts first of all requires a thorough understanding of degradation processes and also a good knowledge of the spatial extension of stable and endangered ecosystems at the regional level.

Although ecosystem processes with their implications for the future progress of desertification have been intensely studied at many field sites in the Mediterranean region (e.g. MEDALUS, 1993), it is not yet clear how findings from field studies at patch-scale can be extrapolated and upscaled to relatively large areas. But it is believed that remote sensing systems can significantly contribute to solve this problem. Satellite remote sensing also provides a way to map degraded (i.e. environmentally sensitive) areas and is virtually the only data source that permits repeated monitoring of land degradation dynamics. Both mapping and monitoring represent important prerequisites for drafting and implementing development plans and policy decisions about the sustained use of Mediterranean land resources.

3.1. Monitoring Land Degradation with Remote Sensing Systems

Operational earth observation satellites with multi-spectral sensor systems (e.g. Landsat-MSS and -TM, SPOT-HRV) have not only high spatial (30 or 20 m) and intermediate spectral resolution (6 or 3 bandpasses in the reflective part of the solar spectrum) but also rather large standard revisit intervals (16–26 days). The optical systems on these satellites register the downwelling solar radiation that has been reflected by the earth's surface and transmitted through the atmosphere. Electronic detectors measure the incident energy in several spectral bandpasses. The resulting current is transformed into a digital count with an analog/digital converter. And this signal is transmitted to a ground receiving station, where the data are processed to form image frames. One of the most important issues is to select land degradation process indicators that generally apply to the Mediterranean as a whole and can be observed with operational (i.e. spaceborne) remote sensing systems. In principle, accelerated water erosion is widely believed to be one of the most important sources of soil degradation, which together with the destruction of vegetation cover and structure, contributes to the potential increase of desertification in Mediterranean Europe. Both of these processes are often connected since the degree of soil degradation in many ways reflects the state of vegetation that covers and conditions it (Perez-Trejo, 1994).

Sparse vegetation communities, which are typical of deserts and semiarid regions, also have importance with respect to the present controversy over changes in global carbon and the resulting effect on climate. Although contributing only slightly to the total carbon pool, these plant communities are perhaps the most sensitive to changes in climate. Therefore, monitoring their dynamic behavior with remote sensing systems might even provide regional evidence of climatic change (Smith *et al.*, 1994). We conclude that the characterization of soil conditions (i.e. the erosional state of soils) and vegetation dynamics constitutes core elements for monitoring the dynamic behavior of Mediterranean ecosystems with remote sensing systems. But although remote sensing provides a convenient source of information, the data collected by these instruments do not directly correspond to the information we need. We must therefore interpret the signal that has interacted with remote objects by the properties of these remote objects (Verstraete, 1994).

3.2. Remotely Sensed Primary Parameters, Thematic Concepts, and Derived Indices

Engineering data on the detector sensitivity (i.e. calibration coefficients) allow us to reconvert encoded image grey values into measured radiances. Radiative transfer calculations can be used to correct for atmospheric effects, such that the surface-reflected radiance is restored from the satellite-measured signal. Dividing this term by the downwelling solar irradiance gives us an important *primary parameter*, which is termed 'bidirectional reflectance' ρ (Figure 1).

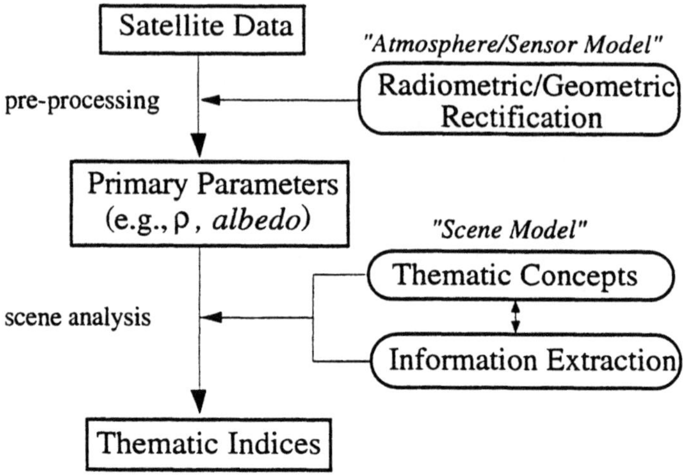

Fig. 1. The conversion of satellite raw data into standardised thematic information layers.

Although of relevance to the global radiation budget, albedo and reflectance changes per se are not direct indicators of land degradation processes, particularly when we consider spatially complex areas like the European Mediterranean. A simple increase in albedo/reflectance (surely characteristic for the desert encroachment in the African Sahel) might here be due to changes in land surface characteristics (e.g. increase of greenhouse areas, maturing cereals, nonphotosynthetic vegetation, clearing of woodlands) that do not necessarily imply negative effects. It is therefore not sufficient to simply map albedo or its changes over time. We need to infer the environmental impact of reflectance/albedo changes by characterizing their *physical nature* in land surface conditions. In other words, we must analyze albedo and reflectance changes in comparison to the multi-spectral characteristics of a range of known surface materials. We thus need a suitable *scene model* for converting multi-spectral reflectance into thematic information as shown in Figure 1. Here the question is not so much to identify a particular surface type, but to characterize it (Verstraete, 1994). The variety of methods that have been proposed range from empirical spectral indices to the design and inversion of physically based models. While the applicability of the approaches depends on the nature and accuracy of the desired information and the availability of resources (i.e. sensor characteristics), an important prerequisite for their operational use is that they satisfy requirements for standardization and portability.

Both the development of suitable indices and their interpretation in the thematic context of land degradation monitoring requires a *conceptual framework* that allows us to draw concise conclusions about land surface conditions. Though these underlying concepts might vary as a function of regional ecosystem characteristics (i.e. depending on physiographic conditions such as parent material and aridity), we request that the results from different regions can be consistently evaluated on a higher hierarchical level, allowing us to assess the system's susceptibility to further degradation by using image-derived but also ancillary information layers. Important conclusions will nevertheless depend on the capability to analyze multi-temporal time series. For this reason, retrospective studies are so important with respect to developing approaches for continuous monitoring of environmental changes (Graetz, 1994).

4. Toward Standardized Approaches

The most important requirement in satellite remote sensing is to conceptualize and streamline the data analysis so we can obtain consistent indicators for land degradation. This requirement implies the development of standardized processing schemes that can be efficiently applied to a variety of landscape types in the Mediterranean basin.

4.1. Radiometric Preprocessing Issues

Today we are in the favorable situation in that several radiative transfer codes are available. If constrained with the same atmospheric parameters, these codes provide consistent results (e.g. Conel *et al.*, 1988). Reflectance factor retrievals from the Landsat TM bands have been achieved with an accuracy of ±0.005 to 0.02 (Holm, 1989; Moran *et al.*, 1992; Wrigley *et al.*, 1992; Markham *et al.*, 1992). And this level of precision can also be accomplished when the atmospheric parameters are estimated from scene data themselves (Hill and Sturm, 1991; Hill and Aifadopoulou, 1990). The major problem in retrieving reflectance factors from satellite data lies in the sensitivity to the absolute radiance calibration of the sensors. But the scientific user community is increasingly aware of this problem. In-flight calibrations for Landsat-TM and SPOT data conducted at various high-reflectance sites are useful for determining at least an absolute radiometric calibration gain and monitoring the sensor degradation with time (e.g. Slater *et al.*, 1987; Thome *et al.*, 1993).

4.2. Conceptual Framework for Analyzing Vegetation and Soil Conditions

Thematic concepts provide the rationale for translating remote measurements of primary parameters (e.g. ρ) into relevant thematic information about vegetation and soil conditions. Ideally, such concepts should be valid for any location in large ecosystems. In reality, however, some adaptations might be needed to account for regional variations.

4.2.1. *Mapping Soil Conditions*

Eroded soils are often recognized through typical soil color changes resulting from removed topsoil. It is nevertheless difficult to define a universally applicable concept that accounts for a variability of soil types and the corresponding sequence of pedogenetic horizons. Our approach refers to basic concepts that consider soil development to be either *progressive* or *regressive* with time (Birkeland, 1990). Under progressive development, soils become better differentiated by horizons, and horizon contrasts intensify. In contrast, regressive pedogenesis refers to the addition of material to the surface at a rate that suppresses soil formation (e.g. eolian dunes, glacial moraines, distal fans) or the suppression of pedogenesis and the truncation of soil horizons by surface erosion. Both progressive and regressive pedogenesis alter the soil surface, and these alterations, because of corresponding color changes, are detectable through the wavelength-dependent variations of ρ (e.g. Baumgardner *et al.*, 1985; Escadafal, 1994).

The intensity of brunification and rubification and the organic matter content of the topsoil material thus provide important diagnostic features for the spectral identification of a majority of undisturbed Mediterranean soils (e.g. cambisols, fluvisols, luvisols, vertisols, rendzinas). Compared to undisturbed Mediterranean soils, soil erosion produces truncated soil profiles with decreasing amounts of iron oxides and organic carbon and increasing proportions of parent material (e.g. lithosols, regosols). Most parent materials differ spectrally from developed soil substrates because of spectral absorption features and increased albedo levels. The resulting concept, which is based on the spectral contrast between developed substrates and parent materials, seems to provide a widely applicable framework for relating spectrally detectable surface phenomena to Mediterranean soil conditions, thereby satisfying an important requirement for successfully applying remote sensing techniques (Hill, 1993; Hill *et al.*, 1994). But the validity of such concepts has to be carefully analyzed in the context of the physiographic conditions under which they should be applied. Modifications might be required, for example, in cases of extreme aridity where soil-forming processes do not permit noticeable amounts of organic components to accumulate (e.g. Escadafal, 1994).

4.2.2. *Mapping Vegetation Abundance*

Vegetation attributes are usually described by structure, dynamics, and taxonomic composition. Of the three, taxonomy is the least important. The classification most compatible with remote sensing relates to the projected foliage cover (PFC or cover) and the life form of the tallest vegetation stratum (Graetz, 1990). But about 75% of the earth's land surface is covered by sparse vegetation that transmits the color of the soil beneath, i.e. the PFC is below 1. In particular, semiarid ecosystems such as in the Mediterranean basin are dominated by such vegetation communities. For these, the soil surface itself should be as much an object of attention as is the vegetation (see Graetz, 1990), and the key issue is therefore to provide accurate estimates of green vegetation abundance, estimates that are not biased by the spectral contribution of background components (i.e. soils and rock outcrops). Attention should also be given to the spectral characteristics of non-green components of plant canopies and associated plant litters, which largely contribute to the reflectance of terrestrial surfaces in semiarid ecosystems (Elvidge, 1990). Although we know that the spectral resolution of earth observation satellite systems is not adequate to consistently differentiate between dry plant components and soils, efforts have to be made to resolve ambiguities from the image context (Smith *et al.*, 1990).

4.3. Advanced Data Interpretation – The Spectral Mixing Paradigm

We need information extraction methods that give largely unbiased estimates for green vegetation cover, permit the identification of soil-related spectral information, and allow sufficient standardization for multitemporal monitoring. Traditional multispectral classification approaches and most vegetation indices are not ideally suited to fulfilling these requirements (Hill *et al.*, in press). Since the inversion of physically based bidirectional reflectance models against satellite data is not feasible with available data sets, we wish to draw attention to suitable semiempirical models. One of the most promising approaches has become known as "Spectral Mixture Analysis" (SMA) (e.g. Adams *et al.*, 1989; Smith *et al.*, 1990). Assuming that most of the spectral variation in multispectral images is caused by mixtures of a limited number of surface materials, SMA attempts to model the multispectral reflectance ρ as a mixture of representative 'prototype' spectra, the so-called "spectral endmembers" (i.e. vegetation, soil, and bedrock components; shade as an illumination component; etc.). Linear mixing within the footprint of a multichannel spectrometer or imaging sensor further assumes that the surface components are large or opaque enough to allow photons to interact with only one component, i.e. the radiative transfer processes are additive. Spectra can then be unmixed by inverting the linear mixing equation

$$R_i = \sum_{j=1}^{n} F_j \cdot RE_{ij} + \varepsilon_i \quad \text{and} \quad \sum_{j=1}^{n} F_j = 1$$

through a least squares regression, while constraining the sum of fractions (i.e. the proportional weights) to one; R_i denotes the reflectance of the mixed spectrum in band i, RE_{ij} the reflectance of the endmember spectrum j in band i; F_j is the fraction of endmember j, and ε_i the residual error in band i. A unique solution is possible as long as the number of spectral components (endmembers) does not exceed the number of bands plus one.

The mixing paradigm provides a type of image enhancement that is not only intense but also physically meaningful (Craig, 1994). Its objective is to isolate the spectral contributions of important surface materials (endmember abundance) before these contributions are edited and recombined to produce thematic maps (Adams *et al.*, 1989). We have successfully adopted this approach to analyze the spectral information content related to the erosional state of soils (Figure 2) but also to derive precise maps of soil conditions and improved estimates of green

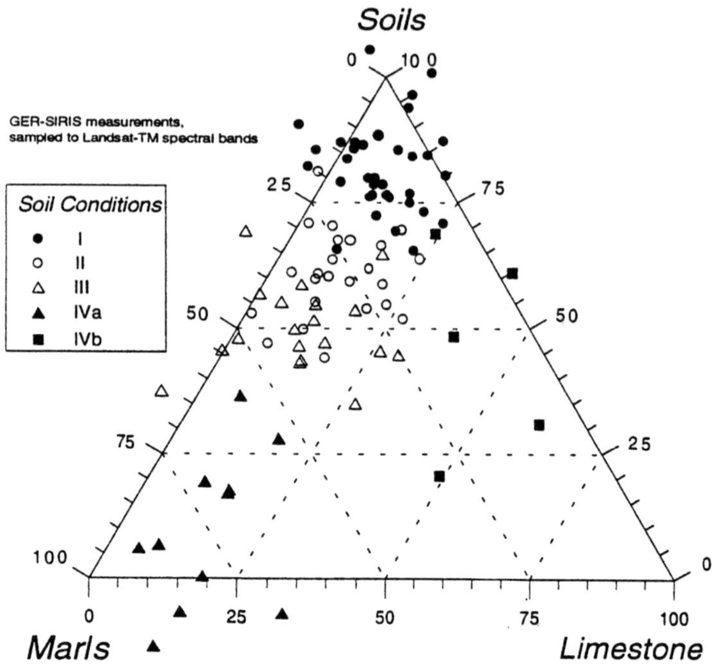

Fig. 2. Ternary plot of about 100 field- and lab-measured soil spectra at Landsat-TM spectral resolution, where the mixing volume is defined by one soil (vertic cambisol) and two bedrock spectra (marls and limestone). Soil conditions range from undisturbed soils (I) to substrates that are increasingly affected by erosion (II-IV) (Hill *et al.*, 1994b).

vegetation abundance from various types of multispectral images (e.g. Hill, 1993; Hill *et al.*, 1994).

An essential part of SMA consists in identifying optimized sets of spectral endmembers that are representative of the major variations of regional soil and vegetation conditions. Though endmembers can be retrieved from the image itself, we prefer to choose them from collections of spectroradiometric field or laboratory measurements, the so-called spectral libraries. The spectral library concept may be questioned with regard to the infinite number of spectral individuals encountered in the field. But there is reason to believe that we can represent the spectral variability in which we are interested by relatively few base categories (i.e. developed soils, fresh or weathered rock outcrops, green vegetation, nonphotosynthetic vegetation) as represented in Figure 2. Such libraries are already being compiled (e.g. Altherr *et al.*, 1991), and the compilation continues in fieldwork in Greece, Spain, France, and Italy.

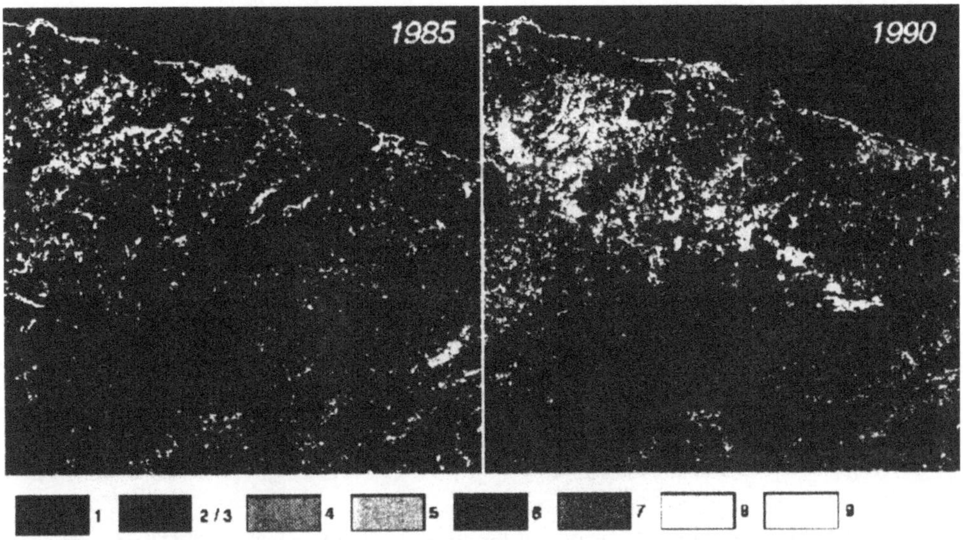

Fig. 3. Peloponese, Greece: Comparison of land surface properties based upon a synoptic evaluation of Landsat-derived information layers on soil conditions and vegetation abundance. The numbers refer to "desertification risk levels" in the sense of varying susceptibility to further degradation, ranking from "stable" (1: vegetation cover more than 50% and soil condition index I) to "irreversible degradation" (9: vegetation cover less than 20% and soil condition index IV).

4.4. Susceptibility Analysis through the Evaluation of Satellite-Derived Information Layers

We consider it very important to further evaluate temporal changes in soil and vegetation conditions in the context of a more ecologically based framework. Again, repeated monitoring requires standardization, mainly because any improvements in the analysis need to be easily incorporated into existing monitoring schemes by reprocessing already available image-derived information layers (e.g. soil condition index, estimated green vegetation cover). As a first attempt we have proposed a simplified technique to combine soil- and vegetation-related information layers (Hill, 1993; Hill *et al.*, 1994). This technique is based on the fact that in a more synoptic perspective an environmental ranking of soil conditions in Mediterranean ecosystems can be defined as a function of green vegetation cover (because of its protective role), thereby distinguishing for example between regrowth/succession on eroded soils and sparse vegetation cover on well-preserved soil resources (Figure 3).

The proposed susceptibility index represents only a first attempt to combine image-derived information layers of soil and vegetation conditions through a strictly formalized procedure. We know that this index should be revised and further developed in close cooperation with ecologists and specialists from the geosciences. New concepts or findings from detailed field studies are easily integrated by updat-

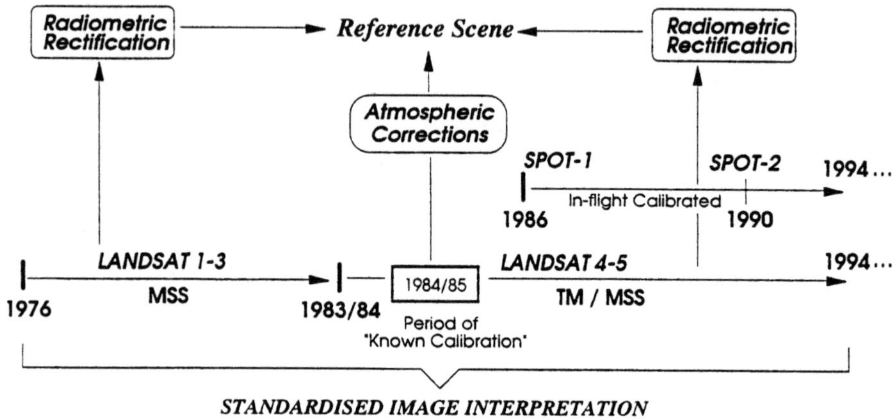

Fig. 4. Radiometric rectification pathways for the analysis of Landsat and SPOT time series.

ing the rule base module such that these findings can be readily applied to new satellite images and archived data products. The incorporation of terrain parameters from digital elevation data (e.g. slope, exposition) appears particularly attractive for obtaining an improved evaluation of ecosystem characteristics, such as water availability and erosion risks.

5. A Satellite Observatory for Monitoring Mediterranean Land Degradation

Finally, we wish to discuss basic elements of a future operational environmental monitoring system for the Mediterranean basin. This system requires a concise definition of the region to be observed but also includes strategies for implementing efficient processing schemes as a function of existing data types, archives, present and future sensor availability, and overpass cycles. Although this proposal focuses on the use of operational earth observation satellite systems (i.e. Landsat and SPOT), we believe that existing and future low-resolution satellites with more frequent coverage (e.g. the NOAA AVHRRs, the planned VEGETATION instrument onboard SPOT-4, MERIS, etc.) and active microwave systems (e.g. ERS-1, ERS-2, RADARSAT) should be incorporated at a later stage. These instruments are particularly required where the temporal dynamics of changes at the landscape level, identified through operational high-resolution systems, need to be understood in a larger spatial context. The data of microwave sensors are also expected to complement the surface spectral reflectance with soil/vegetation parameters retrieved from radar backscatter and to provide all-weather capabilities for the fast assessment of the effects of natural disasters.

5.1. Operational Land Degradation Mapping and Monitoring

Identifying degraded areas in the sense of an environmental inventory provides the fundamental basis for such approaches. To better understand the *processes* of land degradation from their spatial context, it is essential to monitor soil conditions and the disturbance regime of plant communities over time, including their successional recovery. With regard to the available data archives*, we wish to particularly emphasize the importance of retrospective studies that may provide the key not only to understanding the present situation, but also for optimizing our approaches to regular monitoring following the scheme of Figure 4.

Since the retrospective analyses of archived data and regular monitoring are sensitive to the absolute radiance calibration of the sensors, we need practical approaches to minimize the uncertainty about these calibration coefficients. In-flight calibration coefficients for earth observation satellites are available from specifically designed experiments (e.g. Slater *et al.*, 1987; Thome *et al.*, 1993). Though the coefficients are not continuously updated, they are at least valid for specific periods of the sensor's lifetime, allowing us to follow alternative pathways that do not require that each individual scene be radiometrically corrected through radiative transfer calculations (Figure 4). It would be sufficient to apply atmospheric corrections only to reference scenes from so-called periods of known calibration and to routinely adjust the radiometry of earlier or later image acquisitions with reference to pseudo-invariant surfaces in the scenes (Schott *et al.*, 1988). This approach, which we call radiometric rectification, has already been successfully applied to a time series of Landsat TM data (Hill *et al.*, 1994).

5.2. Stratified Sampling for Environmental Monitoring

The criteria for defining the regional extension of Mediterranean-type ecosystems have been a scientific issue (e.g. Di Castri and Mooney, 1973; Di Castri *et al.*, 1981). In principle we have to consider the Iberian Peninsula (except its humid parts in the north), the French *Midi*, the Appennine Peninsula, the Croatian-Albanian coast, Greece, western and southern Turkey, western Syria, Lebanon, Israel, parts of Libya, northern and central Tunisia, and Algeria and Morocco north of the Atlas massifs. Even when we limit ourselves to the Mediterranean member states of the European Union, we should accept that a complete coverage with data from earth observation satellites is not feasible. In particular the repeated monitoring for change detection would largely exceed our financial and data analysis capacities.

Instead, we propose to define a sampling frame of representative sites that account for the major physiographic and agroeconomic variations within the Mediterranean basin. Such a sampling scheme must be based on the path/row reference system of the Landsat and SPOT orbiting patterns complemented by

* Landsat-MSS data for European areas are available from 1976, Landsat-TM coverage started in 1983.

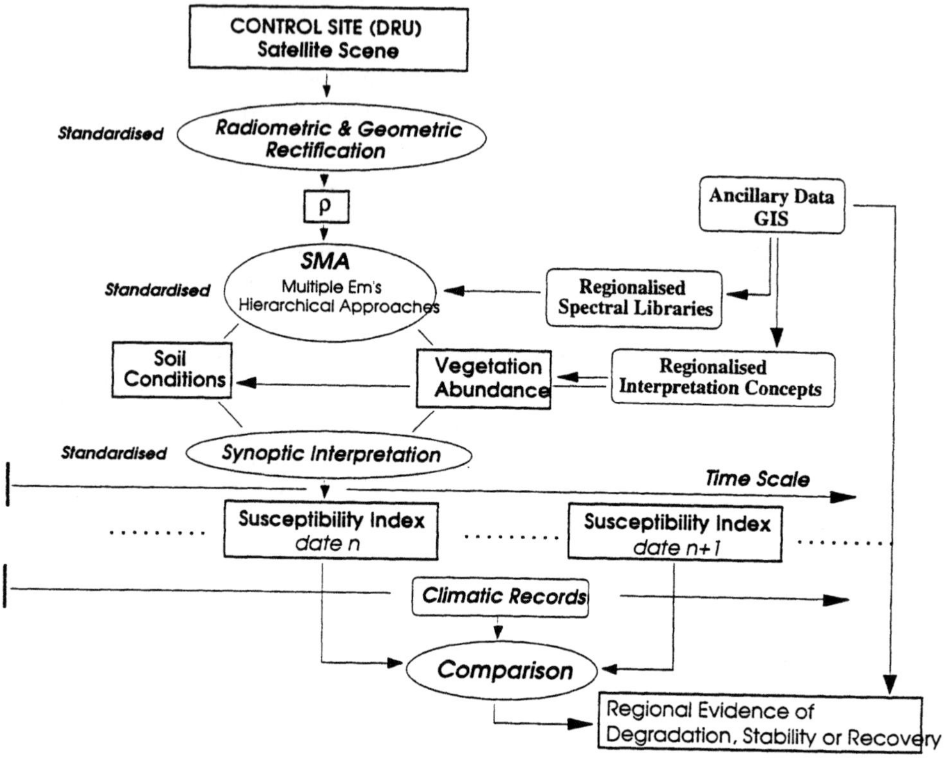

Fig. 5. Processing scheme for deriving standardised land degradation indices from earth observation satellite data.

stratifying the region into zones with homogeneous climatic and edaphic characteristics (ecoregions). This requirement is important because any comparative analysis of land degradation dynamics must account for the physiographic variability within the Mediterranean basin. We expect that these ecoregions can be derived largely from the evaluation of already available cartographic documents and topographic data (e.g. Commission of the European Communities, 1985; Council of Europe and Commission of the European Communities, 1987). Each image frame then represents a measurement for analyzing how changes in climate, drought, land use, and fire regimes generate regional changes in ecosystem processes and patterns that can influence future land degradation. As such, each data layer may be integrated into a nested hierarchy of aerial units with similar responses to desertification (Desertification Response Units), as recently proposed by Imeson *et al.* (in press).

5.3. Standardized Thematic Interpretation

Our proposed scheme consists of several standardized modules characterized by the ellipsoids in Figure 5. Radiometrically rectified satellite images provide the primary parameter ρ for a given control site, and we then use spectral mixture analysis (SMA) to convert the multispectral surface reflectance into soil- and vegetation-related information layers. Although a unique set of standard endmembers has been identified and can be used for different areas of primarily carbonatic rocks, we believe that the spectral unmixing can sometimes be optimized by accounting for regional endmember characteristics (Figure 5). The conceptual framework may also require changes in view of regional conditions, but it looses validity only in areas of extreme aridity.

The synoptic interpretation module uses the intermediate information layers (e.g. soil condition index, projected foliage cover) for computing an index of degradation and susceptibility to land degradation processes (Figure 5). The comparison of susceptibility indices from different years will provide evidence of either degradation, stability, or recovery on a regional scale. But an important objective of the thematic interpretation is to separate the rhythmic phenological changes of growth and senescence from episodic changes introduced by climate or human-induced disturbance. We must therefore also incorporate the climatic records in order to understand whether the meteorological situation in both years was at all comparable (Figure 4). Since the use of abnormally humid or dry years should be avoided, we cannot define a fixed recurrence rate. Instead, we propose a flexible monitoring scheme that guaranties at least one scene per successive 5-year-interval. Moreover, to exclude artifacts resulting from phenological effects and to minimize radiometric distortions that result from bi-directional effects and illumination differences, images from different seasons should not be directly compared.

6. Conclusions

We have summarized important sources and processes of Mediterranean ecosystem degradation to discuss options for monitoring land degradation with remote sensing systems. Unlike meteorological approaches that concentrate on the exchange of matter, momentum, and radiation between the earth's surface and the atmosphere, our approach has emphasized the importance of thematic interpretation pathways for characterizing actual land surface conditions.

Furthermore, we have shown how remotely sensed primary parameters, such as the spectral surface reflectance ρ, can be converted into a standardized characterization of soil conditions and vegetation abundance. In this context, we have introduced the term "thematic concepts" by which we understand the conceptual background for identifying functional links between surface reflectance and vegetation and soil characteristics. Such concepts are mainly based on research in geosciences and ecology.

We require that these concepts not only be valid for local experiments but that they can be applied to extended regions with specific physiographic conditions (e.g. bioclimate, lithology, soil-forming processes). As long as these multiple concepts can be related to regional-scale ecosystems, we even foresee using them in responding to physiographic conditions of particular regions. The data analysis employs linear Spectral Mixture Analysis as a core element. SMA provides an intense but also physically meaningful image enhancement that ideally complements the conceptual framework we have introduced. Selectively editing the resulting fraction images permits vegetation- and soil-related spectral information to be efficiently separated, an important advantage for obtaining more objective estimates of green vegetation abundance and for mapping substrate-related spectral soil properties more independently from the disturbing influence of sparse vegetation cover or soil moisture and illumination differences. SMA further holds the potential of having required processing parameters largely standardized. The inputs from individual analysts can be minimized, which is an advantage for designing operational processing schemes. Although some processing modules that appear suited for standardized data analysis within operational schemes are already available, more efforts are needed, particularly for defining suitable sampling schemes (i.e. selecting monitoring sites representative of large ecoregions), for extending the thematic concepts for an interpretation of remotely sensed primary parameters as a function of ecosystem characteristics, for incorporating ancillary information (climatic records, lithology, topography), and for improving processing modules with respect to the physiographic variability within the Mediterranean basin.

Finally, we wish to emphasize the importance of a retrospective analysis of earth observation satellite images covering almost a 20-year period. The integrated interpretation of the satellite-derived information layers, climatic records, and results from detailed field studies may give us a new perspective on understanding land degradation processes in the European Mediterranean.

References

Adams, J.B., M.O. Smith and A.R. Gillespie. 1989. Simple models for complex natural surfaces: A strategy for the hyperspectral era of remote sensing, in *Proceedings of the International Geoscience and Remote Sensing 1989 Symposium*, July 10–14, Vancouver, Vol. 1, pp. 16–21.

Altherr, M., J. Hill, M. Mehl and P. Pollicini. 1991. NASA multi-sensor airborne campaign: Reflectance spectroscopy Ardèche, June-July 1991. Technical Note no. I.91.120. Joint Research Center, Institute for Remote Sensing Applications, Ispra, Italy.

Birkeland, P.W. 1990. Soil-geomorphic research – A selective overview. *Geomorphology* **3**: 207–224.

Baumgardner, M.F., E.R. Stoner, L.F. Silva and L.L. Biehl. 1985. Reflective properties of soils, In N. Brady (Ed.), *Advances in Agronomy* 38, Academic Press, New York, pp. 1–44.

Commission of the European Communities. 1985. *Soil map of the European Communities*, scale 1:1,000,000. Luxembourg: Office for Official Publications of the European Communities.

Conel, J.E., R.O. Green, R.E. Alley, C.J. Bruegge, V. Carrere, J.S. Margolis, G. Vane, T.G. Chrien, P.N. Slater, S.F. Biggar, P.M. Teillet, R.D. Jackson and M.S. Moran. 1988. In-flight radiometric calibration of the Airborne Visible/Near-Infrared Spectrometer (AVIRIS), in P.N. Slater (Ed.), *Recent advances in sensors, radiometry, and data processing for remote sensing, Proceedings of the International Society for Optical Engineering* (SPIE) **924**: 179–195.

Council of Europe and Commission of the European Communities. 1987. *Map of the natural vegetation of the member countries of the European Community and the Council of Europe*, Office for Official Publications of the European Communities, Luxembourg.

Craig, M.D. 1994. Minimum-volume transforms for remotely sensed data. *Institute of Electrical and Electronics Engineers Transactions on Geoscience and Remote Sensing* **32**(3): 542–552.

Di Castri, F., D.W. Goodall and R.L. Specht (editors). 1981. Mediterranean-type shrublands, *Ecosystems of the World*, 11, Elsevier Scientific Publishing Company, Amsterdam.

Di Castri, F. and H. A. Mooney (editors). 1973. *Mediterranean type ecosystems, origin and structure*. Springer-Verlag, Berlin.

Elvidge, C.D. 1990. Visible and near infrared reflectance characteristics of dry plant materials, *International Journal of Remote Sensing* **11**: 1775–1795.

Escadafal, R. 1994. Soil spectral properties and their relationship with environmental parameters–Examples from arid regions, in J. Hill and J. Mégier (Eds.), *Imaging spectrometry – A tool for environmental observations*, Kluwer Academic Publishers, Dordrecht, pp. 71–87.

Graetz, R.D. 1990. Remote sensing of terrestrial ecosystem structure: An ecologist's pragmatic view, in R.J. Hobbs and H.A. Mooney (Eds.), *Remote sensing of biosphere functioning*, Springer-Verlag, New York, pp. 5–30.

Graetz, R.D. 1994. Empirical and practical approaches to land surface characterisation and change detection, in J. Hill and D. Peter (Eds.), *Remote sensing for land degradation and desertification monitoring in the Mediterranean basin*, Proceedings of a workshop held in Valencia, Spain, 13–14 June 1994. In preparation.

Grenon, M. and M. Batisse (editors). 1989. *Futures for the Mediterranean basin, the Blue Plan*, Oxford University Press, Oxford.

Hill, J. 1993. Monitoring land degradation and soil erosion in Mediterranean environments. *ITC Journal* **4**: 323–331.

Hill, J. and D. Aifadopoulou. 1990. Comparative analysis of Landsat-5 TM and SPOT HRV-1 data for use in multiple sensor approaches. *Remote Sensing of Environment* **34**: 55–70.

Hill, J., J. Megier and W. Mehl. In press. Land degradation, soil erosion and desertification monitoring in Mediterranean ecosystems, *Remote Sensing Reviews*.

Hill, J., W. Mehl, M.O. Smith and J. Megier. 1994. Mediterranean ecosystem monitoring with earth observation satellites, in *Proceedings of the 13th EARSEL Symposium on Remote Sensing – From Research to Operational Applications in the New Europe*, Dundee, Scotland, 28 June-1 July. Springer-Verlag, Budapest, pp. 131–141.

Hill, J. and B. Sturm. 1991. Radiometric correction of multitemporal Thematic Mapper data for use in agricultural land cover classification and vegetation monitoring. *International Journal of Remote Sensing* **12**: 1471–1491.

Holm, R.G., M.S. Moran, R.D. Jackson, P.N. Slater, B. Yuan and S.F. Biggar. 1989. Surface reflectance factor retrieval from Thematic Mapper data. *Remote Sensing of Environment* **27**: 47–57.

Imeson, A.C., L.H. Cammeraat and F. Perez-Trejo. In press. The response of landscape-units to desertification, in J. Thomes and J. Brandt (Eds.), *Mediterranean desertification and land use*, Wiley, New York.

Le Houerou, H.N. 1990. Ecological guidelines to control land degradation in European Mediterranean countries, in J.L. Rubio and R.J. Rickson (Eds.), *Strategies to combat desertification in Mediterranean Europe*, EUR 11175 EN/ES. Office for Official Publications of the European Communities, Luxembourg, pp. 331–360.

Markham, B.L., R.N. Halthore and S.J. Goetz. 1992. Surface reflectance retrieval from satellite and aircraft sensors: Results of sensor and algorithm comparisons during FIFE. *Journal of Geophysical Research* **97**: 18,785–18,795.

MEDALUS. 1993. Mediterranean desertification and land use. Executive summary, abstracted from the first phase of the MEDALUS Project. Bristol, United Kingdom.

Moran, M.S., R.D. Jackson, P.N. Slater and P.M. Teillet. 1992. Evaluation of simplified procedures for retrieval of land surface reflectance factors from satellite sensor output. *Remote Sensing of Environment* **41**: 169–184.

Perez-Trejo, F. 1994. *Desertification and land degradation in the European Mediterranean*. EUR 14850 EN. Office for Official Publications of the European Communities, Luxembourg.

Schott, J.R., C. Salvaggio and W.J. Volchok. 1988. Radiometric scene normalisation using pseudo-invariant features. *Remote Sensing of Environment* **26**: 1–16.

Slater, P.N., S.F. Biggar, R.G. Holm, R.D. Jackson, Y. Mao, M.S. Moran, M. Palmer and B. Yuan. 1987. Reflectance- and radiance-based methods for the in-flight absolute calibration of multispectral sensors. *Remote Sensing of Environment* **22**: 11–37.

Smith, M.O., J.B. Adams and D.E. Sabol. 1994. Mapping sparse vegetation canopies. In J. Hill and J. Megier (Eds.), *Imaging spectrometry – A tool for environmental observations*, Kluwer Academic Publishers, Dordrecht, pp. 221–235.

Smith, M.O., S.L. Ustin, J.B. Adams and A.R. Gillespie. 1990. Vegetation in deserts: I. A regional measure of abundance from multispectral images. *Remote Sensing of Environment* **31**: 1–26.

Thome, K.J., D.I. Gellmann, R.J. Parada, S.F. Biggar, P.N. Slater and M.S. Moran. 1993. In-flight radiometric calibration of Landsat-5 Thematic Mapper from 1984 to present. *Proceedings of Photo-Optical Instrument Engineering Symposium*, 12–16 April, Orlando, Florida, pp. 126–130.

UNEP. 1991. *Status of desertification and implementation of the United Nations plan of action to combat desertification*, United Nations Environment Programme, Nairobi.

Verstraete, M.M. 1994. Scientific issues and instrumental opportunities in remote sensing and high resolution spectrometry, in J. Hill and J. Megier (Eds.), *Imaging spectrometry – A tool for environmental observations*, Kluwer Academic Publishers, Dordrecht, pp. 25–38.

Wrigley, R.C., M.A. Spanner, R.E. Slye, R.F. Pueschel and H.R. Aggarwal. 1992. Atmospheric correction of remotely sensed image data by a simplified model. *Journal of Geophysical Research* **97**: 18,797–18,814.

DESERTIFICATION OF THE EASTERN KAROO, SOUTH AFRICA: CONFLICTING PALEOECOLOGICAL, HISTORICAL, AND SOIL ISOTOPIC EVIDENCE

M. T. HOFFMAN*, W. J. BOND** and W. D. STOCK**
*National Botanical Institute, Private Bag X7, Claremont, 7735, South Africa; **Department of Botany, University of Cape Town, Private Bag X16, Rondebosch, 7700, South Africa

Abstract. The desertification debate in South Africa has benefitted greatly in recent years from the contributions of a wide range of disciplines. In this paper we review the conflicting and supporting evidence for degradation in the eastern Karoo as reported in recent archaeological, historical, and stable carbon isotope studies as it relates to three key aspects of the debate: the precolonial environment, the rate and nature of change, and the relative contributions of humans and climate to the process. First, all studies suggest a greater grassiness at some time in the past, but researchers disagree on the timing of the switch to more shrubby conditions in the eastern Karoo. Second, regional rainfall records for the past 2 decades reveal an above-average rainfall period, and numerous long-term surveys show an increase in grass cover over the same period. These findings question the expanding Karoo hypothesis as well as the argument that the Karoo's carrying capacity has decreased in recent years. Finally, the relative responsibilities of humans and climate in the degradation process remain poorly understood and generally have not formed the focus of investigation.

1. Introduction

The extensive South African desertification literature spans more than a century of active debate. The desertification problem has not only been of interest to academics but has had important financial consequences for the South African state. For example, in the belief that too many livestock have been the root cause of the perceived degradation of South Africa's arid and semiarid rangelands, more than R 45 million was spent in the 1970s alone on implementing the government's stock reduction scheme (Pringle, 1982). Also, the rationale for the current National Grazing Strategy rests on the assumption that the Karoo is a desertified landscape that continues to degrade (Du Toit *et al.*, 1991), with each step in the degradation process being more difficult and expensive to reverse (Milton *et al.*, 1994). More recently, the desertification issue has entered the political debate as the cost of the apartheid years has been broadened to include the degradation of the arid and semiarid southern African environments (Wilson and Ramphele, 1989; Durning, 1990).

But is this the best or only view on which to make policy decisions, spend money, and interpret our history? What if, in fact, views conflict over the nature of precolonial, eastern Karoo environments and the rates of change that have taken place before and since European colonists settled in the region from the late 18th century? What is the best opinion to base our decisions on?

Environmental Monitoring and Assessment **37:** 159–177, 1995.

This review is an attempt to present a broad spectrum of scientific evidence pertaining to the South African desertification debate, particularly for the eastern Karoo. First, we describe the eastern Karoo environment and vegetation dynamics and outline the region's desertification literature. Then we discuss the main arguments as contained in the recent paleoecological, archaeological, and historical publications. Finally, we discuss the recent use of stable carbon isotopes in the soil to address the problem. Throughout the text we have tried to point out conflicting results that have emerged from the different methodologies and have highlighted the spatial and temporal limitations of each technique, suggesting where further work is still needed.

2. The Eastern Karoo Environment

2.1. Climate and Long-Term Rainfall Trends

The bulk of the desertification literature for South Africa is concerned with vegetation changes in the eastern Karoo (Figure 1), an extensive and poorly defined region that straddles the Orange River. The eastern Karoo comprises a broad ecotone between the Grassland and Nama-Karoo biomes (Rutherford and Westfall, 1986) and represents a relatively featureless plain with mountain ranges delimiting its borders to the south and east. The vegetation of the eastern Karoo forms part of Acocks' (1953) False Upper Karoo Veld Type. According to Acocks (1953), the development of this veld type "...constitutes the most spectacular of all the changes in the vegetation of South Africa. The conversion of 32,000 square km of grassveld into eroded Karoo can only be regarded as a national disaster."

Average rainfall for the eastern Karoo is about 362 mm per annum (Figure 2) and decreases from about 447 mm per annum in Bethulie, at the northeast corner of the region, to about 285 and 283 mm per annum at De Aar and Murraysburg at the western and southwestern parts of the region respectively. Rain falls mainly in the summer between October and March although there is a distinct trend for a late summer/autumn pattern towards the west. The mean annual temperature at Middelburg, in the center of the region, is 15.1 ° C. A more complete description of the eastern Karoo climate and environment in relation to the rest of the Karoo region is contained in Cowling *et al.* (1986) and Cowling and Roux (1987).

The long-term rainfall record for the region (Figure 2) reveals that the past 2 decades of the 19th century were generally far wetter than the period from 1903–1973 when at least three separate and extended drought events occurred during the years 1919–1931, 1944–1949, and 1962–1973. Since 1974, however, the region has experienced one of the highest rainfall periods on record with a number of very high rainfall years.

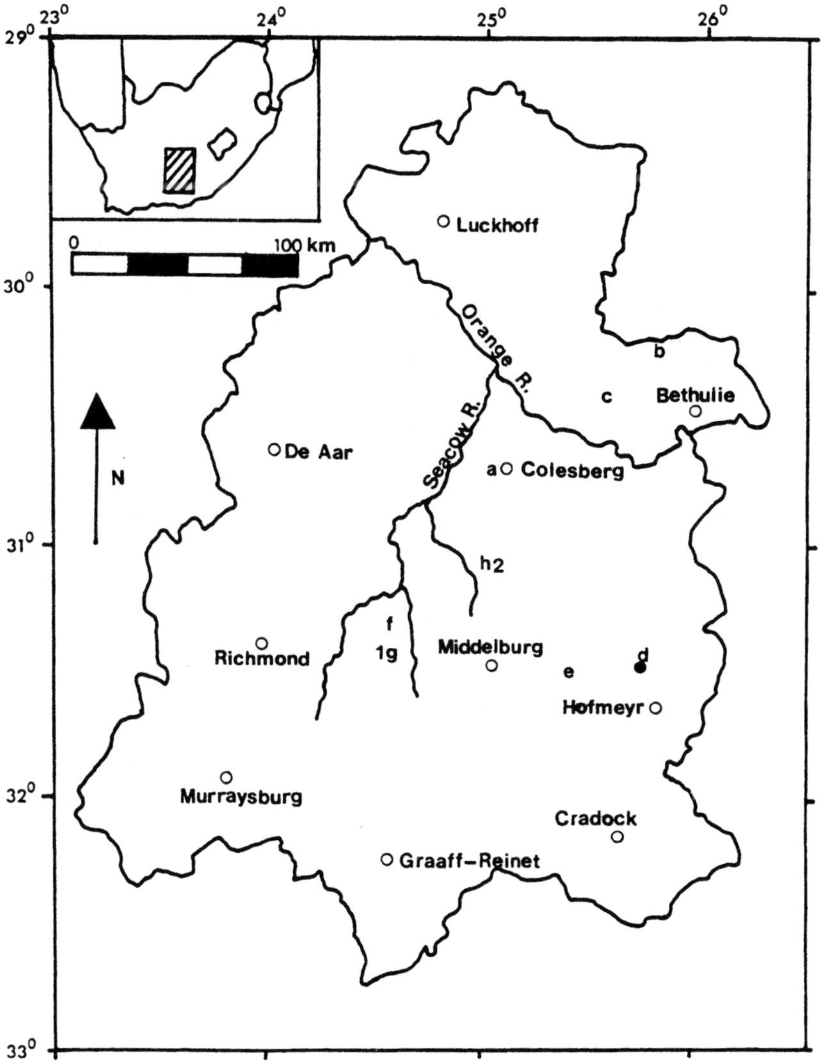

Fig. 1. The eastern Karoo as delimited in this review showing important place names and features mentioned in the text. The location of the stable carbon isotope analysis study sites are shown by the letters: a = Colesberg, b = Hillside, c = Driekuil, d = Hofmeyr, e = Culmstock, f = Wonderheuwel, g = Paardevlei, h = Oppermanskop; key archaeological sites include (1) Abbots Cave and (2) Oppermanskop and the photographic station in Figure 4a-c is shown as a dot.

2.2. Vegetation dynamics

The vegetation dynamics of the eastern Karoo are not well described, and there is disagreement over the relative response of grasses and shrubs to rainfall, grazing, and between-guild competition. One aspect, however, which has received much attention has been the large year-to-year fluctuations in the amount of grass and to

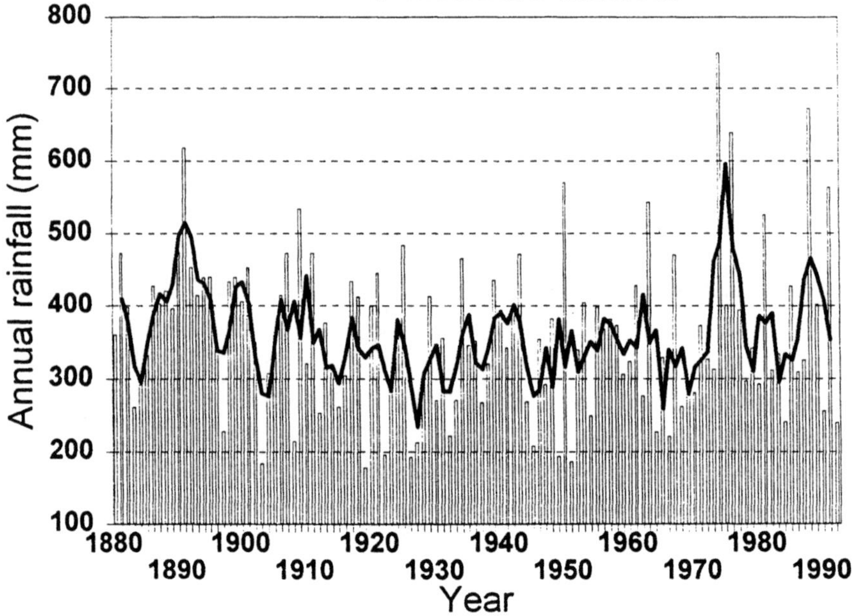

Fig. 2. Mean annual rainfall totals (bars) and 3-year running mean for 16 eastern Karoo stations from 1880–1992. The long-term annual mean for all stations over the period is 362 mm.

a lesser extent shrub cover (Roux, 1966; Novellie and Strydom, 1987; O'Connor and Roux, 1995). While the mechanisms have not been determined experimentally, it is thought that both high annual (Palmer *et al.*, 1990; O'Connor and Roux, 1995) and summer rainfall (Roux, 1966; Novellie and Strydom, 1987; Hoffman *et al.*, 1990) favor the recruitment and growth of especially C_4 grasses. Since summer rain (Sept.-Feb.) and mean annual rainfall are highly co-correlated in the eastern Karoo, it may be difficult without experimentation to establish which variable most influences the response of grasses. For example, in our study area, the Spearman rank correlation coefficients (r_s) for the correlation between summer rainfall and annual rainfall totals for the 16 rainfall stations with records between 1880–1992 (Figure 2) were between 0.60 and 0.89 with a mean for all stations of 0.86 ($p < 0.001$ in all cases).

Hoffman *et al.* (1990) and Milton and Hoffman (1994) have proposed a cyclical model of succession in the eastern Karoo in which the rapid response of grasses to high summer (or annual) rainfall results in a corresponding decline in dwarf shrubs. Grasses are both taller and generally have shallower and more extensive root systems than shrubs. It is suggested that these morphological traits enable grasses to out compete the dwarf shrubs for both light and moisture. Roux (1966) argues, however, for the central role winter precipitation plays on shrub recruitment and growth. O'Connor and Roux's (1995) exhaustive analysis of one of the few long-term data series for the region finds little evidence for either the between-guild competition suggestion proposed by Milton and Hoffman (1994) or for the winter rain hypothesis as determinants of shrub cover suggested by Roux (1966).

There is strong evidence that summer grazing reduces grass cover over periods from 1 to a few year's duration (Roux, 1966; O'Connor and Roux, 1995). But the long-term effect of summer grazing on grass abundance is poorly known because, irrespective of grazing treatment, grasses appear able to recruit following significant rainfall events (Roux, 1966; Novellie and Strydom, 1987). We will return to many of these themes in the dynamics of the vegetation of the eastern Karoo in the following discussion.

3. The Desertification Debate in South Africa

Hoffman and Cowling (1990) and Dean *et al.* (1995) discuss the history of the South African desertification debate, which up until about 1920 consisted of a series of anecdotal comments alluding to the poor condition of the vegetation and the impact of domestic stock on vegetation composition and cover (e.g. Brown, 1875; Hobson, 1896). The first serious scientific investigation of the problem is contained in the comprehensive Drought Investigation Commission's (1923) report. Without negating the effect that the series of severe drought years between 1911–1919 (Figure 2) had had on the vegetation of the Karoo in general, the members of the Commission all warned of the dire consequences of degradation resulting from too many animals on the land. As will be shown later, these warnings were ignored by the farming community as sheep numbers nearly doubled over the next decade.

By the late 1940s the first general scenarios of the nature of precolonial eastern Karoo environments and rates of change were published (De Klerk, 1947; Tidmarsh, 1948). But the most important and lasting impact on the desertification debate in South Africa resulted from John Acocks' (1953) beautifully illustrated color maps showing the nature of precolonial South African environments and the current and future states under different management scenarios. Not only did Acocks present an easily understandable message, but he also predicted a gloomy future for the subcontinent. According to Acocks (1953), unless urgent steps were taken, by the year AD 2050 the unproductive shrubby Karoo vegetation would extend over much of the subcontinent, replacing the more productive grasslands and arid savanna vegetation in turn. The impact of too many animals would be directly responsible for these shifts in arid land boundaries. This message has been repeated time and again ever since, even in the relatively modern ecological literature (Moll and Gubb, 1989). Little original work contributed to the debate until fairly recently.

In recent years, however, interest in the desertification problem has resurged, and current debates focus on three key issues. The first concerns the nature of the precolonial eastern Karoo environment. There is disagreement as to whether the eastern Karoo had more or less perennial grass cover than exists at present. Advocates of the view that the eastern Karoo has been desertified argue that the precolonial environment contained a better perennial grass sward cover than that of today and that the change to a dwarf shrubland was coincident with the colonization of the region by the first European, seminomadic pastoralists or trek-boers (Acocks, 1953; Bond *et al.*, 1994).

The second key issue is centered on the rate of expansion of the dwarf shrublands into the more productive grassland communities of the Grassland biome to the north and east of the Nama-Karoo biome. Those who advocate that the Karoo has become desertified argue for a predictable annual expansion of between about 2 and 5 km per annum (Moll and Gubb, 1989, and many others; see Hoffman and Cowling, 1990). Others find little evidence for this view and argue instead that the long-term changes to the species composition of this region need to be evaluated in the light of the dynamics of component organisms (Dean *et al.*, 1995). In particular, the effect of summer and annual rainfall totals on especially grass cover needs to form a part of the debate.

The third issue addressed by the desertification debate in South Africa concerns the relative responsibilities of overstocking versus drought in influencing ecosystem shifts in grassiness and shrubiness. The desertification proponents point out that no statistically significant evidence shows that rainfall has decreased in the Karoo during the past 100 years and emphasize the role of overstocking in lowering the carrying capacity of the region's rangelands (Dean and Macdonald, 1995).

Recent paleoecological, historical, and ecological sources have contributed greatly to our understanding of these three issues. Each contribution is discussed below.

4. Paleoecological and Archaeological Data

Paleoecologists and archaeologists have used a variety of sources to reconstruct environmental conditions in the eastern Karoo from the Holocene (and earlier in some cases) to the present. Sampson (1994), for example, has used mammal bone, bird bone, and ostrich egg shell deposits in caves in the Seacow River Valley in the central part of the eastern Karoo to reconstruct the nature of precolonial environments. Avery (1991), on the other hand, has used micromammal remains of rodents and insectivores recovered from owl pellets in caves to show how broad habitat classes such as shrubs and grasses may have changed with time. Finally, Scott and Bousman (1990) use fossil pollen in hyrax middens to reconstruct the past vegetation and source of proxy paleoclimatic data.

Whatever the methodology, all authors agree that climatic conditions and vegetation composition of the eastern Karoo have greatly fluctuated during the last millennium. There is also agreement that when high rainfall periods have dominated, grassiness in the sequence has concomitantly increased. All authors imply that the past is characterized by substantial and widely fluctuating shifts in the amount of grass and shrub cover in the landscape with time and that each of these phases may have lasted for periods up to a few centuries.

But this is where the agreement appears to end, and there is conflict around two further issues. The first concerns the timing of the onset of shrubbiness in the eastern Karoo, and the second deals with the effect that overgrazing by European stock may have had on the vegetation of the region, particularly during the 19th century.

With regard to the first issue, Avery (1991) suggests that "The considerable improvement in vegetation cover in the 17th century must have been brought about by an increase in, and greater predictability of, rainfall." Sampson (1994), on the other hand, finds that there was a "....very marked drop in grass pollen between c. [AD] 1600 and 1750..." and that "...this must reflect a decline in grass cover, probably resulting from diminishing rainfall...." Scott and Bousman (1990) concur with Sampson (1994) in suggesting that the onset of shrubby conditions predated the arrival of European stock farmers in the region by stating that "After 300 yr B.P. [i.e. AD 1650] dry shrub vegetation increased, and grassy elements declined..." (See also Bousman and Scott, 1994) Later we will discuss the contribution of the traveller's record to this debate.

The second issue where paleoenvironmental data conflict concerns the effect of European stock farmers on grass and shrub cover, particularly in the 19th century. Sampson (1994), for example, argues that the drop in grass cover in the 1890s, inferred from micromammal deposits, "...equates with the period of record overstocking and must reflect chronic overgrazing by livestock..." Bousman and Scott (1994) agree with this argument, but only in part, by suggesting that "Overgrazing probably contributed to this trend [of less Poaceae and more Compositae], especially in the late nineteenth and early twentieth centuries when stocking rates were high..." They accept, however, that "...it is unlikely to have been the sole cause." Finally, Avery (1991) suggests that "There is no direct evidence from the micromammals to show that stock herding had any effect on vegetation structure up to the beginning of the present century."

Much of the disagreement outlined above may arise from the rather limited spatial extrapolation possible for many of the paleoecological techniques used. For example, the fossil pollen in hyrax middens depends on the location of hyrax middens themselves. The generalization of events from the data from one or a few middens to the eastern Karoo region is problematic, particularly if some of the sites are located at ecotones between atypical vegetation types, as is the case in the Oppermanskop site. While the temporal resolution is excellent (Avery, 1991; Bousman and Scott, 1994; Sampson, 1994), many more palaeoecological and archaeological sites are needed to resolve some of the more important disputes.

5. Historical and Photographic Sources

5.1. The Traveler's Record

The use of the traveler's record as a source of historical information for eastern Karoo vegetation conditions has only been explored for the period 1777–1843. (See Hoffman and Cowling (1990) for details of the traveller's record.) The earliest reports support Avery (1991) in indicating a grassy eastern Karoo during the late 18th century. But from very early on in the 19th century until 1843, there are repeated descriptions of grassless and arid eastern Karoo environments. The wooled sheep industry was not yet well developed, and stock numbers were still relatively low at this time (Thom, 1939). It is therefore likely that the absence of grass was influenced more by the repeated and severe droughts, as reported by Van der Merwe (1938) during the first 3 decades of the 19th century, than by overstocking.

The detailed journals of the early travelers often, but not always, provide excellent site descriptions (Hoffman and Cowling, 1990). Journals, together with the large number of records available, usually present a useful method for interpreting vegetation condition over a large area. But short-term seasonal events are often difficult to uncouple from long-term trends. For the eastern Karoo, at least, the traveller's record is really useful only for the late 18th and 19th century.

5.2. Matched Photographs

Hoffman and Cowling (1990) and Hoffman (1991) used a series of matched photographs of the Karoo, spanning as many as 70 years in some cases, to argue that general ecological conditions in the landscapes photographed in 1989 were without exception either better or not significantly worse than landscapes shown in earlier images. Hoffman and Cowling (1990) warned, however, of the limitations of this technique, as short-term seasonal fluctuations, particularly in grass cover, were likely to bias the interpretations of long-term trends in vegetation composition. Showing large fluctuations in grass cover, Figures 4a–c emphasize the need for an ongoing monitoring program to discriminate significant trends in vegetation change from seasonal and short-term fluctuations.

Perhaps the most important contribution of matched photographic images is to have challenged the entrenched view that the arid dwarf Karoo shrublands are expanding into the more productive highveld grasslands at a rate of between 2 and 5 km per annum. This method also highlights the need to include short-term (1–5 years) dynamic models of vegetation change in any discussion of how the eastern Karoo environment has been altered during the last 300 years and the area's current rate and extent of change. But this method has important limitations, not the least of which is the availability of the images themselves, which, for the Karoo, characteristically span only the last 75 or so years. In addition, it is often difficult

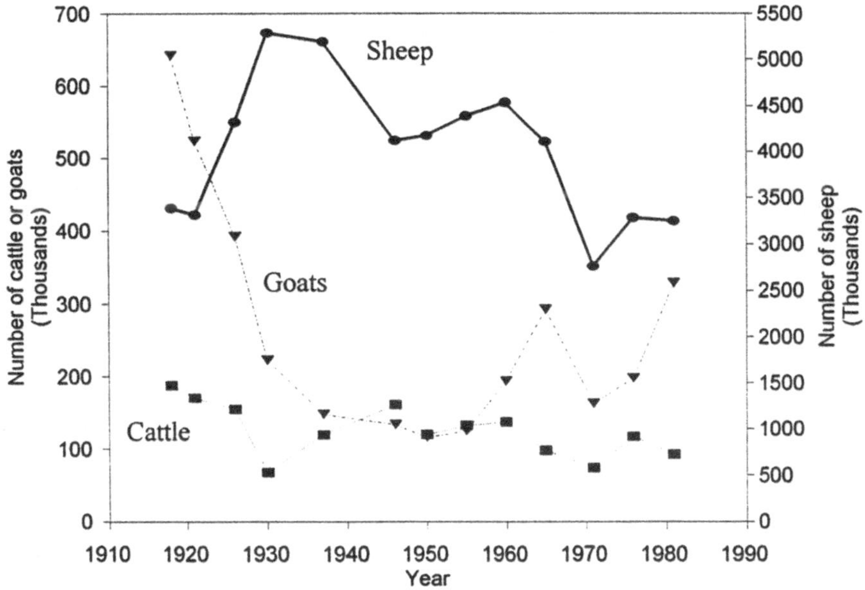

Fig. 3. Changes in total number of cattle, sheep and goats in 18 eastern Karoo magisterial districts between 1918 and 1981.

from just a pair of matched photographs to separate short-term fluctuations from long-term trends in vegetation condition.

5.3. Survey Data

A number of data sets for the eastern Karoo measure vegetation composition changes over periods from a few years (Novellie and Strydom, 1987) to two decades or more (Roux, 1966; Hoffman and Cowling, 1990; Hoffman *et al.*, 1990; Palmer *et al.*, 1990; O'Connor and Roux, 1995). Unfortunately, the two longest relatively unbroken data sets (Roux, 1966; O'Connor and Roux, 1995) do not cover changes during the past 25 years but rely on anecdotal observations and ad hoc measurements only for this later period. But Hoffman and Cowling (1990) show that for a wide range of sites in the eastern Karoo there is little doubt that in 1989 grass cover had increased when compared with surveys in the early 1960s. Again, the influence of short-term seasonal responses of component organisms to the long-term scenario is not well understood.

But the ecological arguments proposed by Hoffman *et al.* (1990) and Milton and Hoffman (1994) suggest that during the generally high rainfall and low stocking period that has characterized the eastern Karoo during the past 2 decades (Figures 2 and 3), grass cover has also increased significantly in a number of years since 1974. This view is confirmed by data from Novellie and Strydom (1987), Hoffman and Cowling (1990), Palmer *et al.* (1990), Bousman and Scott (1994), and O'Connor and Roux (1995) and by observations of M.T. Hoffman since 1985 and the general perceptions of inhabitants of the eastern Karoo. No evidence suggests that grass

Fig. 4a.

Figs. 4(a)–(c). Matched photographs of a site between Steynsburg and Hofmeyr in the eastern Karoo (31°27′S 25°45′E) showing the long-term changes between (4a) September 1925 and (4b) august 1989 and recent short-term changes to (4c) September 1993. In 1989 the vegetation cover was measured between 65–75% but this had declined 25–35% by 1993, only slightly more than the estimated 15–25% cover present in 1925. The major fluctuation in cover between 1989 and 1993 reflects changes in short-lived grasses, especially *Eragrostis lehmanniana, Aristida curvata, Tragus koeleroides* and *Cynodon dactylon.*

cover has decreased in the eastern Karoo during the past 20 years. This general pattern has important consequences for interpreting the results of the stable carbon isotope analysis discussed later.

Survey data provide the most detailed evidence of vegetation change but are generally available only for the last few decades. Moreover, survey data are expensive to maintain, and even state-funded institutions have not managed to sustain an ongoing monitoring program of eastern Karoo semiarid shrublands. Perhaps the most important contribution of long-term vegetation surveys is that their data can be incorporated into spatially and temporally explicit models in which grazing and climatic parameters can also be incorporated.

5.4. HISTORICAL CHANGES IN STOCKING RATES

South Africa possesses an excellent agricultural census record of livestock numbers, on a magisterial district basis, particularly since 1918. Dean and Macdonald (1994) have compared the changes in the number of Large Stock Units (LSU)

Fig. 4b.

Fig. 4c.

per 100 ha for all magisterial districts in the Karoo during the 20th century. They conclude that the majority of districts show a decline in stock numbers when the numbers of the past (1911–1930) are compared with those of the present (1971–1981). They attribute this decrease primarily to a decline in the carrying capacity of these semiarid and arid rangelands and argue that market forces and state policy have had little influence on the trends.

Our analysis of the changes in the number of sheep, goats, and cattle in the eastern Karoo from 1918 to 1981 concurs with Dean and Macdonald (1994) in showing the decrease in the numbers of animals supported by eastern Karoo rangelands this century (Figure 3). This decrease, however, has not been uniform with time, and interpretation of the complex pattern is difficult. The change in livestock numbers is characterized initially by a dramatic increase in especially sheep numbers between 1921 and 1930, a levelling off to 1937, and then a step-wise drop in two major steps until the last published agricultural census in 1981.

The increase in sheep numbers between 1921 and 1930, concomitant with the sharp decline in goat numbers, suggests that eastern Karoo farmers during this period completely ignored the dire warnings of the Drought Investigation Committee (1923) report and reacted instead to the market forces of the time. Beinart (1994) has suggested that the drive to increase sheep flocks during this period could have been prompted by the drop in wool prices from 13.5 pennies per kg in 1920 to 2 pennies per kg in 1932 (Department of Agricultural Economics and Marketing, 1961) and the obvious need to have more animals to generate the same income from the land.

The drop in sheep numbers in the 1940s and again in the 1960s appears related to drought, but it is difficult to understand why sheep numbers have not risen since the 1971 record century low. Dean and Macdonald (1994) argue that the carrying capacity of the land has declined to a level that can no longer support as many animals as before. While this argument may well hold for parts of the grassless, southern Karoo shrublands, it does not appear to be the only explanation for events in the eastern Karoo. Since 1974 a number of very high rainfall years have been recorded in this region (Figure 2), and the response of the vegetation has been equally dramatic, as discussed above and as shown by Novellie and Strydom (1987), Hoffman and Cowling (1990), Palmer *et al.* (1990), and O'Connor and Roux (1995).

It seems implausible that this significant measured increase in primary productivity, including that of palatable forage in the eastern Karoo during the past 2 decades (Hoffman and Cowling, 1990), can translate to a reduced carrying capacity for the region. It is, therefore, unlikely in the eastern Karoo that the low stock numbers of the period 1971–1981 can be attributed to a drop in the region's carrying capacity. Instead we suggest that the state-subsidized stock reduction scheme has effectively limited animal numbers on the range and should be granted more weight in Dean and Macdonald's (1994) general explanation for historical changes in stock numbers in the Karoo.

To emphasize this point we conclude with Pringle's (1982) description of the stock reduction scheme as "the greatest campaign in South African conservation history." Pringle (1982) states:

> The target area [of the stock reduction scheme] was the Karoo. The aim was to get the grass back. The method was to pay farmers not to farm. And a mind-boggling amount of money was set aside to cover the costs for five years. No farmer was forced to join the scheme – he joined if he wanted to, and 4095 wanted to. The carrying capacity of every farm was assessed in terms of stock units, the farmer reduced his stock by a third of this quota, and every year was paid the income lost by not farming this third.

6. Evidence from Soil Isotopes

The most recent addition to the desertification debate has analyzed stable carbon isotopes from the organic matter fraction in the soil profile at 11 Karoo sites (8 of which are located in the eastern Karoo region outlined in this study) (Figure 1) to describe the relative grassiness or shrubbiness of these environments through time (Bond *et al.*, 1994). This technique is possible only because C_3 shrubs and C_4 grasses have slightly different photosynthetic systems. As plant material from shrubs and grasses enters the organic matter pool and is translocated down the soil profile, older material accumulates at depth, while more modern organic material is found closer to the surface. One can then measure the stable carbon isotope ratios of the organic carbon within different soil layers and discern whether the organic material is derived predominantly from grasses (δ^{13}C values of between –10 to 14‰) or from shrubs (δ^{13}C values of between –26 to –30‰).

Bond *et al.* (1994) sampled the soils down to 0.6 m, in some cases removing successive 0.05 m layers for analysis (Table I), taking care in selecting study sites to avoid areas with clear mammal disturbance or evidence of digging. After this recently published work a few of the samples have been dated by C14 dating techniques, allowing for a greater comparative analysis of the findings of the study. Problems with the interpretation of the dates, however, hamper a detailed understanding of historical events in the eastern Karoo.

At Wonderheuwel (Table I), C14 dates at different depths down the profile confirmed one of the central assumptions of the method, namely that deeper parts of the soil profile contain older organic material than samples closer to the surface. For example, the dates for the 0.03–0.035 m layer at Wonderheuwel were recorded as 700 ± 80 yrs BP, while for the 0.02–0.025 m layer values obtained were 410 ± 70 yrs BP. At Wonderheuwel, as at five other sites sampled, the 0.05–0.01 m layer contained relatively large amounts of recent organics dated to the last 30 to 40 years only.

Two important conclusions from the work of Bond *et al.* (1994) are of relevance here. The first is their suggestion that the precolonial eastern Karoo environment

TABLE I

Carbon isotope ratios ($\delta\ ^{13}C$ in ‰) of vegetation and soil samples at 0.05 m intervals at 11 sites in the Karoo. Symbols refer to study site locations shown in Figure 1

Vegetation or soil depth (m)	Sites Colesberg (a)	Hillside (b)	Driekuil (c)	Hofmeyr (d)	Culmstock (e)	Wonder-heuwel (f)	Paardevlei (g)	Opper-manskop (h)	Beaufort West (i)	Victoria West (j)	Tierberg (k)
Vegetation	–21.59	–16.46	–21.00	–19.26	–21.64	–24.01	–25.06	–21.58	–22.93	–20.45	–23.32
Soil depth											
0–0.05	–17.13	–15.62	–19.11	–18.31	–19.66	–20.04	–19.84	–21.71	–20.98	–20.68	–21.91
0.05–0.01	–14.16	–13.68	–16.72	–16.50	–16.72	–18.08	–18.02	–19.89	–20.61	–20.09	–21.56
0.01–0.015	–13.31	–12.43	–14.05	–15.83	–16.19	–17.39	–18.71	–19.88	–20.00	–19.97	–21.92
0.015–0.02	–12.13	–12.58	–13.28	–15.71	–16.50	–16.70	–18.78	–19.22	–19.30	–19.73	–21.56
0.02–0.025	–11.64	–12.32	–13.75	–15.29	–16.61	–16.39	–18.08	–19.00	–19.11	–19.38	–20.96
0.025–0.03	–11.51	–12.74		–15.38	–16.37	–15.54		–18.85	–18.93	–19.09	–20.95
0.03–0.035	–11.5	–12.14		–15.26	–14.82	–15.08		–17.80	–18.93	–19.88	–20.97
0.035–0.04	–11.56	–12.71		–15.63	–14.48	–15.53		–16.75	–18.63	–18.65	
0.04–0.045	–10.98	–12.44		–16.00	–14.53	–14.53		–16.52		–19.00	
0.045–0.05	–10.98	–13.01		–16.00		–14.53				–19.00	
0.05–0.055				–15.32		–14.42				–19.93	
0.055–0.06				–15.32		–14.42				–19.93	

should be characterized as a broad but relatively stable transition zone between the more arid dwarf shrublands to the northwest, west, and southwest, and the highveld grasslands to the north, northeast, and east (Table I). Sites close to the Grassland biome boundary, such as Hillside, Colesberg, and Driekuil, show a relatively strong grass signature ($\delta^{13}C$ values of between –10 to –14‰) in the deeper part of the soil profile below about 0.02 m. This signal becomes progressively weaker and shifts to a more shrubby pattern at the sites further and further away from the Grassland biome. The three sites outside of the eastern Karoo as delimited here (Beaufort West, Victoria West, and Tierberg) all indicate a dominance of shrubs, even at depths, suggesting that they have probably never supported a dominant grass sward for any length of time in the past.

Central to the argument of Bond *et al.* (1994) that the eastern Karoo environment has been a stable transition zone is that below 0.02–0.025 m there are no large shifts in the C_3 and C_4 signatures at any of the sites. This fact suggests a relatively stable mix of grasses and shrubs in the landscape over time. This hypothesis fundamentally differs from two alternative views. It differs from the perennial eastern Karoo grassland idea, favored by Acocks (1953) and others (e.g. Roux and Vorster, 1983). If their view were correct, then all sites in the eastern Karoo area should have shown a stronger grass signature at depths below about 0.02 m. This, however, is not the case. This difference is important because management options for the region are all aimed at recovering a pristine grassland state that never existed for much of the eastern Karoo.

The stable isotope evidence also differs from the archaeological and ecological models, which argue for significant and lengthy fluctuations in shrub and grass cover dependent largely on shifts in rainfall conditions and not for stable mixes of grasses and shrubs in a broad transition zone (Avery, 1991; Bousman and Scott, 1994; Sampson, 1994). For the archaeological view to hold, the stable isotope signature should show large fluctuations from C_3 to C_4 signatures between successive layers, at least for samples from the suggested precolonial period, below about 0.02 m. Again, the data from Bond *et al.* (1994) do not support this view.

The second conclusion from the stable isotope study of relevance to this review is the suggestion of Bond *et al.* (1994) that the greatest shift from a grassy to a shrubby physiognomy has occurred during the most recent period as stable isotope values become increasingly more negative higher up the soil profile (Figure 5). If the radio carbon dates for the upper soil layers are to be interpreted fairly literally, then Bond *et al.* (1994) argue for a decrease in grass cover in the eastern Karoo in the last several decades. This position supports Acocks' (1953) spatial models of a recently degraded Karoo but is contrary to the photographic and survey evidence outlined above. This position also contradicts the ecological models for the region that suggest an increased grass response following high rainfall events such as the region has experienced since the mid-1970s (Figure 2).

One possible way to resolve this conflict is to separate the fine organic carbon fraction from the coarse fraction in the upper soil layers and to measure the

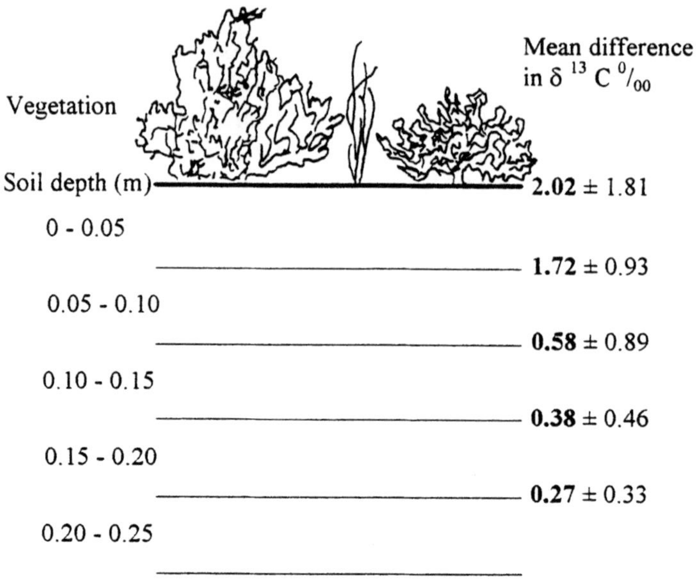

Fig. 5. The mean difference in δ^{13}C (‰) values between the vegetation sample and the surface soil layer and between each successive soil layer thereafter for 11 sample sites in the Karoo.

TABLE II

Isotope ratios (δ ^{13}C ‰) of different soil organic matter (SOM) size fractions in the 0–0.05 m surface soil layer at 10 sites in the Karoo

Study site	Coarse SOM fraction (younger) (>63 μm)	Fine SOM fraction (older) (<63 μm)	Difference	Increase in:
(a) Colesberg	−19.70	−18.17	−1.53	Shrubs
(b) Hillside	−13.54	−15.46	+1.92	Grasses
(c) Driekuil	−21.60	−17.24	−4.36	Shrubs
(d) Hofmeyr	−20.91	−17.19	−3.72	Shrubs
(e) Culmstock	−21.18	−18.12	−3.06	Shrubs
(f) Wonderheuwel	−19.12	−19.23	+0.11	Grasses
(g) Paardevlei	−21.11	−19.25	−1.86	Shrubs
(h) Oppermanskop	−22.82	−20.92	−1.90	Shrubs
(i) Beaufort West	−22.72	−19.82	−2.90	Shrubs
(j) Victoria West	−23.29	−20.70	−2.59	Shrubs

carbon isotope ratios for each of these elements separately as done by Stock *et al.* (1993). Theory suggests that the coarse fraction represents younger organic material whereas the fine material is older. Analysis of the eastern Karoo material

(Table II), however, suggests that the coarse fraction does not contain more C_4 material than the finer fraction. For most sites, in fact, opposite is true, and the general trend appears to be a shift to increased shrub cover in the more recent period. The results of this analysis thus appear inconsistent with the changes in the eastern Karoo as reflected in the recent historical and ecological record.

Finally, it is difficult to reconcile the differences between the stable carbon isotope data and other lines of evidence. Perhaps stable isotopes integrate events over much longer time periods than other techniques, and differences may reflect different scales of measurement rather than real differences in the composition and dynamics of eastern Karoo environments. Nevertheless, stable isotope analyses hold much promise for resolving many of the long-standing desertification debates in the region. Their spatial resolution is better than most other techniques since sample locations do not depend on specific positions in the landscape or associated archaeological material. But many complications to the dating of soil organic carbon have not yet been resolved, and the temporal resolution of the technique appears rather coarse.

7. Conclusions

The South African desertification debate has benefitted enormously from the wide range of methodologies that have been used in recent years to understand the problem. Considerable progress appears to have been made, but the three core areas of conflict outlined earlier have still not been satisfactorily resolved.

First, with regard to the physiognomy of precolonial eastern Karoo environments, there appears general agreement that they were more grassy at some time in the past. But no consensus exists as to either the extent or stability of this grassiness in the region. Also, the timing of the onset of more shrubby conditions is disputed, with some authors suggesting that this onset predated the arrival of European stock farmers. The date for the shift from more grassy to shrubby conditions is important since it may facilitate a finer partitioning of the relative responsibilities of climate and livestock for the perceived degradation of the region. If the dating problems of soil organic carbon can be adequately resolved, then the stable carbon isotope work outlined above may provide one of the best ways out of this difficulty.

Second, no recent evidence has supported the argument for a predictable annual expansion of the Karoo into the highveld grasslands. Instead, the importance of seasonal and short-term fluctuations in grass and shrub cover in any degradation model for the region now appears widely recognized (Dean *et al.* 1995).

Finally, the area where progress appears to have been weakest is in understanding the effects of humans and climate, separately and in combination, on key degradation processes of the region. Carefully constructed long-term experiments may give us our best answers to these important questions.

Acknowledgements

The Department of Environment Affairs provided additional financial assistance to M.T. Hoffman. The carbon isotope work was funded by the Division of Forest Science and Technology of the CSIR. We wish to thank our colleagues, including Margaret Avery, William Beinart, Guy Midgley, Tony Palmer, Dan Sonnenberg, Julia Lee-Thorp, and Royden Yates for either helping to collect data or providing valuable comments on some ideas expressed here. We also thank Nicky Allsop and Karen Wienand for providing soil analyses, John Vogel for providing the carbon 14 dates, and the Computing Centre for Water Research and the Agricultural Research Council for providing the long-term rainfall records.

References

Acocks, J.P.H. 1953. Veld types of South Africa. *Memoirs of the Botanical Survey of South Africa* **28**: 1–128.

Avery, D.M. 1991. Micromammals, owls and vegetation change in the eastern Cape Midlands, South Africa, during the last millennium. *Journal of Arid Environments* **20**: 357–369.

Beinart, W. 1994. Environmental destruction in sheep farming areas of South Africa: Soil erosion, animals and pastures over the longer term. Unpublished discussion paper. Department of Historical Studies, University of Bristol, United Kingdom.

Bond, W.J., W.D. Stock and M.T. Hoffman. 1994. Has the Karoo spread? A test for desertification using carbon isotopes from soils. *South African Journal of Science* **90**: 391–397.

Bousman, B. and L. Scott. 1994. Climate or overgrazing: The palynological evidence for vegetation change in the eastern Karoo. *South African Journal of Science* **90**: 575–578.

Brown, J.C. 1875. *Hydrology of South Africa*, Henry King and Co., London.

Cowling, R.M. and P.W. Roux (editors). 1987. The Karoo biome: A preliminary synthesis. Part 2, Vegetation and history. *South African National Scientific Programmes Report* **142**: 1–133.

Cowling, R.M., P.W. Roux and A.J.H. Pieterse (editors). 1986. The Karoo biome: A preliminary synthesis. Part 1, Physical environment. *South African National Scientific Programmes Report* **124**: 1–115.

Dean, W.R.J., M.T. Hoffman, M. Meadows and S.J. Milton. 1995. Desertification in the semi-arid Karoo, South Africa: Review and reassessment. *Journal of Arid Environments* **30**: 247–264.

Dean, W.R.J. and I.A.W. Macdonald. 1994. Historical changes in stocking rates of domestic livestock as a measure of semi-arid and arid rangeland degradation in the Cape Province, South Africa. *Journal of Arid Environments* **26**: 281–298.

De Klerk, J.C. 1947. Pastures of the southern Orange Free State a century ago and today. *Farming in South Africa* April: 347–354.

Department of Agricultural Economics and Marketing. 1961. *Handbook of agricultural statistics: 1904–1950*. Government Printer, Pretoria.

Drought Investigation Commission. 1923. *Final Report of the Department of Agriculture and Forestry*. Cape Times for Government Printer, Cape Town.

Durning, A. 1990. *Apartheid's environmental toll*. Worldwatch Publication Vol. 95. Worldwatch Institute, Washington.

Du Toit, P.F., A.J. Aucamp and J.J. Bruwer. 1991. The National Grazing Strategy of the Republic of South Africa: Objectives, achievements and future challenges. *Journal of the Grassland Society of Southern Africa* **8**(4): 126–130.

Hobson, S. 1896. Deterioration of the veld. *Agricultural Journal* **9**(19): 491–492.

Hoffman, M.T. 1991. Is the Karoo spreading? *Veld and Flora* **77**(1): 4–7.

Hoffman, M.T., G.D. Barr and R.M. Cowling. 1990. Vegetation dynamics in the semi-arid eastern Karoo, South Africa: The effect of seasonal rainfall and competition on grass and shrub basal cover. *South African Journal of Science* **86**: 462–463.

Hoffman, M.T. and R.M. Cowling. 1990. Vegetation change in the semi-arid eastern Karoo over the last two hundred years: An expanding Karoo – fact or fiction? *South African Journal of Science* **86**: 286–294

Milton, S.J. and M.T. Hoffman. 1994. The application of state-and-transition models to rangeland research and management in arid succulent and semi-arid grassy Karoo, South Africa. *African Journal of Range and Forage Science* **11**: 18–26.

Milton, S.J., W.R.J. Dean, M.A. Du Plessis and W.R. Siegfried. 1994. A conceptual model of arid rangeland degradation. *Bioscience* **44**: 70–76.

Moll, E.J. and A.A. Gubb. 1989. Southern African shrublands, in C.M. Mackell (Ed.), *The biology and utilization of shrubs*, Academic Press, New York, pp. 145–175.

Novellie, P. and G. Strydom. 1987. Monitoring the response of vegetation to use by large herbivores: An assessment of some techniques. *South African Journal of Wildlife Research* **17**: 109–117.

O'Connor, T.G. and P.W. Roux. 1995. Vegetation changes (1949–1971) in a semi-arid, grassy dwarf shrubland in the Karoo, South Africa: Influence of rainfall variability and grazing by sheep. *Journal of Applied Ecology* **32:** 612–626.

Palmer, A.R., C.G. Hobson and M.T. Hoffman. 1990. Vegetation change in a semi-arid, succulent dwarf shrubland in the eastern Cape. *South African Journal of Science* **86**: 392–395.

Pringle, J. 1982. *The conservationists and the killers. The story of game protection and the Wildlife Society of Southern Africa*, TV Bulpin and Books of Africa, Cape Town.

Roux, P.W. 1966. Die uitwerking van seisoensreënval en beweiding op gemengde Karooveld. In *Proceedings of the Grassland Society of Southern Africa* **1**: 103–110.

Roux, P.W. and Vorster M. 1983. Vegetation change in the Karoo areas. *Proceedings of the Grassland Society of Southern Africa* **18**: 25–29.

Rutherford, M.C. and R.H. Westfall. 1986. The biomes of southern Africa – an objective categorization. *Memoirs of the Botanical Survey of South Africa* **54**: 1–98.

Sampson, C.G. 1994. Ostrich eggs and Bushman survival on the north-east frontier of the Cape Colony, South Africa. *Journal of Arid Environments* **26**: 383–399.

Scott, L. and C.B. Bousman. 1990. Palynological analysis of hyrax middens from southern Africa. *Palaeogeography, Palaeoclimatology, Palaeoecology* **76**: 367–379.

Stock, W.D., W.J. Bond and D. Le Roux. 1993. Isotopic evidence from soil carbon to reconstruct vegetation history in the south-western Cape Province. *South African Journal of Science* **89**: 153–154.

Thom, H.B. 1936. *Die geskiedenis van skaapboerdery in Suid-Afrika*, Swetz en Zeitlinger, Amsterdam.

Tidmarsh, C.E.M. 1948. Conservation problems of the Karoo. *Farming in South Africa* **23**: 519–530.

Van der Merwe, P.J. 1938. *Die noordwaartse beweging van boere voor die Groot Trek (1770–1842).* Pretoria Nasionale Pers Beperk, WP van Stockum & Zoon, Den Haag.

Wilson, F. and M. Ramphele. 1989. *Uprooting poverty: The South African challenge*, David Philip, Cape Town.

GOVERNMENT POLICY INFLUENCES ON RANGELAND CONDITIONS IN THE UNITED STATES: A CASE EXAMPLE *

JERRY L. HOLECHEK* and KARL HESS, JR.**

*Department of Animal and Range Sciences, Box 30003, New Mexico State University, Las Cruces, NM 88003, USA; **Cato Institute, 1000 Massachusetts Avenue, NW, Washington, D.C. 20001–5403, USA

Abstract. Since the early 1970s the Bureau of Land Management (BLM) and the Forest Service have emphasized the monitoring approach in managing their rangelands. After a 5-year period of intensive monitoring, stocking rate and other management actions are typically adjusted, depending on whether a definite downward or upward trend in range condition is observed. Examining the Afton Allotment on BLM land in southcentral New Mexico, we demonstrate serious flaws in the monitoring approach and other BLM grazing policies. Monitoring reflects past management but does not consider the future. Perverse incentives for permittees to maximize permit value rather than sustain the forage base still remain in place. The "Range Reform '94" proposals by the U.S. Department of the Interior do not address these incentives and other flaws in public land policy. We would reform federal grazing land policy by coupling grazing fees to grazing intensity. Ranchers choosing conservative sustainable grazing intensities would pay low fees, whereas those choosing heavy grazing intensities would be charged higher rates. We would also implement other reforms: making livestock grazing permits transferable to other uses, discontinuing subsidies, eliminating restrictions on nonuse, offering incentives for public land investment, and establishing a biodiversity fund.

1. Introduction

Since the late 1970s management of federal rangelands has been heavily centered on monitoring changes in trend in ecological condition. Under this approach stocking rate and other management practices are adjusted at 5-year intervals in response to the outcome of monitoring. The rationale behind the monitoring approach is to base management changes on the trend in condition of grazing allotments.

Although the monitoring approach is supported by many range professionals, it also has its critics. The monitoring approach is costly, time intensive, and reflects past rather than present conditions. Moreover, vegetation trend data can be unreliable. Environmentally concerned individuals have charged that monitoring-based management allows government agencies to delay management decisions on rangelands undergoing degradation where management changes will conflict with permittees.

The Afton Allotment in southcentral New Mexico, controlled by the Bureau of Land Management, provides an opportunity to evaluate the effectiveness of the

* This paper was supported by the New Mexico Agricultural Experiment Station, Las Cruces and was part of project 1–5–27417.

Environmental Monitoring and Assessment **37:** 179–187, 1995.

monitoring approach. Intensively monitored since 1982, this allotment has been the source of considerable controversy among environmental groups, the permittee, and the Bureau of Land Management. The situation is somewhat unusual in that the conflict has focused on interpretation rather than the reliability of monitoring data. We will examine the history of the Afton Allotment and then discuss its implications for federal rangelands policy.

2. History

The Afton Allotment (No. 3056) in southcentral New Mexico was created in 1981 as a result of a grazing permit transfer to the current permittee. On September 16, 1982, the grazing preference on the Afton Allotment was set at 1,284 animal unit months or 126 cattle yearlong (animal unit years). This decision also initiated a rangeland monitoring program that was to continue for 5 years. At the end of this period a new preference could be established in response to the results of the monitoring data. As a result of land acquisition and a fence change on September 11, 1984, the size of the allotment was increased to 10,594 ha, and preference was changed to 130 animal unit years (AUYs) with 20 more temporary AUYs. About 80 percent of the Afton Allotment lies within the Aden Lava Flow Wilderness Study Area established on November 15, 1980.

Upon completion of the Southern Rio Grande Environmental Impact Statement in 1981, the Afton Allotment was placed in the Improve (I) category. This classification was based on the ecological condition rating of the allotment, lack of permanent water sources, the potential for improvement, and the opportunity for economic returns from investment.

During the 1982–1986 period the allotment was stocked at a light to conservative average rate of 77 ha/AUY. Research by Holechek (1991) and Holechek *et al.* (1994) showed that this level of stocking (about 25–35% use of forage species) resulted in a major improvement in range condition and forage production over a 25-year period on an experimental area similar to the Afton Allotment. During the 1982–1986 period the permittee also developed several waters to improve livestock distribution. All groups generally agreed that the permittee demonstrated a genuine commitment to improving land condition. But she undoubtedly was also motivated by the prospect of increasing her equity in the allotment by obtaining a higher preference number of AUYs at the end of the monitoring period.

3. Monitoring Results

During the 1982–1986 period, the key species-key area approach discussed by Cook and Stubbendieck (1986) and Holechek *et al.* (1989) was used at five locations to collect data on vegetation, ecological condition, vegetation trend, forage utilization,

TABLE I

Livestock use, range ecological condition, forage production, and precipitation data for the Afton Allotment in southcentral New Mexico in the 1982–1993 period

	Year											
	1982	1983	1984	1985	1986	1987	1988	1989	1990	1991	1992	1993
Actual cattle animal unit years (AUY)	106	138	155	150	150	150	150	168	170	185	190	202
Stocking rate (ha/AUY)	100	77	68	70	70	70	70	63	62	57	56	52
Precipitation Afton Allotment (cm/year)[1]	20.8	27.61	24.43	40.97	30.18	–	–	35.08	21.01	31.39	–	–
Precipitation NMSU Ranch (cm/year)[2]	18.69	24.49	17.98	35.56	32.16	29.59	23.14	30.05	21.77	24.13	43.82	23.29
Forage utilization (%)[3]	23	25	40	18	20	38	36	31	29	26	30	–
Fall forage standing crop (kg/ha)	151	–	–	–	612	387	628	495	550	722	630	340
Fall black grama standing crop (kg/ha)[4]	49	–	–	–	98	63	42	101	24	119	149	125
Ecological range condition score	49	–	–	–	42	43	41	48	52	54	53	48

[1] Long-term average precipitation is 21.72 cm.
[2] Long-term average precipitation is 21.31 cm.
[3] Utilization data were collected in late spring and reflect use of forage produced in the previous year.
[4] Black grama is a perennial grass that is considered to be a key forage species for cattle and a dominant component of the climax vegetation.

forage standing crop, and precipitation. A summary of these data is provided in Table I.

Although the Afton Allotment showed no real change in ecological condition in the 1982–1986 period, perennial grass and total forage production more than doubled due to favorable precipitation and conservative stocking. Monitoring data collected on the Afton Allotment by the BLM was generally considered sound by concerned environmental groups, the permittee, and New Mexico State University's range science faculty.

4. Application of Monitoring Results

Because of the increase in forage on the Afton Allotment the permittee requested that the preference AUYs be increased from 130 to 250 AUYs in 1987. This request was based on consultation with the Range Management Task Force of New Mexico State University and personnel in the New Mexico Department of Agriculture. Because 80% of the allotment lay within a wilderness study area, the outcome of monitoring and later management changes were of particular concern to environmental groups. The BLM decided to keep the preference AUYs for the allotment at 130. This decision was appealed by the permittee and went to court in April 1989.

A hearing was held and a decision was issued based in part on technical testimony and a report developed by Jerry Holechek, a range science faculty member at New Mexico State University. His approach involved calculating the stocking rate for the Afton Allotment using BLM monitoring data and a stocking rate procedure (Holechek, 1988) based on range research. The court recommendation was to immediately grant the permittee an increase of 20 AUYs preference and 20 AUYs temporary. Depending on the results from monitoring over the next 5 years, an additional 52 AUYs preference could be granted at BLM's discretion. We estimate that court costs for the defendant and plaintiffs were about $50,000 each.

From 1987 to 1991 forage production and ecological condition improved on the Afton Allotment. Rainfall during the summer growing season was nearly 20% above average in this period. On 1 September 1993, the permittee increased the number of AUYs consumed on the allotment from 190 to 202. In the spring of 1994 she was issued a proposed decision granting a preference increase of 20 AUYs for a total of 170 AUYs because the allotment showed improvement in the 1987–1991 monitoring period (see Table I).

Starting in the summer of 1992 drier conditions prevailed on the allotment with outright drought in 1994. By the summer of 1994 heavy use of primary forage plants (mesa dropseed, black grama) and deterioration in ecological condition were evident in several parts of the allotment. But according to the permittee, cattle numbers were not reduced and remained at 202.

5. Problems with the Monitoring Approach

In our opinion the Afton Allotment demonstrates a serious flaw in the monitoring approach to federal rangeland management. This flaw centers on the fact that present policy fails to reward permittees for responsible stocking rate decisions and may actually encourage them to implement ecologically harmful stocking rate decisions. In effect, this policy flaw decouples the monitoring process from on-the-ground management and facilitates land degradation.

For example, starting with acquisition of the Afton Allotment permit in 1981, the permittee voluntarily stocked her allotment at a light to conservative rate to facilitate range improvement and enhance allotment carrying capacity. Monitoring during this period confirmed low livestock utilization and an increase in available forage. In response to the apparent positive return to management, the permittee requested that the BLM grant a temporary increase in authorized stocking for the 1985–87 period to determine through monitoring if an increase in allotment preference was warranted. In a decision issued by the BLM on 17 January 1986, the District Manager for the Las Cruces District denied the request. Among the reasons given for the decision was the failure of the permittee to graze the allotment at full authorized use:

> Over the past four years your actual use, based on your information, has been less than authorized use. This could raise the question as to whether or not an increase in grazing use is really needed.

The BLM also denied the permittee's later request for a permanent increase in preference. This denial resulted in the 1989 administrative hearing.

Following the court's 1989 ruling to increase preference on the Afton Allotment and to engage in a 5-year monitoring program to assess stocking increases up to the 202 AUY ceiling, a new set of incentives were presented to the permittee. Because of the apparent penalty for taking past nonuse for range improvement and her fear of instability in the public grazing land policy environment, the permittee felt compelled to increase cattle numbers on her allotment as quickly as possible to prove to the BLM that the Afton Allotment could handle the court-specified maximum preference of 202 AUYs.

New incentives have injected uncertainty and unpredictability into the management of the Afton Allotment, highlighting the failure and frustration of sophisticated land-based monitoring. By voluntarily destocking the Afton Allotment in the early years of her permit, the permittee believed she could increase preference and enhance the market value of her grazing permit. She had an economic incentive to graze conservatively. That incentive, however, was largely removed when she found that extensive nonuse for range improvement actually jeopardized her ambitions to expand authorized use.

During the 1992–1994 dry period, the permittee felt she must stock her allotment at the maximum rate or she would lose the opportunity to be granted a preference of 202 AUYs. Clearly, her management strategy had shifted from one of conservative grazing to one of intensive grazing in the aftermath of the 1989 court decision. She determined that her economic interest, as defined by her grazing permit, now lay more in protecting and increasing licensed numbers of livestock on her allotment rather than sustaining the forage base. Twelve years of intensive, state-of-the-art monitoring on the Afton Allotment are culminating in policy-driven, drought-exacerbated overgrazing.

6. Policy Changes to Encourage Better Management

We believe that federal grazing land policies should be changed to encourage rather than discourage conservative grazing, particularly during drought periods. An overwhelming amount of research reviewed by Holechek *et al.* (1989), Vallentine (1990), Heitschmidt and Stuth (1991), and Holechek (1993) shows that vegetation residues remaining after grazing are critical to sustaining soil, forage, watershed, and wildlife components on grazing lands. Validated, peer-reviewed stocking rate procedures (Holechek and Pieper, 1992) by Holechek (1988) and Troxell and White (1989) will sustain grazing land resources without severely compromising financial returns.

6.1. Incentives for Conservative Stocking

We would encourage conservative stocking rates on federal grazing lands by linking grazing fees to the level of forage use. Ranchers choosing conservative stocking rates (see guidelines in Holechek *et al.*, 1989 and Vallentine, 1990 for individual range types) would receive low rates ($2.00/animal unit month), whereas those choosing heavy stocking rates would be charged accordingly (up to $12/animal unit month). Under this type of program ranchers would essentially be charged for what they remove from the system.

Under our program ranchers would have the responsibility of developing and implementing a management plan for their allotment. Costs of monitoring the effectiveness of the plan would be sustained by the rancher and concerned environmental groups using private certified range consultants. Our experience with the Afton Allotment reveals that reliable monitoring data can be obtained for $0.25–$0.37/ha/year. This type of land depending on cattle prices will give net returns ranging from $0.25 to $2.50 ha/year (Holechek, 1991, 1992). Ranchers whose monitoring results achieved management goals could have their tenure of grazing privileges extended for up to 30 years. Restrictions on grazing privilege tenure and high grazing fees could be imposed on ranchers who were willfully degrading their allotments. If parties disagreed over monitoring results, the BLM, the rancher, or environmental groups could implement a separate monitoring evaluation using another certified consultant.

6.2. Incentives for Establishing a Grass Bank for Drought

We would encourage policy changes to allow ranchers to set aside 20% of their grazing lease as a forage reserve for drought. The incentive here would be to charge low grazing fees ($1/AUM) and greatly extend grazing privilege tenure (30–50 years) to ranchers who put part of their allotment in a grass bank. We believe that both conservative stocking and the grass bank could be applied on public lands without major enforcement costs or problems. Proper education on the financial benefits of conservative stocking coupled with discontinuing the emergency feed subsidy administered by the U.S. Department of Agriculture should result in compliance by the majority of ranchers. Market forces would be allowed to punish those who failed to comply.

6.3. Transferable Grazing Permits

Ranchers on federal lands can derive income from their grazing permits only by pasturing livestock. The laws and policies that mandate such use lie at the heart of the rancher-environmentalist conflict. These regulations deny the nonranching public access to managing public lands. Discontinuing livestock grazing to enhance recreation, build big game populations, or protect biological diversity may be desirable to the public at large, but to the rancher it means the loss of income and the loss of perceived "forage rights."

By changing the laws and policies that restrict forage use to the production of livestock and by making grazing allotment permits marketable to nonranchers, we could remedy the major problems that plague federal grazing lands. First, the forage rights that ranchers defend would be tradeable on the open market and convertible to any number of uses. Ranchers holding grazing permits could diversify their operations by allocating forage to paying uses other than livestock. This incentive would particularly apply to ranchers who voluntarily choose to conservatively stock their allotments. Ranchers could then derive income from using their allotted forage to improve riparian conditions, expand big game populations, or restore endangered species.

Environmentalists who want to make rangelands cattle free or simply to reduce livestock numbers to more acceptable levels could do so more quickly and less expensively than is now feasible by political or judicial means. By purchasing forage rights from ranchers on a willing buyer-willing seller basis, they could pursue their conservation goals without conflict. In some situations environmentalists might find the tool of conservation easements a practical option on federal lands. Instead of purchasing all of a rancher's forage rights, they could purchase the grazing privileges along a critical riparian zone or in an upland site with unique environmental value.

States, cities, and towns could acquire allotment permits, particularly when erosion control, wildlife, or recreation values exceeded those of livestock. Agencies, private organizations, and individuals seeking to protect endangered species such as the desert tortoise could buy allotment permits and apply the forage rights to species recovery. With this approach over-obligated grazing privileges could be purchased by almost anyone at a fraction of the cost that would otherwise be imposed on taxpayers because of litigation or intensive rangeland management.

6.4. Discontinuing subsidies

In our opinion all subsidies to permittees, such as returning half of grazing fees from federal lands for range betterment, should be discontinued. Historically, federal range reclamation subsidies have tended to encourage bad land practices by bailing out land users who have neglected their lands or stubbornly stuck to bad land practices (Hess, 1992; Holechek and Hess, 1994). In our opinion the worst of the subsidies is the emergency feed program administered by the U.S. Department of Agriculture Agricultural Stabilization and Conservation Service. Under this subsidy ranchers can receive 50% of the cost of extra feed in drought from the Federal Government. Ranchers on federal rangelands can qualify for this subsidy, which encourages the stocking of federal rangelands at excessive levels during dry periods when the range is easily damaged. In New Mexico ranchers qualify for this subsidy about 5 out of every 10 years and receive payments on average of nearly $4,000 a piece. In 1990, a year of well-above-average precipitation, this subsidy was used by ranchers throughout the state on both private and federal rangeland (Torell and Word, 1991).

6.5. Eliminating restrictions on nonuse

Current restrictions on the taking of grazing nonuse on federal lands should be immediately eliminated. In an era when real livestock prices adjusted for inflation are at historically low levels (oversupply) and recreational demands are rapidly increasing (Holechek and Hawkes, 1993), it makes little sense to require that federal land grazing permits be exercised or transferred to another applicant.

6.6. Protection of investment in public lands

Permittees, whether ranchers or environmentalists, must have assurances that their investments in federal lands are protected and that the benefits derived from those investments can be captured and enjoyed by the responsible parties. This does not mean that a portion of benefits cannot or should not be allocated to the general public. It only means that stewardship should not be expected to sustain itself on goodwill alone. For people to invest of themselves over the long run, they must have access to the fruits of their labors.

6.7. Biodiversity Trust Fund

We call for the creation of a biological diversity trust fund financed from public land user fees, administered at the local level, and available to all Americans through competitive grants. Not all resources on federal lands have an economic value. Biological diversity, for example, has no price tag and therefore cannot be stewarded effectively through market mechanisms. But by taking a substantial proportion of user fees generated through the market process, we can create a source of funding for critical federal land resources.

We believe the biodiversity trust fund would round out the program of market reforms outlined above. Such funds would empower ranchers and environmentalists alike to play a major role in restoring and protecting federal lands. Environmentalists, for example, could use trust funds to buy out forage rights for site-specific protection of species and habitat. Likewise, ranchers could use trust funds to expedite species recovery and habitat restoration on their allotments or to fund land practices that have benefits not measurable in dollars and cents.

References

Cook, C.W. and J. Stubbendieck (editors). 1986. *Range research: Basic problems and techniques*, Society for Range Management, Denver, Colorado.

Heitschmidt, D.D. and J. Stuth (editors). 1991. *Grazing management: An ecological perspective*. Timber Press, Inc., Portland, Oregon.

Hess, K., Jr. 1992. *Visions upon the land*. Island Press, Covelo, California.

Holechek, J.L. 1988. An approach for setting the stocking rate. *Rangelands* **10**: 10–14.

Holechek, J.L. 1991. Chihuahuan Desert rangeland, livestock grazing, and sustainability. *Rangelands* **13**: 115–120.

Holechek, J.L. 1992. Financial aspects of cattle production in the Chihuahuan Desert. *Rangelands* **14**: 145–149.

Holechek, J.L. 1993. Managing stocking rates to achieve range resource goals. In *Managing Livestock Stocking Rates on Rangeland Proceedings*. Texas A&M University, College Station, Texas.

Holechek, J.L. and J. Hawkes. 1993. Desert and prairie ranching profitability. *Rangelands* **15**: 104–109.

Holechek, J.L. and K. Hess. 1994. Free market policy for public land grazing. *Rangelands* **16**: 63–67.

Holechek, J.L. and R.D. Pieper. 1992. Estimation of stocking rate on New Mexico rangelands. *Journal of Soil and Water Conservation* **47**: 116–119.

Holechek, J.L., R.D. Pieper and C.H. Herbel. 1989. *Range management principles and practices*. Prentice-Hall, Englewood Cliffs, New Jersey.

Holechek, J.L., A. Tembo, A. Daniel, M.J. Fusco and M. Cardenas. 1994. Long term grazing influences on Chihuahuan Desert rangeland. *Southwestern Naturalist* **39**: 392–349.

Torell, L.A. and W. Word. 1991. *Range livestock cost and return estimates for New Mexico, 1990*. New Mexico Agricultural Experiment Station Research Report 659. Las Cruces.

Troxell, T.R. and L.D. White. 1989. *Balancing forage demand with forage supply*. Report B-106. Texas A&M University Extension Service, College Station, Texas.

Vallentine, J.F. 1990. *Grazing management*, Academic Press, New York.

INTEGRATED RESPONSE PLOT DESIGNS FOR INDICATORS OF DESERTIFICATION *

R. O. KUEHL*, R. P. BRECKENRIDGE** and M. PANDA*

*The University of Arizona, Tucson, AZ 85721 and Lockheed Environmental Systems and Technologies, Las Vegas, Nevada 89119, USA; **Idaho National Engineering Laboratory, P.O. Box 1625, Idaho Falls, Idaho 83415–2213, USA

Abstract. The improvement of land management practices on lands susceptible to desertification requires information on the status and condition of the existing resources as well as any change occurring in the resource condition over time. The Environmental Monitoring and Assessment Program (EMAP) of the U.S. Environmental Protection Agency has developed a statistical survey design for monitoring the condition of ecological resources on large spatial scales. EMAP-Rangelands used a uniformity sampling study in 1993 to evaluate response plot designs for three categories of indicators (soils, vegetation, and spectral reflectance) to be used for monitoring ecological condition of a site. The response plot design study was developed to integrate on-site measurements for the three indicator categories. The study was conducted on the Colorado Plateau in southern Utah in three rangeland resource classes (grassland, desertscrub, and conifer woodland) of differing productivity levels in an attempt to develop a common plot design for all three resource classes. Basic measurement units were developed to facilitate integration of data collection. Preliminary spatial analysis of the sampling study found considerable differences in variation patterns among the study sites and measurement categories for the indicator classes used by EMAP-Rangelands. Evidence of substantial trends in the indicator measurements on monitoring sites relative to regional trends leads to the conclusion that nonstationary spatial models for biological processes on a monitoring site may be needed to fulfill the requirements for developing plot designs and indicator criteria.

1. Introduction

Occurring in arid and semiarid areas, the process of desertification is the result of human activities or climatic change (Hellden, 1991; Primack, 1993). One of the main difficulties in assessing and trying to control desertification is trying to determine the threatened areas before they reach a threshold beyond which huge amounts of resources are needed to correct the problem area. This problem is compounded by the different schools of thought within rangeland science. Should these areas be managed following a more traditional range succession-retrogression model prompted by Clementsian ecology or following a state and transition model that looks at changes resulting from excessive erosion conditions or from major changes in weather such as drought (Westoby *et al.*, 1989). To address these concerns and collect data for assessing issues related to desertification, assessment

* The U.S. Environmental Protection Agency, through the Office of Research and Development, funded the research described here. This paper has been subjected to the Agency's peer and administrative review and has been approved as an EPA publication. The U.S. Government has the right to retain a nonexclusive, royalty-free license in and to any copyright covering this article.

Environmental Monitoring and Assessment **37:** 189–209, 1995.

and survey methods are needed that accurately characterize the resource and can give managers information in a timely manner. The assessments need to be made at a scale large enough to capture the variability of the site and represent the condition of the resource within similar ecological classes of the region. This paper addresses the variability of arid resources in the context of regional monitoring. The plot design study reported here is important to desertification because without understanding the patterns and site-specific processes that govern change at a site, scientists cannot make associations between human- or climate-induced change and select management options appropriately.

Surveys are a good way to estimate the status of natural resources. Traditionally surveys have been applied to resources where commodities were removed to estimate the amount of merchantable wood volume by tree species and diameter class (Schreuder and Czaplewski, 1993), to evaluate crop production, or to gain data on the condition of water or soil resources. In the late 1980s the Environmental Protection Agency (EPA) and collaborating federal agencies, universities, research institutes, and private organizations established the Environmental Monitoring and Assessment Program (EMAP) to survey the status and determine trends in the condition of the ecological resources in the United States. EMAP was divided into resource groups to focus the science on different areas. Indicators of resource condition were tailored to the ecological unit and to resources being surveyed (e.g. fish harvest in lakes, vegetation structure and composition on rangelands, or soils quality for forests).

The primary objective of EMAP is to provide information, with known confidence, on status and trends in the condition of the Nation's ecological resources as well as estimates, with known confidence, of their extent. EMAP developed a statistical sampling design for the survey of natural resources as a basis for statistical inferences about the condition and extent of ecological resources. The design furnishes a probability sample of ecological resources that forms the foundation for statistical estimation with known confidence levels and the framework to detect changes and trends in ecological indicators. The monitoring plan and survey design are crafted with regional or national assessment as goals (e.g. making statements about changes in desertification or regional shifts in vegetation communities). Thus monitoring and assessment are on a much different scale than that required to address issues associated with localized perturbations of the environment.

The EMAP-Rangelands resource group (EMAP-Rangelands) used a uniformity sampling study to evaluate response plot designs for a group of preselected indicators, spectral properties, soil properties, and vegetation composition and abundance. These indicator categories of arid ecosystem condition were previously tested during the summer of 1992 in the southeastern Utah portion of the Colorado Plateau (Kepner *et al.*, 1994). These indicator categories were selected through workshops and peer reviews (Breckenridge *et al.*, 1993, in press; Mouat *et al.*, 1992; Kepner and Fox, 1991) and appear to meet all the criteria for indicator development as suggested by Hunsaker and Carpenter (1990), such as being applicable and interpretable on a regional scale. Thus these indicator categories are of high priority for EMAP-Rangelands evaluation. Although they appear to demonstrate the highest potential or capability for diagnosing ecosystem change (i.e. the ability to be merged with other data sets to make integrated assessments of ecosystem condition at the regional level), they must first be further field tested for confirmation of regional diagnostic ability before their incorporation into long-term implementation.

2. Study Objectives and Rationale

The objective of the 1993 EMAP-Rangelands pilot study was to determine the optimum sampling support area and plot size for measuring the indicator attributes for vegetation, soils, and spectral properties in EMAP-Rangeland's extensive resource classes of the Colorado Plateau biogeographical region, where the *sampling support area* is that area at the monitoring site that provides an adequate sample representation of EMAP-Rangelands subpopulations. The extensive EMAP-Rangelands resource classes included in the 1993 study were the desertscrub, grassland, and woodland resource classes in the Colorado Plateau biogeographical region (Kepner *et al.*, 1993).

The development of indicators of the condition of rangeland ecosystems depends on the quality of measurements needed to quantify the indicators. Effective ecological interpretation of information gathered on monitored sites requires measurements that adequately describe the biological communities at the sites. To make these measurements consistent with the biology of the EMAP-Rangelands resource class requires a sample plot configuration, within a sampling support area, that is sufficient to capture the characteristics (i.e. degree of variability) of the site's biological communities.

A distinction must be made between the sample plot design and the sample survey design that will be used for the EMAP-Rangelands monitoring survey. The monitoring sites for EMAP-Rangelands are point samples in the context of the EMAP survey design (Overton *et al.*, 1990; White *et al.*, 1992). Point samples for measurements such as temperature do not need a support area around the point to acquire the measurement. But point samples for characteristics such as vegetation cover and soil properties require observation in some area surrounding the point to describe the characteristic identified by the point sample.

The plot design at a site may be considered a "response" design for indicator measurement at a monitoring site, whereas the collection of monitoring sites constitutes the survey design. The measurements on the sample plot at a site will be used as one observation of the indicator in the survey design. The survey sample is complete only when that indicator observation is included with the observations collected at all other sites combined by resource class within the landscape or region. The observations from the survey design are those used for regional and landscape interpretations.

Interpretation of ecological condition at EMAP-Rangelands monitoring sites requires an adequate representation of the vegetation and soil communities at a site. The ecological concept of minimal area (Dietvorst *et al.*, 1982) is usually defined as the smallest area on which the species composition of a plant community is adequately represented. Two renderings of the minimum area concept have been described, the *biological* and *methodological*. *Biological minimum area* refers to the smallest area on which the species composition of a plant community is adequately represented or, equivalently, to the size of stand required to be well developed. *Methodological minimum area* within a stand refers to the size of sample plots needed to complete at least an adequate description of that stand (Barkman, 1989).

The methodological minimum area has been further divided into qualitative and quantitative parts. The qualitative area is that plot size above which the number of species does not increase at all or increases only insignificantly within the same stand. The quantitative area is distinguished as that above which the quantitative shares of all species do not change significantly (Barkman, 1989). For example, Barkman (1989) observed that the gain in information is less each time the sample plot area is doubled, particularly with respect to the cost of sampling. Thus EMAP-Rangelands concluded that a pragmatic definition of methodological minimum area should be that plot size whose further enlargement produces an insufficient gain in information for the added cost.

Researchers have used several criteria to determine the minimum plot size and shape needed to adequately describe an area. Species area curves, similarity analysis, frequency area curves, species representation, and pattern representation have had various applications. The Braun-Blanquet cover abundance scale (Bonham, 1989) is commonly used as a measurement to evaluate the minimum area. Historically, monotonic relationships have been observed between the size of a sampled area and the criteria used to determine the methodological minimum area such as species area curves and similarity measurements (Dietvorst *et al.*, 1982; Barbour *et al.*, 1980). Monotonic relationships between size of sampled areas and variances of measured variables have also been observed in a variety of settings, including biomass of grasses and forbs (Wiegert, 1962), tree volume (Tardif, 1965), basal area (Bormann, 1953), agricultural experimental yield trials (Smith, 1938), surveys for plant disease incidence (Proctor, 1985), and agricultural acreage and yield sample surveys (Cochran, 1977).

One standard method to assess efficient plot sizes and configuration considers an empirical model developed by Smith (1938), which relates variance to size of plot. The method has been used for decades to evaluate plot sizes for agricultural field experiments and for natural systems (Brummer *et al.*, 1994). A parameter estimated for the Smith model from the plot data is used in conjunction with a cost function to obtain an efficient size of sample plot. Extensions and modifications of the Smith variance model include terms to account for variation between plants, (Pearce, 1955; Freeman, 1963) and plot shape, (Reddy and Chetty, 1982). The use of spatial correlation analysis in place of or in conjunction with the Smith model by Modjeska and Rawlings (1983) included a generalized method of plot optimization. Brewer and Mead (1986) extensively reviewed models that have been developed by themselves and others for analyses of field crop experiments to evaluate the efficiencies of plot designs.

Other methods to determine optimum plot size include the tangent to the species area curve used by Hyder *et al.* (1963) for plant frequency data in sagebrush-bunchgrass communities and sample size requirements for specific degrees of precision for confidence interval estimation with cost constraints (Bormann, 1953; O'Regan and Arvanitis, 1966).

The measurement of variables for indicators on EMAP-Rangelands monitoring sites is intended to reflect the status of biotic and abiotic communities within the sampled resource community. Thus the sampling support area must be large enough to adequately characterize indicator variables for the plant and soil communities under consideration.

The literature previously cited on plot size has been concerned primarily with vegetation communities. In addition, plot size in natural systems most often refers to quadrat for which the entire quadrat area is censused for vegetation community characteristics, such as biomass and species abundance or composition. The agricultural studies have concentrated on plot size studies for biomass, grain, or fruit yield in comparative experimental trials.

The uniformity sampling trial is a commonly used method to provide data for several types of studies related to sampling area. Uniformity trials for agricultural crops use harvest data from a field planted to a single cultivar divided into small contiguous areas known as basic units. These data are used to find optimum plot sizes from among aggregations of the basic units.

Contiguous rows of quadrants are used in ecological studies to evaluate the occurrence of contagious patterns. Increasing quadrat sizes are built up by blocking adjacent quadrants in pairs, fours, eights, and so forth, and the variances of different block sizes are related to block size in a graph to detect different scales as peaks in the graph (Greig-Smith, 1952).

Contiguous quadrants are measured over a designated area in ecological studies to determine biological minimum areas (Barkman, 1989). The contiguous quadrants are combined into larger quadrants of different sizes, and various measures are calculated from the nested sets of quadrants to ascertain biological and methodological minimum areas.

The EMAP-Rangelands survey requirements for sampling plots differed from previous studies in several facets. The sampling plots have to serve as monitoring sites for a suite of indicator measurements with the potential for repeated visits to the site over decades. A complete census of the plots is neither feasible nor desirable under these circumstances. Plots must be set up with minimal disturbance

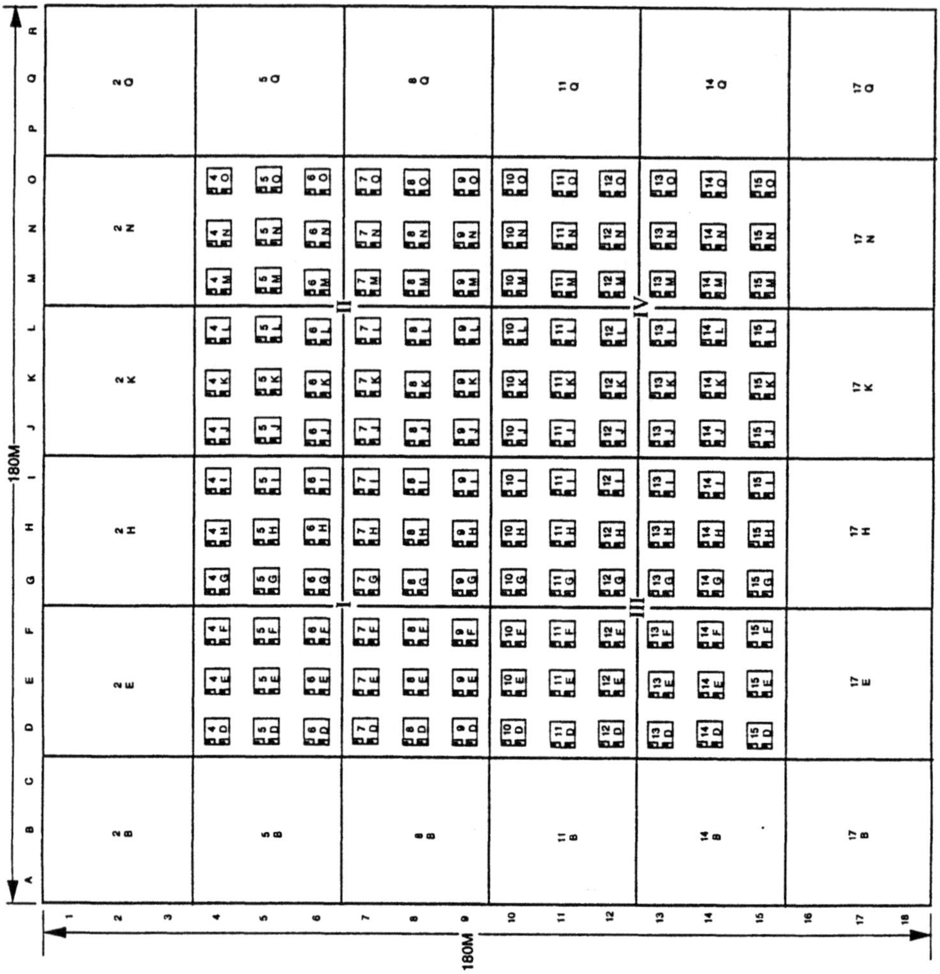

Fig. 1. Macroplot Design for soils, vegetation, and spectral properties indicators.

to the measurement areas. The number of plots or replicate samples at the site for monitoring purposes will be the number required for specified confidence estimates in the survey. Thus questions of replication blocking with different treatments on the plots within blocks as in experimental trials are not relevant to the survey.

Finally, for indicator attributes in this study EMAP-Rangelands is concerned not only with vegetation but also with soils and spectral measurements, which provide a much stronger preponderance of evidence for assessing desertification. Thus EMAP-Rangelands must construct a sample plot design that integrates the sampling support area and optimum plot size required for the measurements made for the three indicator categories.

3. Approach

The field sampling areas required for this study were coordinated and integrated by three indicator groups. Large macroplots estimated to be larger than the indicator sampling support area were established in relatively homogeneous resource classes of either desertscrub, woodland, or grassland. The macroplots were the basis for a uniformity sampling trial at each site. Each macroplot (Figure 1) was subdivided into a rectangular array of basic sampling units. The size of basic units was such that a sample measurement could be made for each of the selected measurements in the three indicator categories in each of the biomes.

This study evaluated five combinations of resource class and productivity levels. Productivity was defined as the potential mean annual vascular vegetation biomass as described in existing ecological (range) site descriptions (USDA, 1976; USDI, 1990). Within a biogeographic province, vegetation community physiognomy and cover characteristics may differ significantly between naturally low production sites and high production sites. Physiognomy and cover characteristics, however, are expected to exhibit greater similarity between communities with similar production potential when subjected to similar environmental and anthropogenic stresses.

Arid environments in the United States typically produce natural communities that range from a mean annual productivity of less than 178 kg/ha to more than 890 kg/ha because of the natural variance in the interrelationships between soil, climate, and vegetation community. Ecological site correlation procedures developed by the Bureau of Land Management (Leonard *et al.*, 1992) suggest natural breaks at various productivity levels. Grouping of sites less than 445 kg/ha (low), 445 to 890 kg/ha (medium), and greater than 890 kg/ha (medium-high) are used for comparison. Also described are other criteria, such as dominant species and community composition, that relate to resource class characteristics at the larger scale considered by EMAP.

Study sites for 1993 were selected to represent readily observable differences in resource class and productivity combinations, but not every possible combination was represented because of time and cost constraints. The desertscrub resource class is the most extensive vegetation class in the Great Basin biogeographic region. Low, medium, and (medium) high-producing sites were selected to represent this resource class. A medium-producing grassland and high-producing pinyon-juniper site were also selected. Low-producing grassland and pinyon-juniper sites are rare in the Great Basin biogeographic province.

The size of the macroplot for the 1993 plot study (Figure 1) was selected to be more than large enough to provide an adequate sample thought to characterize the local scale of variation at the site. Literature and professional judgment were used to select the macroplots for soils, vegetation, and spectral indicators, but they all had to be of a size that could be supported with existing funding and staff. The final macroplot size (180 m × 180 m) was adjusted to maximize integration of data between the indicator groups and to ensure that all data could be collected within

1 week. The overall size of the soil macroplot was selected to be 180 m by 180 m to allow adequate soil samples to characterize variability.

Previous research (Campbell, 1978; Warrick *et al.*, 1986; Wilding and Drees, 1983) suggests that a spacing of 30 m would result in spatially independent samples. Thus the 180 m by 180 m macroplot contained 36 basic units (30 m by 30 m). A total of 36 other 10 m by 10 m basic units were nested within the 180 m by 180 m macroplot to have spatially dependent basic units in order to measure spatial correlations.

Soil properties have different orders of variation and spatial correlation. To measure the variation and spatial correlation, observations and samples were taken at different densities. The 10 m spacing was chosen to estimate the spatial correlation of surface properties that vary more at a small scale and have higher coefficients of variation. These surface properties were organic carbon, total nitrogen, and hydraulic conductivity (Mausbach *et al.*, 1980; Wilding and Drees, 1983). Properties that are variable at a larger scale and have coefficients of variation from 15–30% were sampled at 30 m intervals. These properties were the texture and structure of the A horizon. Properties that are less variable and have coefficients of variation of less than 15%, such as the soil color and thickness of the A horizon and soil classification, were sampled at 60 m intervals.

Spectral reflectance properties of biotic and abiotic objects tend to be wavelength dependent, and determining the relationships and the support area is critical for characterizing or discriminating the object. The plot size selected for the spectral part of this study was based on the requirement to collect both ground-based and satellite spectral measurements to evaluate their relationship with estimated vegetation and soil features on the site. The spectral properties plot requirements were developed from findings of other researchers (Gholz, 1982; Waring *et al.*, 1978) about the relationship between ecosystem structure (e.g. biomass, LAI) and functional properties (e.g. net primary productivity).

Satellite sensors (Landsat Thematic Mapper) typically collect data that is integrated or mixed into small segments (pixels) representing the spectral properties of an area about 30 m by 30 m on the ground. The ground spectral data is collected using a portable spectrometer at a smaller scale, typically 30 cm when the instrument is held 1 m above the ground. Ground spectral measurements were spatially coincident with vegetation measurements (collected in 5 m by 5 m quadrats) and combined so that the relationship between vegetation composition and abundance and surface spectral attributes could be determined at the basic unit of 10 m by 10 m. Ground spectral measurements were collected for the entire macroplot of 120 m by 120 m so that several satellite pixels could be combined to develop a spectral signature for the entire macroplot.

Vegetation composition, structure, and abundance are measured because of their importance as indicators of changes in arid and semiarid ecosystems. Several sampling options were considered, and a decision was made to use a modified Daubenmire cover class method within a quadrat along a line transect. Similar methods are used in other inventory, monitoring, and evaluation programs conducted by the USDA Forest Service in the intermountain West (USDI, 1985; O'Brien and Van Hooser, 1983).

For plant sampling, rectangular plots have been found to yield better results than other shapes. A rectangle with sides in a 1:2 ratio works well and was selected for grass and shrub quadrats. Tree quadrats were increased to 5 m by 5 m to account for the larger spacing between the pinyon-juniper plants and modified to a square to be integrated into overall macroplot design. Vegetation samples were coordinated with spectral and soils sampling so that relationships between the data could be evaluated. Vegetation line transects of 120 m were used, and plots were located every 5 m (two 1 m $\times$ 2 m quadrats were within each basic unit) to provide adequate coverage to characterize the variability of the more widely spaced organisms in arid ecosystems.

Centered within the soil macroplot, the macroplot for vegetation and spectral properties was 120 m by 120 m. The macroplot size for each indicator group was selected to provide enough replication to supply adequate degrees of freedom to statistically evaluate the spatial variability of the resources across the site.

4. Results and Discussion

The observations for plot design analysis are spatially related whether they come from ecological studies of natural systems, agricultural experiments, or the current study on plot design for EMAP-Rangelands monitoring.

The success of any method of analysis on spatially related data depends considerably on some assumptions about statistical properties of spatial data. More specifically, assumptions are required about the spatial process under study. These properties are discussed in detail by Ripley (1981), Haining (1990), and Cressie (1991).

For the uniformity sampling study the data $Y(s_1), Y(S_2), \ldots, Y(s_n)$ collected at spatial locations $s_i = \{x_{1i} x_{2i}\}$ for $i = 1, 2, \ldots, n$ are considered observations from some biological process at the scale of the macroplot somewhat larger than 1 ha at any of the study locations.

One of the desired features of the data for plot design analysis is that of stationarity in the spatial setting. Informally, stationarity refers to variational properties of the data that do not change throughout the sampled region of interest. More formally, a second order stationary process, $Y(s)$, will have a constant mean, $E[Y(s)] = \mu$ for all locations s in the region of interest, and a constant covariance function

$$Cov[Y(s), Y(t)] = E[Y(s) - \mu][Y(t) - \mu]$$

for the distance s to t anywhere in the region of interest for a constant direction. If the covariance depends only on the distance and not the direction, the process $Y(s)$ is considered isotropic, a common assumption in many spatial statistics studies.

TABLE I
Spatial correlation matrix for total grass-forb cover on a desertscrub site

	k										
l	−5	−4	−3	−2	−1	0	1	2	3	4	5
0	−0.18	0.19	0.00	−0.08	0.18	1.00	0.18	−0.08	0.00	0.19	−0.18
1	−0.11	−0.06	0.22	0.08	−0.03	0.03	0.52	−0.02	−0.13	−0.08	0.06
2	0.16	−0.25	−0.22	0.43	−0.02	−0.02	−0.01	−0.06	−0.04	−0.15	−0.12
3	−0.19	0.24	0.03	−0.09	0.00	0.40	0.11	−0.11	−0.06	−0.01	−0.11
4	0.04	0.09	0.01	0.14	0.00	−0.06	0.77	−0.01	−0.16	−0.09	0.01
5	0.41	−0.14	−0.12	0.61	0.12	−0.02	−0.03	0.02	−0.15	−0.08	−0.09

TABLE II
Standardized mean square errors for total grass-forb cover on a per-unit basis for plots a × b units in size on the grassland site

	b					
a	1	2	3	4	5	6
1	1.00	1.23	1.39	1.57	1.71	1.77
2	1.36	1.84	2.15	2.50	2.78	2.89
3	1.68	2.46	2.96	3.48	3.92	4.11
4	2.04	3.08	3.76	4.48	5.08	5.34
5	2.39	3.72	4.61	5.53	6.29	6.62
6	2.76	4.35	5.43	6.56	7.47	7.87

A number of methods may be used to explore the stationarity or nonstationarity, as the case may be, of the process, $Y(s)$, with the spatially observed data. For the uniformity sampling trials, the exploration included estimates of spatial correlations, variograms, trend across the sites, and variances for a variety of plot configurations. But nonstationarity of the covariance and of the mean are difficult to jointly evaluate, and completely satisfactory methods are not fully developed (Handcock and Wallis, 1994).

The initial analyses conducted on the 1993 plot data included the estimate of spatial correlations along with variances for sample plot configurations at the five sites for all of the indicator measurements. The computational algorithm for the covariances, correlations, and plot variances follow those of Modjeska and Rawlings (1983).

Spatial correlations, $r(l,k)$, are illustrated with total grass-forb cover measures on a desertscrub site (Table I) where $r(l,k)$ is the correlation between observations separated by l rows and k columns in the 12×12 array of data. Because of symmetry of correlations, $r(l,k) = r(-l,-k)$ and $r(l,-k) = r(-l,k)$.

Under many of the common stationary models one would expect to see some patterns in the correlation matrix. But no regular patterns were apparent with these correlations, and in general we found no regular patterns for most of the indicator measurements and sites.

Standardized mean squares, $STMSE(ab)$, for sampling plots of size a rows by b columns computed for total grass-forb cover on the grassland macroplot (Table II) differed considerably among plot shapes. The mean square $MSE(ab)$ for any configuration was divided by the mean square for the unit size plot, $a = 1$ and $b = 1$, to achieve the standardized value. The relative efficiency of two plot shapes of the same size can be evaluated by their ratios. Consider plots of six basic units, for example, $STMSE(61) = 2.76$, $STMSE(32) = 2.46$, $STMSE(23) = 2.15$, and $STMSE(16) = 1.77$. The ratio of standardized mean squares for a 6×1 plot to 1×6 plot is 2.76/1.77 = 1.56, indicating the 1 row by 6 column plots are 56% more efficient than the 6 row by 1 column plots. Likewise, the efficiencies of 3×2 and 2×3 plots relative to the 6×1 are 1.12 and 1.29 respectively. These efficiencies show that plots were more efficient as they became elongated in the west-to-east direction. Upon viewing the first row and first column of Table II, one can see that any long and narrow plot in the north-to-south direction (i.e. $2 \times 1, 3 \times 1, \ldots, 6 \times 1$) is less efficient than a long and narrow plot in the west-east direction, (i.e. $1 \times 2, 1 \times 3, \ldots, 1 \times 6$). Thus the plots elongated in the west-to-east direction captured considerably more variation on the site than those oriented in the north-to-south direction. Similar differences in efficiencies can be seen for other plots of same size but different configurations. Again, similar results were observed for all measurements at all sites.

Because nonstationary processes can produce trends in the spatial extent of the data, trends across rows and columns were evaluated for the data on the 12-row by 12-column sampling grid. Patterns of row effects and column effects in two-way arrays have commonly been analyzed with traditional analysis of variance methods. Underlying the two-way analysis of variance is the additive decomposition

$$\text{data} = \text{constant} + \text{row} + \text{column} + \text{residual}$$

and trends in the two-way array are revealed by the magnitudes and signs of the row and column effects.

The row and column effects for the current study were evaluated with a median polish (Tukey, 1977) on the two-way array. The median polish resists the influence of outliers and reduces their effect on the evaluation of row and column contributions to the variation in a two-way table.

The row effects from the median polish for total vascular plant cover in the north-to-south direction on a desertscrub site (Figure 2) show definite trends, as does the plot of column effects in the west-to-east direction on the site (Figure 3). Other variables on the sites also exhibited row and column patterns.

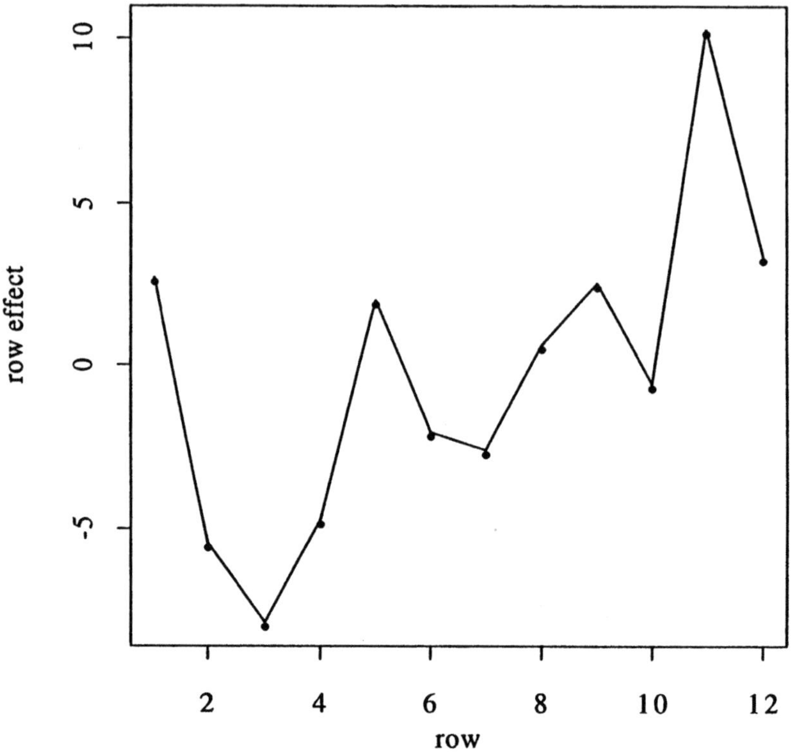

Fig. 2. Row effects for total vascular plant cover.

The stability of the mean and variance on the macroplots can be evaluated with a moving window similar to the moving average technique used for time series. The mean and relative variation, percent coefficient of variation (%CV), for the normalized difference vegetation index (NDVI) was computed in 2×2 moving windows of basic units across the macroplot. The moving average for NDVI on a desertscrub plot (Figure 4) appears to be fairly stable, but the %CV of NDVI measurements (Figure 5) are quite divergent across the macroplot, suggesting the possibility of a nonstationary variance structure.

For all variables at all sites these nonstationarity diagnostics for the most part exhibited patterns that discourage the use of the simple stationary spatial models most often used for plot design studies. In addition, the patterns or behavior of the diagnostics showed no consistency among sites or among variables.

If the response plot design were of concern only for a specific site, the nonstationarity is less troublesome. The orientation of transects or shapes of sampling plots can be determined conveniently from the standardized mean squares and costs as suggested by Modjeska and Rawlings (1983). Of course, each site in that case would have its unique sampling requirements.

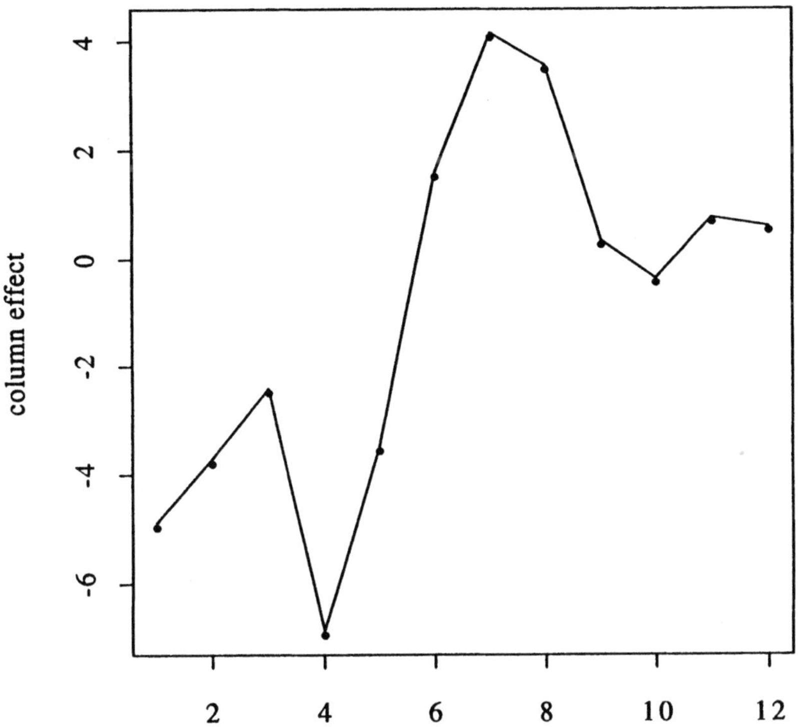

Fig. 3. Column effects for total vascular plant cover.

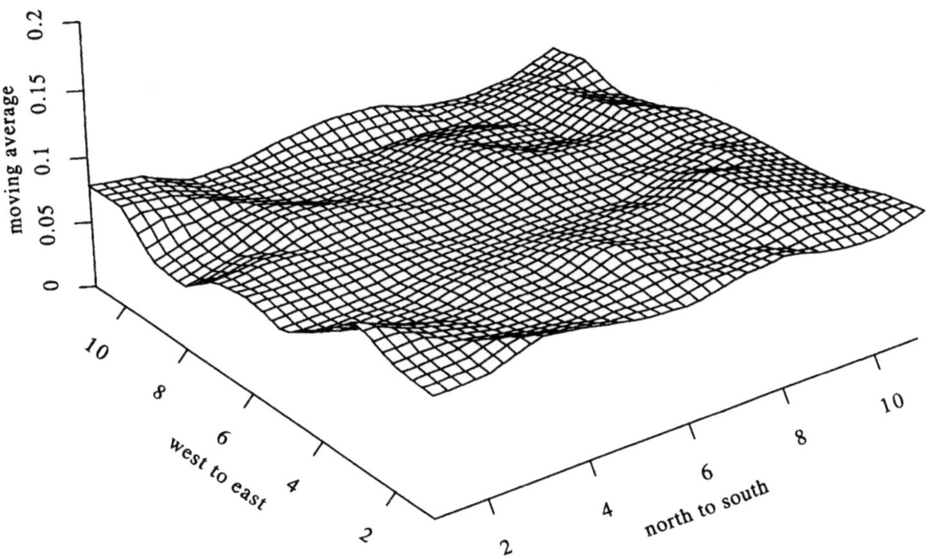

Fig. 4. Moving avarage for NDVI in a 2 × 2 window.

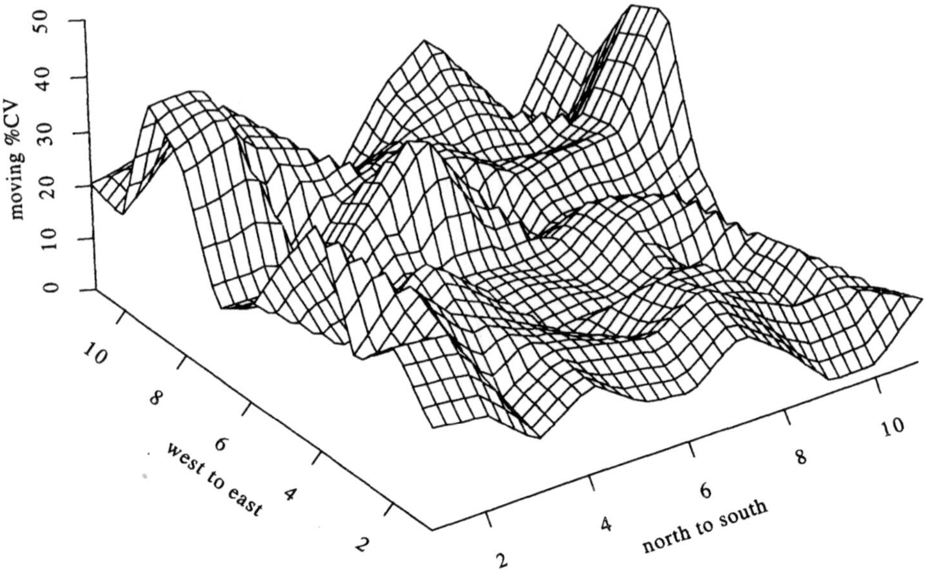

Fig. 5. Moving coefficient of variation for NDVI in a 2 × 2 window.

TABLE III

Averages of site means and coefficients of variation for basic units from uniformity sampling

Variable	Mean	% Coefficient variation
Vegetation Cover (n = 144)		
Total vascular plant	23.3	61
Total shrub	15.5	60
Total grass-forb	3.7	142
Dominant shrub	13.7	69
Dominant grass-forb	1.7	282
Total tree (1 site)	12.6	77
Spectral Reflectance (n = 144)		
NDVI	0.16	22
SAVI	0.11	21
Albedo	0.19	21
Soils (n = 36)		
Organic carbon	0.54	40
Nitrogen	0.04	43
C/N ratio	10.75	61
Water content	5.42	18

But the objective for EMAP-Rangelands requires developing a sampling strategy that applies generally to all monitoring sites. Thus the requirements for stationarity to evaluate the covariance structure across sites and variables become an important consideration for the plot design development.

Historically high sampling variability has been experienced on rangeland sample sites, resulting in large sample-size requirements for precision in the estimates of means and other parameters. The average site means and coefficients of variation from the basic units at the five sites for some of the measurements (Table III) differ considerably in variation among the different measurements for indicator classes in indicating that sample size requirements will vary considerably. For some indicator measurements the needed sample information appears to increase with the detail required in the indicator. The variation for species-level information, such as dominant species cover, tends to be greater than that for total vegetation cover, thus requiring larger sample sizes for the same level of information. Likewise, carbon-to-nitrogen ratios in soil samples tended to be more variable than the individual carbon or nitrogen samples.

If biological processes are operating at a scale equal to or smaller than the sampled area in such a way to produce trends with local minima and maxima on the site, then the usual sample statistics are inadequate to characterize the site. The sample variances fail to separate the variation in trends across the site from the natural sampling variation about the trend. And the site means may be inadequate to describe the site characteristics. Thus these simple stationary statistical models are insufficient to characterize the range site.

Under the supposition that biological and geological or soil processes are operating at the site scale, the data can be modeled as $Y(s) = \mu(s) + Z(s)$ where $\mu(s)$ is the deterministic mean structure or trend over the spatial extent of the sampled plot and $Z(s)$ is the stationary correlated error process.

The challenge with this broader view model for the range site is to achieve a sensible decomposition of the variation into its trend and random components. The component $\mu(s)$ models the large scale variation on the site as a consequence of biological processes in operation at the scale of the sampled site. The $Z(s)$ component models the random sampling errors about the mean structure, $\mu(s)$. This representation is analogous to that used in time series analysis to model trend and error processes.

Since the model recognizes spatial variability at the large scale through $\mu(s)$ and at the small scale through $Z(s)$, the objective for an analysis of the model is to estimate the trend on the site, $\mu(s)$, and subsequently, from the estimates of the residuals, estimate the correlations and variances associated with the stationary process, $Z(s)$.

Interpolated surface plots of the 144 observations for total vascular plant cover (Figure 6), total grass-forb cover (Figure 7), and total shrub cover (Figure 8) on a desertscrub site reveal substantially irregular surfaces. The nonstationary mean model presumes that the roughness of the surface plot for total vascular plant

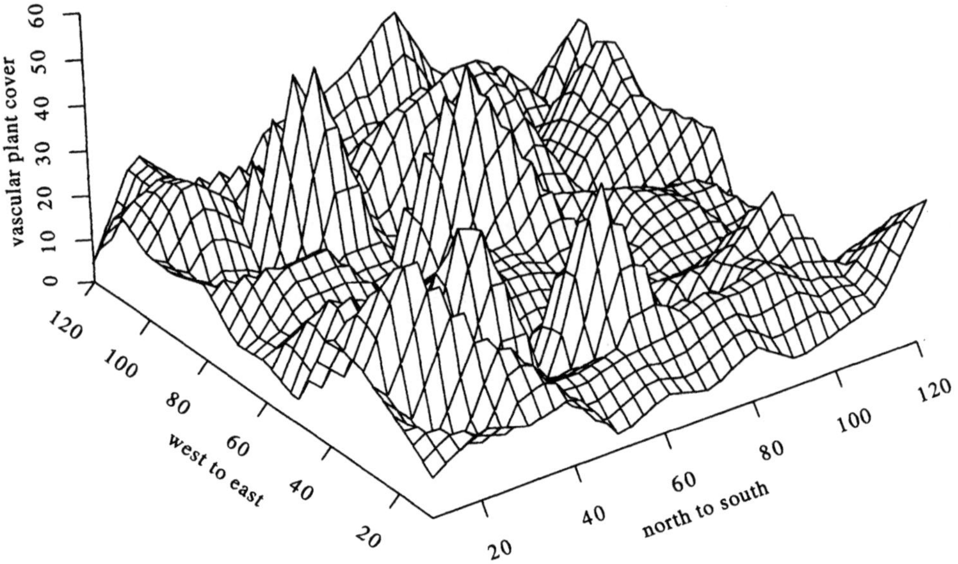

Fig. 6. Total vascular plant cover.

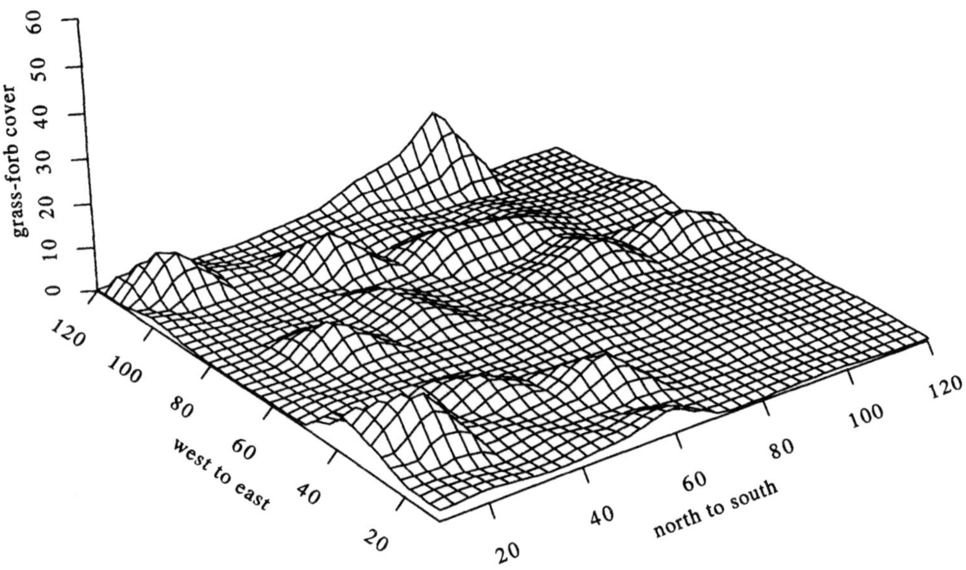

Fig. 7. Total grass-forb cover.

and shrub cover and the obvious trend in the grass-forb cover results from more than simple random sampling variation. Specifically this model presumes that the surfaces can be explained as the aggregate of an unknown deterministic mean structure plus a correlated error structure. Providing an informative description of the range site through the large-scale trend $\mu(s)$ and the attendant correlated error

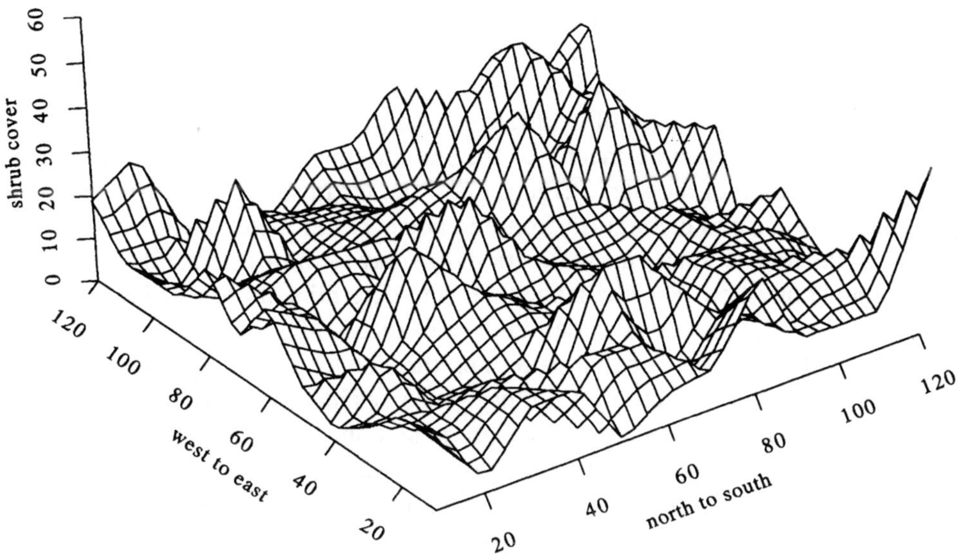

Fig. 8. Total shrub cover.

structure for $Z(s)$ is a requirement placed on model estimation. The desired result is a parsimonious partition of the variation between the parameterization for $\mu(s)$ and a suitably stationary error structure $Z(s)$. Parameterization of the mean function $\mu(s)$ and the subsequent estimates for the stationary error process are subjects of much current research in spatial statistics. Ripley (1981) and Cressie (1991) extensively discuss some of the approaches.

The challenge posed by this investigation requires a decision as to the need of nonstationary models for these local-scale monitoring plots when inferences are ultimately based on the results of these point samples over a much broader regional landscape. One consideration in making that decision is the range of variation experienced at the two scales. For example, the range in basic unit total vascular plant cover values was typically 2% to 70% at any one of the sites. The averages of the five sites ranged from 17% to 29%. Because the local scale of variation considerably exceeds the larger scale amount, the nonstationary model at the local site may be a reasonable choice.

Another consideration in deciding to use a nonstationary model is whether the potential exists for local processes to occur at these monitoring sites. For example, the total vascular plant cover (Figure 6) is an aggregation of the total grass-forb cover (Figure 7) and total shrub cover (Figure 8). If the substantial differences in spatial patterns for vegetation cover of different types observed over the 1 ha sites are a function of interactions among and within the different plant communities or between the plant and soil communities over the site, then local processes are in effect, and the nonstationary model is appropriate. An overview of criteria and indicators of rangeland health by the National Research Council (1994) Committee on Rangeland Health recognized many of the biotic and abiotic processes on range sites as indicators of rangeland condition.

Additionally, the existence of processes on the site may substantially affect the criteria used as indicators of ecological condition for the monitoring site. For example, are simple sample means of the measurements adequate to judge the ecological condition of the site? Perhaps not if the mean varies across the site and the signal is difficult to distinguish from the noise. Other metrics derived from an estimate of the nonstationary mean function $\mu(s)$ may be more reasonable indicators of ecological condition. The challenge then is to determine those metrics.

The scale of the processes under study will dictate the scale of measurement distances and the observed spatial pattern on the site. The challenge posed here is to determine the scale of activity for those processes identified for the monitoring process and then match the scale of measurement to the scale of the process for the indicator under consideration in order to have an effective monitoring indicator.

Finally, the measurement tools or methodologies may figure prominently in the modeling process. For example, the use of quadrats, line transect intercepts, or point frames for collecting vegetation data may greatly influence the spatial patterns of the measurements because the cost and measurement errors differ among the methods. Thus the choice of methodology may considerably affect decisions on the most useful metrics for monitoring.

5. Conclusions

Evaluating desertification requires a suite of indicators and methodologies that of necessity must be integrated into an overall monitoring program. The EMAP-Rangelands pilot study addressed several facets of this integration process at the field-sampling level, including the logistical requirements for multiple indicator classes, integrating measurement methodologies, plot traverse, and patterns of responses for the measurements to evaluate plot design strategies. This paper discussed the exploration of spatial relationships in the measurements and the consequences of these relationships for plot design strategies in the monitoring program.

Preliminary analyses clearly found the need to expand current thinking on strategies for plot design and indicator criteria. To properly assess the condition of an arid area, the plot design needs to account for the larger spatial variability associated with natural resources on rangelands. This plot design study showed that some of the measurements exhibit a pattern of lower variability in a north-to-south orientation than east to west. To properly capture this variability, areal cover designs appear to have advantages over long and narrow transects for properly characterizing a site.

The complexities of indicator measurement patterns as they vary among monitoring sites beg the question of more complex site models that incorporate nonstationary mean structure over the monitoring sites. These changes from current practices in modeling site-specific characteristics in turn promote new thinking about the type of metrics needed for indicator criteria to judge ecological condition. The scale of measurement necessarily must match the scale of the process being used as an indicator, and measurement tools or methodologies may figure prominently in the response plot design development.

The selection of indicators for evaluating desertification on rangelands must reflect the biotic and abiotic processes that are producing observed patterns on a sample site. An oversized plot design study is a useful approach for evaluating the scale of processes and their patterns and should be employed in different ecological resource classes (i.e. grassland versus desertscrub) to ensure the resource is accurately assessed before making management decisions.

References

Barbour, M.G., J.H. Burk and W.D. Pitts. 1980. *Terrestrial plant ecology*, Benjamin/Cumming, Menlo Park, California.

Barkman, J.J. 1989. A critical evaluation of minimum area concepts. *Vegetatio* **85**: 89–104.

Bonham, C.D. 1989. *Measurements for terrestrial vegetation*, Wiley, New York.

Bormann, F.H. 1953. The statistical efficiency of sample plot size and shape in forest ecology. *Ecology* **34**: 474–487.

Breckenridge, R.P., W.G. Kepner and D.A. Mouat. 1993. *Monitoring for assessing sustainability of arid ecosystems*. Paper presented at the 25th International Symposium, Remote Sensing and Global Environmental Change. Graz, Austria, 4–8 April, 1993.

Breckenridge, R.P., W.G. Kepner and D.A. Mouat. In press. A process for selecting indicators for monitoring conditions of rangeland health. *Journal of Environmental Monitoring and Assessment*.

Brewer, A.C. and R. Mead. 1986. Continuous second order models of spatial variation with application to the efficiency of field crop experiments. *Journal of the Royal Statistical Society* A **149**: 314–348.

Brummer, J., J. Nichols, R. Engel and K. Eskridge. 1994. Efficiency of different quadrat sizes and shapes for sampling standing crop. *Journal of Range Management* **47**: 84–89.

Campbell, J.B. 1978. Spatial variation of sand content and pH within single contiguous delineations of two soil-mapping units. *Soil Science Society of America Journal* **42**: 460–464.

Cochran, W.G. 1977. *Sampling techniques*. 3rd Ed. Wiley, New York.

Cressie, N. 1991. *Statistics for spatial data*, Wiley, New York.

Dietvorst, P., E. van der Maarel and H. van der Putten. 1982. A new approach to the minimal area of a plant community. *Vegetatio* **50**: 77–91.

Freeman, G.H. 1963. The combined effect of environmental and plant variation. *Biometrics* **19**: 273–277.

Gholz, H. 1982. Environmental limits on above ground net primary production, leaf area, and biomass in vegetation zones of the Pacific Northwest. *Ecology* **63**: 469–481.

Greig-Smith, P. 1952. The use of random and contiguous quadrats in the study of structure of plant communities. *Annals of Botany* **16**: 293–316.

Haining, R.P., 1990. *Spatial data analysis in the social and environmental sciences*, Cambridge University Press, New York.

Handcock, M.S and J.R. Wallis. 1994. An approach to statistical spatial-temporal modeling of meteorological fields. *Journal of the American Statistical Association* **89**: 368–378.

Hellden, U. 1991. Desertification – Time for an assessment? *Ambio* **20**(8): 372–383.

Hyder, D.N., C.E. Conrad, P.T. Tueller, L.D. Calvin, C.E. Poulton and F.A. Sneva. 1963. Frequency sampling in sagebrush-bunchgrass vegetation. *Ecology* **44**: 740–746.

Hunsaker, C.T. and D.A. Carpenter (editors). 1990. *Ecological indicators for the Environmental Monitoring and Assessment Program*. EPA 600/3–90/060. U.S. Environmental Protection Agency, Office of Research and Development, Research Triangle Park, North Carolina.

Kepner, W.G. and C.A. Fox (editors). 1991. *Environmental Monitoring and Assessment Program. Arid ecosystems strategic monitoring plan*. EPA 600/4–91/018, U.S. Environmental Protection Agency, Washington.

Kepner, W. G., R.O. Kuehl, R.P. Breckenridge, J.R. Baker, D. O'Leary, J.M. Lancaster, S.G. Leonard, D.A. Mouat, T.G. Reinsch, M.A. Weltz, D.W. Sutton, R.L. Tidwell, A.C. Neale. G.E. Byers and R.L. Slagle. 1993. *Environmental Monitoring and Assessment Program Arid Ecosystems 1993 implementation plan Colorado Plateau plot design pilot study*. EPA/620/R-93/016. U.S. Environmental Protection Agency, EMSL-LV, Las Vegas, Nevada.

Kepner, W.G., R.O. Kuehl, R.P. Breckenridge, J.M. Lancaster, S.G. Leonard, D.A. Mouat, T.G. Reinsch, K.B. Jones, A.C. Neale, T.B. Minor and N. Tallent-Halsell. 1994. *Environmental Monitoring and Assessment Program Arid Ecosystems 1992 pilot report*. EPA/620/R-94/015. U.S. Environmental Protection Agency, EMSL-LV, Las Vegas, Nevada.

Leonard, S.G., G. Staidl, J. Fogg, K. Gebhardt, W. Hagenbuck and D. Prichard. 1992. *Riparian area management: Procedures for ecological site inventory – with special reference to riparian-wetland sites*. U.S. Department of the Interior, Bureau of Land Management, Service Center, Denver.

Mausbach, M.J., B.R. Brasher, R.D. Yeck and W.D. Nettleton. 1980. Variability of measured properties in morphologically matched pedons. *Soil Science Society of America Journal* **44**: 358–363.

Modjeska, J.S. and J.O. Rawlings. 1983. Spatial correlation analysis of uniformity data. *Biometrics* **39**: 373–384.

Mouat, D.A., C.A. Fox and M.R. Rose. 1992. Ecological indicator strategies for monitoring arid ecosystems, in D.H. McKenzie, D.E. Hyatt and V.J. McDonald (Eds.), *Ecological indicators* (vol. I and II), Elsevier Applied Science, New York.

National Research Council. 1994. *Rangeland health: New methods to classify, inventory, and monitor rangelands*. Committee on Rangeland Classification, Board on Agriculture. National Academy Press, Washington.

O'Brien, T. and D.D. Van Hooser. 1983. *Understory vegetation inventory: An efficient procedure*. Research paper INT-323. U.S. Department of Agriculture Forest Service, Intermountain Forest and Range Experiment Station, Ogden, Utah.

O'Regan, W.G. and Arvanitis, L.G. 1966. Cost-effectiveness in forest sampling. *Forest Science* **12**: 406–414.

Overton, W.S., D. White and D.L. Stevens, Jr. 1990. *Design report for EMAP, Environmental Monitoring and Assessment Program*. EPA/600/3–91/053. U.S. Environmental Protection Agency, ERL, Corvallis, Oregon.

Pearce, S.C. 1955. Some considerations in deciding plot size in field trials with trees and bushes. *Journal of the Indian Society of Agricultural Statistics*. **7**: 23–26.

Primack, R.P. 1993. *Essentials of conservation biology*, Sinauer Associates, Sunderland, Massachusetts, pp. 129–130.

Proctor, C.H. 1985. Fitting H.F. Smith's empirical law to cluster variances for use in designing multistage sample surveys. *Journal of the American Statistical Association* **80**: 294–300.

Reddy, M.N. and C.K.R. Chetty. 1982. Effect of plot shape on variability in Smith's variance law. *Experimental Agriculture* **18**: 333–338.

Ripley, B.D. 1981. *Spatial statistics*, Wiley, New York.

Schreuder, H.T. and R.L. Czaplewski. 1993. Long-term strategy for the statistical design of a forest health monitoring system. *Environmental Monitoring and Assessment* **17**: 81–94.

Smith, H.F. 1938. An empirical law describing heterogeneity in the yields of agricultural crops. *Journal of Agricultural Science* **28**: 1–23.

Tardif, G. 1965. Some considerations concerning the establishment of optimum plot size in forest survey. *Forestry Chronicle* **41**: 93–102.

Tukey, J.W. 1977. *Exploratory data analysis*, Addison-Wesley, Reading, Massachusetts.

U.S. Department of Agriculture (USDA). 1976. *National range handbook*. USDA Soil Conservation Service, Washington.

U.S. Department of the Interior (USDI). 1985. *Rangeland monitoring trend studies.* Technical Reference 4400–4. USDI Bureau of Land Management, Service Center, Denver.

U.S. Department of the Interior (USDI). 1990. *National range handbook.* BLM Manual Handbook H-4410–1. USDI Bureau of Land Management, Washington.

Waring, R.H., W.H. Emmingham, H.L. Gholz and C.C. Grier. 1978. Variation in maximum leaf area of coniferous forests in Oregon and its ecological significance. *Forest Science* **24**: 131–139.

Warrick, A.W., D.E. Myers and D.R. Nielsen. 1986. Geostatistical methods applied to soil science, in A. Klute (Ed.), *Methods of soil analysis. Part 1. Physical and mineralogical methods*, 2nd ed., *Agronomy* **9**: 53–82. American Society of Agronomy, Madison, Wisconsin.

Westoby, M., B. Walker and I. Noy-Meir. 1989. Opportunistic management for rangelands not at equilibrium. *Journal of Range Management* **42**: 265–274.

White, D., A.J. Kimberling and W.S. Overton. 1992. Cartographic and geometric components of a global sampling design for environmental monitoring. *Cartography and Geographical Information Systems* **19**: 5–22.

Wiegert, R.G., 1962. The selection of an optimum quadrat size for sampling the standing crop of grasses and forbs. *Ecology* **43**: 125–129.

Wilding, L.P. and L.R. Drees. 1983. Spatial variability and pedology, in L.P. Wilding, N.E. Smeck and G.F. Hall (Eds.), *Pedogenesis and soil taxonomy: Concepts and interactions*, Elsevier Applied Science, New York, pp. 83–116.

DESERTIFICATION OF SUBTROPICAL THICKET IN THE EASTERN CAPE, SOUTH AFRICA: ARE THERE ALTERNATIVES?

GRAHAM I. H. KERLEY*, MICHAEL H. KNIGHT** and MAURITZ DE KOCK***
**Terrestrial Ecology Research Unit, Department of Zoology, University of Port Elizabeth, PO Box 1600, Port Elizabeth 6000, South Africa; **Scientific Services, National Parks Board, PO Box 110040, Hadison Park 8306, South Africa; ***Dohne Agricultural Research Institute, Private Bag X27592, Greenacres 6057, South Africa*

Abstract. The Eastern Cape Subtropical Thicket (ECST) forms the transition between forest, semiarid karroid shrublands, and grassland in the Eastern Cape, South Africa. Undegraded ECST forms an impenetrable, spiny thicket up to 3 m high consisting of a wealth of growth forms, including evergreen plants, succulent and deciduous shrubs, lianas, grasses, and geophytes. The thicket dynamics are not well understood, but elephants may have been important browsers and patch disturbance agents. These semiarid thickets have been subjected to intensive grazing by domestic ungulates, which have largely replaced indigenous herbivores over the last 2 centuries. Overgrazing has extensively degraded vegetation, resulting in the loss of phytomass and plant species and the replacement of perennials by annuals. Coupled with these changes are alterations of soil structure and secondary productivity. This rangeland degradation has largely been attributed to pastoralism with domestic herbivores. The impact of indigenous herbivores differs in scale, intensity, and nature from that of domestic ungulates. Further degradation of the ECST may be limited by alternative management strategies, including the use of wildlife for meat production and ecotourism. Producing meat from wildlife earns less income than from domestic herbivores but is ecologically sustainable. The financial benefits of game use can be improved by developing expertise, technology, and marketing. Ecotourism is not well developed in the Eastern Cape although the Addo Elephant National Park is a financial success and provides considerable employment benefits within an ecologically sustainable system. The density of black rhinoceros and elephant in these thickets is among the highest in Africa, with high population growth and the lowest poaching risk. The financial and ecological viability of ecotourism and the conservation status of these two species warrant expanding ecotourism in the Eastern Cape, thereby reducing the probability of further degradation of ECST.

1. Introduction

Acocks (1975) discussed two thrusts of desertification in South Africa, predicting the large-scale expansion of karroid shrublands eastward into the grasslands of the Orange Free State and south eastward into the thickets of the Eastern Cape. He named overgrazing by domestic herbivores as the cause of this desertification.

The concept of desertification has promoted considerable debate in South Africa, particularly for the shrubland/grassland interface (Roux and Theron, 1987; Hoffman and Cowling, 1990a; Dean and MacDonald, 1993). In contrast, the degradation of the Eastern Cape Subtropical Thicket (ECST) (Figure 1) has been more generally accepted (Hoffman and Cowling, 1990b; Stuart-Hill, 1992). This vegetation type is recognized as being vulnerable to such degradation due to overgrazing, and there is little evidence for the recovery of severely degraded ECST. Although the

Environmental Monitoring and Assessment **37:** 211–230, 1995.

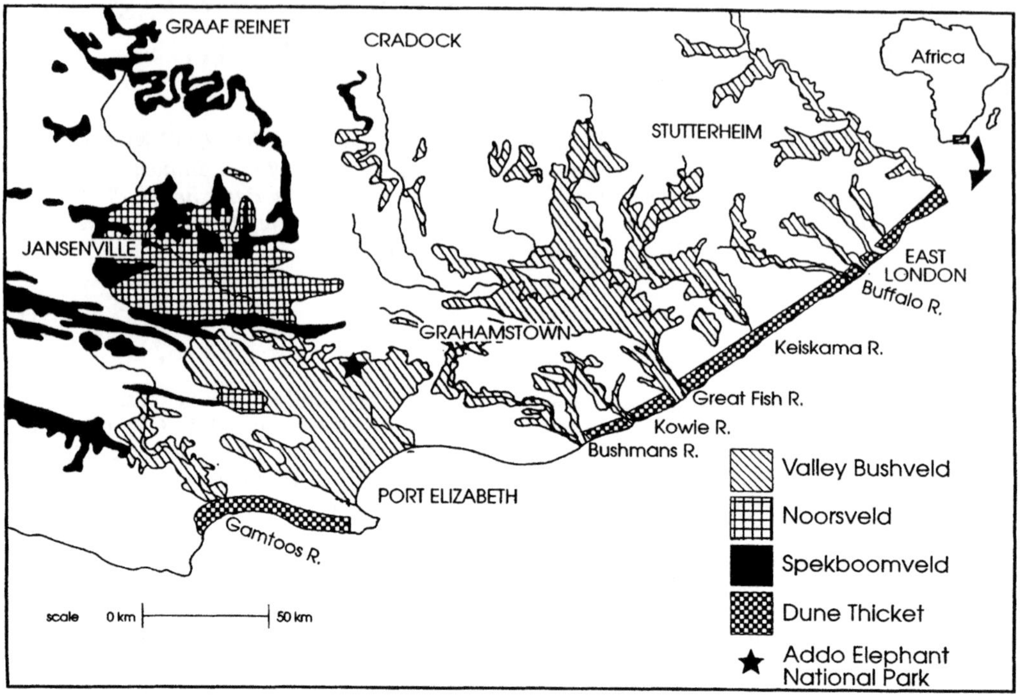

Fig. 1. Map of the Eastern Cape, South Africa, showing the distribution of the principal forms of Subtropical Thicket (from Hoffman and Everard, 1987 and Lubke *et al.*, 1986).

degradation of ECST can largely be ascribed to extensive pastoralism, few attempts have been made to bring about more sustainable land use practices.

The aims of this paper are to review the status, dynamics, and degradation of ECST and to explore sustainable alternate uses for it. This study assesses both the financial and sustainability implications of traditional and alternate uses of this thicket, focusing on the Sundays River drainage system with its Kaffrarian Succulent Thicket (Everard, 1987) form of ECST.

2. Subtropical Thicket

The southeastern seaboard (Eastern Cape, as defined by Bruton and Gess, 1988) of South Africa is a climatically and physiognomically complex region. Rainfall seasonality varies from summer to winter maxima, average annual totals range from less than 200 mm to more than 2000 mm, and temperature regimes vary from subtropical to steppe (Kopke, 1988). Rising to the Great Escarpment via a series of lesser ranges, the coastal plains are deeply incised by rivers that drain into the Indian Ocean. Of these rivers, the Sundays, Great Fish, Great Kei, and Mbashe are the major river systems (Nicol, 1988). This region's vegetation reflects this complexity, with four major phytochoria meeting here: Afromontane, Cape,

TABLE I

Classification of Eastern Cape Subtropical Thickets, according to veld types (Acocks, 1975) or hierarchical vegetation classes (Cowling, 1984; Lubke *et al.*, 1986; Everard, 1987). Figure 1 reflects Lubke *et al.*'s (1986) delineation

Acocks (1975)	Cowling (1984)	Lubke *et al.* (1986)	Everard (1987)
Valley Bushveld	Subtropical Transition Thicket: Order: Kaffrarian Succulent Thicket Order: Kaffrarian Thicket Order: Kaffrarian Thicket	Subtropical Thicket: Valley Bushveld	Subtropical Transitional Thicket: Order: Kaffrarian Succulent Thicket suborder: Xeric Succulent Thicket suborder: Mesic Succulent Thicket Order: Kaffrarian Thicket suborder: Xeric Kaffrarian Thicket suborder: Mesic Kaffrarian Thicket
Spekboomveld		Subtropical Thicket: Spekboomveld	
Noorsveld		Subtropical Thicket: Noorsveld	
Coastal Forest	Subtropical Transitional Thicket: Order: Kaffrarian Thicket	Subtropical Thicket: Dune Thicket	Subtropical Transitional Thicket: suborder: Xeric Kaffrarian Thicket

Tongoland-Pondoland, and Karoo-Namib (White, 1983). These phytochoria form eight vegetation classes (Lubke *et al.,* 1986).

Subtropical Thicket (Figures 1 and 2a) makes up about 25% of the area of the Eastern Cape (Lubke *et al.*, 1986) and is generally confined to areas with less than 850 mm annual rainfall. Acocks (1975) considered ECST to be of Karoo origin because of the high component of dwarf shrubs and succulents. But Huntley (1984) and Rutherford and Westfall (1986) consider these thickets to be within the Savanna Biome. Cowling (1984), Lubke *et al.* (1986) and Everard (1987, 1991) emphasized the transitional nature of the thickets, suggesting that they form the transitional vegetation type between subtropical forest, Afromontane forest, fynbos, Karoo, and grassland vegetation. Cowling and Holmes (1991) also stressed the transitional nature of the ECST, pointing out that the different strata tended to have different biogeographical origins. Thus the tree or overstory shrub stratum is derived from tropical forest communities, whereas the understory or open-habitat species are of karroid affinity.

Acocks (1975) coined the term "Valley Bushveld" to describe the dense thickets of the Eastern Cape (Table I). As the name suggests, this vegetation type is largely confined to moderately deep, well-drained fertile soils of the hot, semiarid, deeply incised river valleys of the coastal region. This type is rarely found higher than 1000 m above sea level. The vegetation of Valley Bushveld is dominated by phanerophytes, occasionally reaching heights of about 5 m but typically around 3 m high (Rutherford and Westfall, 1986). The vegetation has a great variety of growth forms, including succulents, deciduous and evergreen shrubs, lianas, herbs, geophytes, and grasses (Figure 2a). The high incidence of spinescence combined with the dense growth creates a virtually impenetrable, spiny thicket (Everard 1987; Hoffman and Everard, 1987).

Two other thicket veld types tend to be found inland of Valley Bushveld. Spekboomveld is the term used for largely succulent thickets dominated by the spekboom (*Portulacaria afra*). Noorsveld refers to the lower thicket dominated by soet noors (*Euphorbia coerulescens*). Lubke *et al.* (1986) recognized an additional thicket type, Dune Thicket, which is less succulent and confined to low coastal regions. These four thicket types (Table I) are generally recognized as forming ECST (Lubke *et al.*, 1986; Hoffman and Everard, 1987).

Veld types are functional definitions of areas of uniform agricultural potential (Acocks, 1975) and thus do not differentiate between unrelated vegetation types with similar agricultural potential (Lubke *et al.*, 1986). Everard (1987) therefore adopted a syntaxonomic and structural approach to classifying the Valley Bushveld veld type of the ECST and recognized two further classes of ECST (including the Kaffrarian Succulent Thicket of the Sundays River Valley), each with two forms (Table I). The occurrence of these forms appears to be a function of rainfall and altitude (Palmer and La Cock, 1991). Further classification of this and the remaining thicket veld classes is hampered by a lack of sampling, which reflects the poor ecological understanding of ECST (Hoffman and Everard, 1987).

Fig. 2a.

Fig. 2b.

Fig. 2c.

Figs. 2(a)–(c). Examples of Subtropical Thicket (Valley Bushveld form) in various stages of degradation. (a) Undegraded thicket within an ungrazed railway reserve. *Aloe ferox* in the foreground with *Portulacaria afra* in the background. (b) Fenceline contrasts between pastoral management units. The area to the right is heavily degraded, bottom left is moderately degraded and top left is in moderate condition. (c) Characteristic umbrella-shaped tree (*Pappea capensis*) resulting from goat browsing. Contrast the open nature of the background with (a) above (a and b taken 100 m apart).

As expected from the wealth of plant growth forms mentioned above and the transitional nature of ECST, diversity is high, and these may be among the most species-rich formations of woody plants in southern Africa (Hoffman and Everard, 1987). Cowling (1983) ascribes this high diversity in part to the temporal separation of fruit production between species and zoochorous dispersal of seeds to small disturbance patches. Similarly, the relatively low endemism within ECST can probably be ascribed to the transitional nature of the thickets (Everard, 1987).

Although endemism within ECST may be relatively low on a subcontinental scale, on a regional scale ECST has the highest endemism (30%) within the Eastern Cape (Lubke *et al.*, 1986). Furthermore, this endemism is not uniform, being highest among the succulent shrubs and geophytes (Hoffman and Cowling, 1991; Moolman and Cowling, 1994).

3. Dynamics of ECST

Hoffman and Everard pointed out in 1987 that very little is known about the natural dynamics of ECST. Fire plays a minor role in ECST, both because of the low incidence of lightning (Manry and Knight, 1986) and the low flammability of the largely succulent vegetation. Drought is relatively common in the Eastern Cape although Stuart-Hill and Aucamp (1993) showed that ECST in a relatively undegraded condition maintains its forage production during drought and a large proportion of ECST plants have water-storage organs (Hoffman, 1989), possibly ameliorating the impacts of all but the most severe droughts.

Cowling (1983) suggested that these communities were largely bird driven through dispersal of seed from fleshy fruits held high in the canopy to localized disturbance patches suitable for germination and recruitment. Mammals are more important seed dispersers than birds in Dune Thicket (Castley, 1992) and may therefore also be important in other forms of ECST. Factors affecting the fate of seeds have been poorly investigated although small-mammal seed predation is limited (Castley, 1992).

Portulacaria afra is one of the few species for which data exist on seed viability. The seeds of this species have initially high viability that declines rapidly within a few months (Whiting, 1991), with vegetative recruitment via ramets being more important.

The high incidence of ramet production within ECST communities, combined with the general paucity of seedlings, suggests that genets (recruitment via seeds) may not be important in ECST dynamics (Midgley, 1991; Midgley and Cowling, 1993). Midgley (1991) suggested a model for the maintenance of ECST by the low death of individuals and small-scale gaps being filled by ramets. This model, however, conflicts with Cowling's (1983) suggestion of the importance of zoochory in ECST dynamics. Midgley's model suggests that ECST communities are vulnerable to large-scale disturbances.

ECST is nutritious and largely evergreen (Aucamp, 1976; Seydack and Bigalke, 1992; Stuart-Hill and Aucamp, 1993; Haschick, 1994), with a mean wet plant biomass of 18 kg/m^2 (Penzhorn *et al.*, 1974). ECST supports a diversity of vertebrate herbivores, ranging from the diminutive blue duiker (*Philantomba monticola*) (4.5 kg) to the African elephant (*Loxodonta africana*) (Smithers, 1983). The spiny nature and ability of the plants to sprout after being browsed led Midgley (1991) to suggest coevolution between ECST plant communities and their associated herbivores.

A specific example is Stuart-Hill's (1992) proposal that *P. afra* is adapted to browsing from above by elephants, with recruitment via the "skirt" of branches that bend down and root where nodes touch the ground. When heavily browsed from the side by small herbivores, *P. afra* does not form this skirt of rooted branches and thereby cannot develop ramets. The relative importance of megaherbivores in ECST is emphasized by the fact that the elephant and black rhinoceros (*Diceros bicornis*) make up 85% of the biomass of vertebrate herbivores in the Addo Elephant National Park, contributing 78.4% and 7% respectively (Stuart-Hill, 1992).

Midgley (1991) and Stuart-Hill (1992) suggested that large herbivores would have been the major patch disturbance agents in ECST in precolonial times. These impacts would have been in the form of herbivory, trampling, and dung deposition. Elephants increase woody plant density in ECST, possibly through the coppicing of trees broken down by feeding or through increased germination of seeds in soil churned by trampling or in soil enriched by elephant dung (Stuart-Hill, 1992). Germination does increase at sites enriched by rhinoceros dung (La Cock, 1992).

Indigenous herbivores have largely been replaced by domestic herbivores within ECST in the last 200 years. Megaherbivores, in particular, have suffered from the impact of modern man. Black rhinoceros were extirpated from ECST in 1858 (Skead, 1987) and have been reintroduced into a few small conservation areas since the late 1960s (Hall-Martin and Penzhorn, 1977). Elephants were reduced to 11 individuals in the Addo Bush by 1930 (Penzhorn *et al.*, 1974), with two other populations established recently. Thanks to the impenetrability of ECST, the smaller herbivores fared slightly better, and kudu (*Tragelaphus strepsiceros*), bushbuck (*T. scriptus*), grysbok (*Raphiceros campestris*), duiker (*Sylvicapra grimmia*) and bushpig (*Potamochoerus porcus*) are still relatively common outside conservation areas. The consequences of the loss of megaherbivores from ECST are probably profound but are not well understood.

4. Desertification of ECST

The arrival of modern pastoralists and their domestic herbivores has had wider implications for ECST than the loss of megaherbivores. Introducing large numbers of goats and cattle has been associated with overgrazing and degradating ECST (Hoffman and Everard, 1987) although few studies have addressed this issue.

Acocks' (1975) prediction that overgrazing by livestock leads to an increase in karroid shrubs has been shown only for the xeric extremes (Noorsveld) of ECST. But Hoffman and Cowling (1990b) clearly demonstrated a reduction in total plant cover, a replacement of dominant species by less palatable or alien species, and a decrease in cover of perennials with an increase in the cover of annuals associated with heavy grazing in the more mesic forms of ECST. Current pastoral practices lead to a loss of endemic species (particularly endemic succulents and geophytes), which in view of pervasive use of ECST for pastoralism could lead to the extinction of these species (Hoffman and Cowling, 1991; Moolman and Cowling, 1994). Hoffman and Cowling (1990b) hypothesized that the decrease in cover and shift from perennials to annuals would lead to a decrease in productivity. They concluded that ECST plant communities are relatively stable within natural environmental fluctuations but are not resilient to the sustained grazing typical of current pastoralism.

Although overgrazing and degradation are often thought to lead to soil erosion in ECST (e.g. Stuart-Hill and Aucamp, 1993), Hoffman and Cowling (1990b) found little evidence for changes in soil characteristics in their heavily grazed sites. Palmer *et al.* (1988), however, found higher organics, moisture, and Mg and Ca levels in soils under clumps of thicket than in open, grassy areas between the thicket clumps. They attributed this difference to the protection of the soils within the thicket from erosion. La Cock (1992) found similar results and demonstrated high levels of Al in open (∼ degraded) areas adjacent to thicket.

Desertification implies an irreversible loss of secondary productivity (United Nations, 1977). Degraded ECST does have a reduced ability to support herbivores (browsers and grazers) without further degrading the vegetation (i.e. allowing the vegetation to recover to the starting condition before the next offtake by herbivores). ECST in pristine condition varies less than the degraded form in its ability to support herbivores under varying rainfall. Finally, although ECST in an intermediate condition can support more grazers (which use the ephemeral sward produced in degraded ECST) than pristine ECST, this use is not sustainable because the resource base can totally collapse in drought years (Stuart-Hill and Aucamp, 1993). Thus Stuart-Hill and Aucamp (1993) concluded that degraded ECST is an unstable forage base for pastoralism for which pristine ECST is the optimum. They also pointed out that, although allowing ECST to degrade may fulfill a short-term objective (e.g. increase the number of grazers), degradation sharply reduces future management options because degraded ECST does not appear to regenerate. Stuart-Hill and Aucamp's (1993) findings therefore support the contention that ECST degradation does represent desertification.

ECST degradation is particularly apparent at browsing contrasts, where the impacts of domestic herbivores are visually obvious. Thus contrasts between farms are relatively common in ECST with dense vegetation on one side of a fence and degraded shrublands on the other side (Figure 2b). A similar contrast is caused by the foraging height limitations of goats, which can effectively browse up to 1.7 m (Haschick, 1994). Thus, umbrella-shaped trees (Hoffman and Everard, 1987), where virtually all vegetation within the browsing range of goats has been consumed, are characteristic of degraded ECST (Figure 2c).

Why a vegetation type that has apparently evolved under high levels of herbivory (Midgley, 1991) should be so vulnerable to a specific form of herbivory is not clear. Stuart-Hill's (1992) model to explain the lack of recruitment of *P. afra* does not appear to apply to other coppicing species, and the paucity of seedlings in degraded ECST appears to apply to most shrub species. Allometric relationships, however, show that the metabolic requirements per unit body mass of elephants are lower than those of smaller species (Meissner, 1982). Thus replacing elephants with a similar biomass of goats (e.g. Stuart-Hill, 1992) would significantly increase the consumption needed to maintain the biomass of herbivores (Pienaar *et al.*, 1966). In addition, it is likely that elephant populations were originally migratory, with intermittent impacts within any particular area of ECST. The current provision of water and high stocking rates of domestic herbivores suggests more sustained browsing pressure on ECST.

ECST plant species may require a specific suite of environmental conditions for germination and recruitment (Stuart-Hill, 1989), and these conditions may not have been met within the relatively short timeframe that ecologists have been investigating the dynamics of these thickets (e.g. Midgley and Cowling, 1993). The extreme of this hypothesis is that ECST represents a relic vegetation type from a more amenable climatic condition (Palmer, 1990). This hypothesis is complicated by the possibility that successful recruitment may occur only in suitable microsites within ECST and that, once degraded, the microclimates suitable for recruitment are lost (La Cock, 1992). The suggestion that Al levels in degraded ECST soils may be high enough to be toxic for seedlings (La Cock, 1992) needs to be investigated.

Another hypothesis (although not exclusive) is that the megaherbivores are indeed key species for dispersing and germinating seeds. This hypothesis suggests that recruitment should have been ongoing within the Addo Elephant National Park, where elephants were never eliminated. Finally, Stuart-Hill's (1992) ideas on the nature of herbivory suggest that the high levels of herbivory lower down in the canopy may be the issue. Pastoral practices may therefore lead to seedling mortality (by being eaten) and may not allow recruitment (La Cock, 1992) whereas in the past seedlings may have been able to grow larger and establish successfully before suffering the attention of elephants, the dominant browsers.

Whatever the causes of the lack of regeneration of degraded ECST, this is a major concern because of the apparent impossibility of running an economical pastoralism operation at the conservative stocking rates of domestic herbivores needed to maintain ECST in an undegraded condition (Stuart-Hill, 1992). Thus current farming practices are unsustainably "mining" the "fodder capital" of these thickets. This situation is exacerbated once ECST has been degraded due to the reduced productivity and increased variation in productivity of degraded thicket (Stuart-Hill and Aucamp, 1993).

TABLE II

Calculated gross margins for game and goat utilization on the J.C. Steyn farm calculated per small stock unit (SSU: metabolic livestock equivalents). Values are expressed as both South African Rands and US Dollars (US $1 = R3.57)

			R/SSU	US$/SSU
Game				
Gross Income	Sales from meat		81.00	22.69
	Income from hides		3.08	0.86
		Total	**84.08**	**23.55**
Variable Costs	Transport		11.49	3.22
	Ammunition		0.82	0.23
	Other costs		0.17	0.05
		Total	**12.48**	**3.50**
Gross margin for game			**71.60**	**20.06**
Goats				
Gross income	Sales from meat and hides		93.24	26.12
		Total	**93.24**	**26.12**
Variable Costs	Veterinary services and medicines		9.64	2.70
Supplementary fodder	Supplementary fodder		7.00	1.96
	Miscellaneous		0.83	0.23
		Total	**17.47**	**4.89**
Gross margin for goats			**75.77**	**21.22**

Thus desertification of ECST results from overgrazing by the replacing of indigenous herbivores with domestic species, leading to changes in species-specific foraging behavior and sustained herbivory, coupled with overestimates of carrying capacity and economic constraints on the landowner. This process represents a positive feedback system. Degraded thickets lack regeneration and have reduced phytomass, productivity, and species richness. These thickets are further overgrazed and degraded until replaced by an ephemeral weed community that may have virtually no production in drought years. This process appears to be irreversible.

The only way of breaking this desertification feedback loop is preventing overgrazing, although reducing goat stocking rates is not economically feasible (Stuart-Hill, 1992). Alternative management paradigms are urgently needed to allow the sustainable use of ECST and prevent further desertification.

TABLE III

Farm income (less running expenses) calculated for game and goat operations on the 3,000 ha J C Steyn farm. Animal numbers are expressed as both small stock units (metabolic livestock equivalents) and the numbers of animals. Values are expressed as both South African Rands and US Dollars (US $1 – R3.57)

Operation	Stocking rate (small stock units)	Stocking rate (animals)	Rands	US $
Game	296	114	21,196.00	5,937.25
Goats	746	465	56,524.42	15,833.17

5. Game Exploitation in ECST

Game utilization has often been suggested as a feasible option in ECST (e.g. Stuart-Hill, 1992) but must be evaluated by its the economic and ecological implications in relation to farming with domestic herbivores. The Genl. J.C. Steyn Prison Farm (33° 24′ S, 25° 16′ E) on the west bank of the Sundays River runs a 3,000 ha pastoralism operation in thicket vegetation of high ecological quality (Stuart-Hill, 1992). Although the prison (a captive market) uses the products (goat and game meat), the Department of Agriculture uses this operation as a research project, monitoring productivity, income, and expenditures. The goats are stocked conservatively following Department of Agriculture recommendations. The game is exploited by annually cropping a conservative 30% of the population (largely kudu) estimated from helicopter game counts (Fairall, 1994).

We compared the economic benefits of farming goats and harvesting game. To simplify the comparison, we limited it to the difference between variable costs and gross income. Evaluating fixed costs for the J.C. Steyn farm is difficult because of cross-subsidizing from other state functions. But this simplification is justified because the property values for the two operations are the same. A major difference between the two enterprises is the capital cost of fencing, but the expensive perimeter fence required for game management is offset by the costs of internal fences for goat management. These data are presented as income per small stock unit (Meissner, 1982) to allow extrapolation to other sized properties for which the carrying capacity of goats and game is known, as well as total income for the 3,000 ha J.C. Steyn farm.

The gross marginal income per small stock unit of game (US $20.06) is virtually the same as for goats (US $21.22) (Table II, US $1 = South African R3.57). But when calculated for the 3,000 ha property, goat farming represents a major advantage over game utilization (Table III), goats (US $15,833) producing nearly three times more income (after deducting operating expenses) than game (US $5,535).

Superficially then one may conclude that game farming is at a major economic disadvantage when compared to goat farming. This disadvantage can be attributed to the fewer game animals maintained than goats (Table III). Goats probably compete with and competitively exclude game species (Stuart-Hill and Danckwerts, 1988; Haschick, 1994). Thus removing the goats may increase game numbers and income. Despite the economic advantages of goat farming, even conservative stocking rates of goats lead to degradation of ECST (Stuart-Hill and Aucamp, 1993), and the income from goat farming is therefore not sustainable. (Note the need to provide supplementary fodder in Table II.) In contrast, where megaherbivores are excluded, there is no evidence for degradation of ECST being used by indigenous herbivores (Stuart-Hill, 1992; Moolman and Cowling, 1994). Stuart-Hill (1992) found a peak in floristic and habitat diversity when ECST was being used by the medium-sized indigenous herbivores. Therefore, for sustainability the preferable grazing use of ECST consists of indigenous medium-sized herbivores.

The question then arises as to what strategies will increase the financial feasibility of game utilization in ECST? Hobson *et al.* (1994) suggested that game farming does have the potential for higher profitability than stocking goats but is hampered by a lack of technology, expertise, and market structures. Natural herbivore (excluding megaherbivores) populations are relatively low in ECST compared to the artificial stocking rates that can be achieved with goats. So to increase income, larger farming units are needed (Hobson *et al.*, 1994). Research is needed to determine the effects of farm size on the economics of game farming.

Another strategy is to process the products of game farming to increase their value. A variety of options can be taken. According to J. Westcott (personal communication), a venison dealer representing the Camdeboo Meat Processors Pty., processing meat to produce biltong (dried meat) and game sausage – both local delicacies with a high market value – more than doubles the meat's value. Considering the added costs of processing the meat (feasible to operate as a farm business), the J.C. Steyn game farming operation could generate US $9,290 (after running expenses have been deducted) from processed meat sales, markedly reducing the differential between game and goat farming (Table III).

Hunting is a growing and lucrative form of game use, and a number of ECST species (particularly kudu and bushbuck) are in demand for recreational hunting for both meat and trophies. Introducing trophy hunting increases the value of harvesting game by 14% (Els, 1994). Ideally, trophy hunting should be combined with the processing of meat to maximize income. Tourism is another option for landowners in ECST (see next section) and is compatible with hunting. Hunting tends to be restricted to the cooler winter season, and during the remainder of the year the property can be used for tourism. We have no data for the value of farm-based tourism in ECST, although landowners are taking this option.

This assessment of the value of game use shows that considerable income can be sustainably derived from the ECST. These figures, however, are only a first approximation and do not include a variety of economic factors such as property prices and debt levels of farmers or socioeconomic factors such as standards of living of property owners and laborers. The sustainability of land use is an emerging demand that society is placing on landowners (Hanekom and Liebenberg, 1994), and options that allow the sustainable use of ECST will become important to both landowners and society in general.

6. Ecotourism in ECST

The few ecotourism operations within the thickets of the Eastern Cape may in part result from the sheer impenetrability of ECST, the feature that has allowed many species to survive into recent times. Ironically this feature also militates against the desirability of these thickets as ecotourism destinations. This problem is exemplified by a quote from the *Eastern Province Herald* (1919) when the Addo elephants were being systematically eradicated: "There is, it may be noted, no case made out for the preservation of the herd for the pleasure of animal lovers, no man in his senses would venture to enter the Addo Bush to sight the wild elephant since he would have to crawl within a few yards before he could even dimly see the outline of an animal, and then if he were scented his further interest would be blotted out in a charge."

Fortunately, public perceptions of the value of our biota have changed considerably. One can now experience the vegetation and animals of ECST without risking one's life, and the demand for this experience is increasing. The question is whether using ECST for conservation and ecotourism is financially feasible and sustainable? The Addo Elephant National Park (AENP) is the best known of such operations. It is also the only conservation area within ECST where megaherbivore (elephant) populations have survived (despite the sentiments expressed in the above newspaper quote) and will be used as a model to investigate this issue.

The AENP (33° 31′ S, 25° 24′ E) was proclaimed in 1931 to protect the remnants of a much larger population of elephants that had found refuge in the dense thickets of the Addo Bush. Unfortunately, these animals conflicted with farmers in the area, and in 1919/1920 about 120 elephants were shot in a concerted eradication campaign. Only 16 elephants remained, but their safety was finally assured when the herd was enclosed within an elephant-proof fence in 1955 (Hall-Martin, 1980; Hoffman, 1993). A population of Cape buffalo (*Syncerus cafer*) also found refuge in the Addo Bush, and later the AENP, together with a range of other thicket herbivores. Black rhinoceros were reintroduced into the AENP in 1961 (Hall-Martin and Penzhorn, 1977).

Sixty-five km northeast of Port Elizabeth, the AENP is well developed, with 60 km of tourist roads, five artificial waterholes that are foci for gameviewing, a 20-site camping area, and overnight accommodations for 116 guests in 32 well-appointed chalets. A $\sim$ 8 km hiking trail (in an elephant-proof botanical reserve) provides opportunities for tourists to safely explore the vegetation with instructive

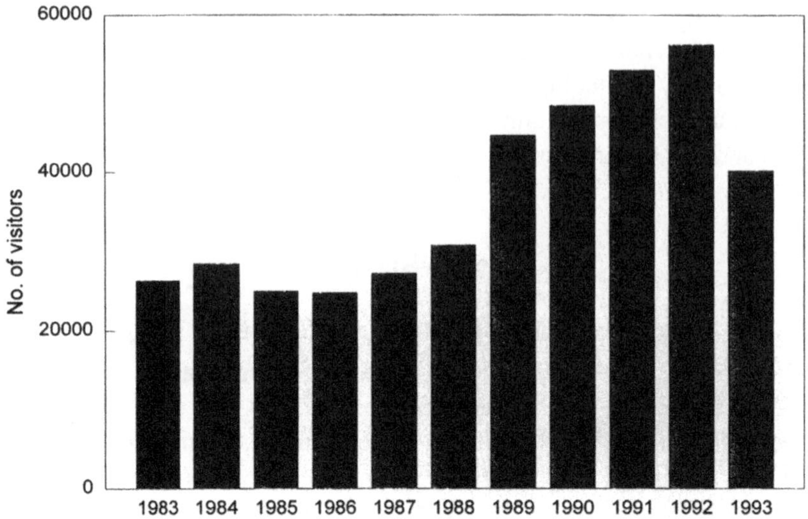

Fig. 3. Annual visitor figures for the Addo Elephant National Park, South Africa, for the period 1979 to 1993. (Decrease in 1993 is a response to the political events before the election in South Africa.)

leaflets for guidance. A floodlit waterhole in front of the main camp and restaurant provides night game viewing for the less adventurous, and night game drives with experienced guides are extremely popular.

The AENP receives about 50,000 visitors a year, of whom 38% spend the night in the park. And visitor numbers are growing despite the political isolation and turmoil in South Africa during the past decade (Figure 3). That 28% of visitors are foreigners shows that the AENP is recognized as an international tourism destination. The number of foreign visitors is expected to increase rapidly with the political settlement in South Africa. Turnover from tourist fees and accommodations, shop sales, and game sales in the 1992/1993 financial year totaled R2.4 million (US $665,000). Excluding the costs (and values) of fixed assets (the same assumptions as for the goat and game operations above) but including the costs of salaries, the AENP recorded a profit exceeding R112,000 (US $30,000) for that year. The AENP has a staff of 57 and an annual payroll of R866,000 (US $234,000), nearly double the number employed at four times the average income of a comparable pastoral operation (about 30 employees on 12,000 ha, each earning R3,697 per annum – 1991 figures from Antrobus *et al.* (1994) – with salaries extrapolated at 15% per annum).

Another level of analysis suggests that the AENP has been subsidized through property purchases. But property values are increasing (Antrobus *et al.*, 1994), and other capital assets are also appreciating. Black rhinoceros numbers are increasing by two individuals a year, and at the current value of R200,000 (US $54,000) per black rhinoceros, this increase represents a considerable growth in assets. A comprehensive study of the economic and employment benefits and opportunities of ecotourism in ECST is needed to confirm the benefits revealed by the present findings. Furthermore, the benefits derived from the AENP need to be measured on a range of scales varying from local communities (e.g. employment, curio sales) through regional (e.g. ripple effects for tourist operators) and national levels (e.g. foreign exchange).

7. Are Megaherbivores Sustainable?

What impacts does megaherbivore ecotourism have in ECST? The elephants were originally confined in an elephant-proof enclosure of 2,270 ha in the AENP in 1954. Although the enclosure prevented conflicts with surrounding landowners, who had suffered during the earlier forays of the unrestrained elephants outside the park, problems with the impact of the elephants within the enclosure were to be expected. Within 20 years the elephant population had trebled to 60 animals (2.6 elephants/km^2). The biomass of vegetation within the elephant enclosure was found to be about half that in adjacent areas. And although species diversity did not significantly differ, at least one species, a tree aloe, *Aloe africana*, was absent from the enclosure (Penzhorn *et al.*, 1974). Penzhorn *et al.* (1974) recommended a maximum density of 0.4 elephants/km^2 and warned that the herd might need to be culled. Between 1977 and 1992 the elephant enclosure was expanded to more than 10,000 ha. The herd has increased to 195 at present, and densities have consistently been above the conservatively estimated recommended levels. Yet the current biomass of vertebrate herbivores in the AENP is well below the predicted sustainable agricultural stocking rates for the area (Stuart-Hill and Aucamp, 1993).

Although elephant herbivory does reduce canopy cover and lead to an increase in unpalatable shrubs (Novellie, 1988; Hall-Martin and Barratt, in press; Stuart-Hill, 1992) and a loss of endemic small succulent shrubs and geophytes (Moolman and Cowling, 1994) and mistletoes (Midgley and Joubert, 1991), these impacts are less severe than impacts of goats stocked at similar or lower biomasses (Stuart-Hill, 1992; Moolman and Cowling, 1994). The vegetation of the Kaffrarian succulent thicket, therefore, appears to be adapted to elephant browsing with limited degradation if elephant densities are kept below two elephants/km^2 (Hall-Martin and Barratt, in press).

Conserving the elephants is one of the objectives of the AENP, and seeing the elephants is a high priority for visitors to the park. Park management is therefore pressured to maintain high populations of elephants, in conflict with the conservation of endemic plants and the dangers of degrading the Kaffrarian succulent thicket to the point of desertification. Several authors have argued that expanding the area available to elephants (with botanical reserves for the endemic succulents and geophytes) is preferable to culling elephants (Stuart-Hill, 1992; Moolman and Cowling, 1994; Hall-Martin and Barratt, in press). In the short term such expansion does seem feasible, particularly with the recently achieved linkage between the AENP and the 34,000 ha Zuurberg National Park. But this linkage provides only a limited area of ECST within Zuurberg National Park, and further ECST habitat will have to be purchased.

8. Megaherbivores as Flagship Species in ECST

The megaherbivores may be playing a pivotal role in ECST. Besides the possibility that they may be keystone species and hence essential to the maintenance of ECST dynamics, they may also be the key to halting desertification of ECST. The charismatic nature of both elephant and rhinoceros and their plight throughout Africa in the face of habitat encroachment and rampant poaching have resulted in these species being advocated as "flagship" species by conservation organizations as a means of furthering public support and awareness of larger environmental issues (Western, 1987). That tourists are attracted to the AENP largely to see the elephants emphasizes this point, as does the estimated US $25 million being generated in viewing elephants in Kenya (Brown and Henry, 1993). The solitary and cryptic nature of the black rhinoceros makes it less valuable as an attraction but valuable from a conservation angle.

ECST appears to be an ideal habitat to provide refuge for these two species. Poaching, the major threat throughout the rest of their range, is nonexistent in the Eastern Cape, possibly reflecting ECST's impenetrability. The densities and population growth of black rhinoceros and elephant in the AENP are among the highest in the world. The AENP black rhinoceros population currently represents the fifth largest free-ranging population of the east African ecotype *D. bicornis michaeli*. Animals from this population will be used to restock other areas, while the current population is replaced with the rarer indigenous ecotype *D. bicornis bicornis*. The importance of ECST in black rhinoceros conservation will further increase (Hall-Martin and Knight, 1994). Two other populations of black rhinoceros have recently been established in ECST (in 1990 in the Andries Vosloo Kudu Reserve and 1993 at Shamwari), highlighting the increasing recognition of the importance of these animals in ECST and conversely the importance of ECST for these animals.

9. Conclusions

Desertification does occur in ECST and will continue as long as pastoralism is the dominant form of land use. Farmers are under financial pressure to exploit ECST, using domestic herbivores for short-term economic benefits (Birch, 1991). Our current ecological understanding of ECST is inadequate to restore degraded thickets, and restoration is unlikely to be financially feasible. The worst-case scenario is that

ECST represents a relic vegetation type (Palmer, 1990), in which ECST cannot be restored and desertification will be irreversible.

The threats to ECST are not confined to overgrazing: large areas have been cleared for crop production and urban expansion (Hoffman and Everard, 1987). ECST is a conservation priority in the Eastern Cape for both floristic (threatened species, endemicity, and diversity) (Hoffman and Everard, 1987) and faunal (particularly megaherbivores) components. Already 9% of the Valley Bushveld form of ECST has been irreversibly degraded (La Cock *et al.*, 1990). The conserved status of ECST has recently increased to 10% although conservation has focused on the Sundays and Fish River catchments (13% conserved) at the expense of the Kaffrarian thicket to the east (3% conserved) (La Cock *et al.*, 1990). Spekboomveld has only 1.8% conserved and has been severely affected by overgrazing (Hilton-Taylor and Le Roux, 1989). But the status of the Noorsveld is the greatest concern since these thickets are under severe grazing pressure and have no conservation areas (Hilton-Taylor and Le Roux, 1989). Increasing the area conserved is therefore a priority within ECST.

Most of ECST, however, is privately owned, and attempts at preventing further desertification within ECST have to be directed through these landowners. Pastoralism with domestic herbivores does not appear to be economically or ecologically sustainable, while game utilization is sustainable but produces less income. The short-term financial rewards of goat farming cannot be justified in any system committed to sustainability. The best form of ECST use is ecotourism, with megaherbivores as the major drawing card. Ecotourism is profitable, ecologically sustainable, and desirable from a conservation perspective. It also offers considerable employment opportunities. The major drawback of ecotourism is the need for large areas to sustain megaherbivores. A solution would be landowners pooling their properties into collective ecotourism operations, which could be combined with game utilization and limited goat pastoralism. The prospects for expanding the ecotourism industry are good, particularly as ECST is one of the few areas in Africa where big game can be viewed without the attendant risk of malaria.

It is unlikely that all of ECST could be converted to ecotourism. But a shift in the emphasis in the land use practices from pastoralism to ecotourism, game utilization, and pastoralism mixtures would increase the overall sustainability of land use in ECST. A critical in-depth evaluation of the economic, social, and ecological implications of the use of ECST is therefore urgently needed. The present analysis suggests that desertification of the ECST is as much a socioeconomic as an ecological problem. Sustainability is becoming increasingly important to society (e.g. African National Congress, 1994; Hanekom and Liebenberg, 1994), and increasing effort will be focused on finding socioeconomic solutions to what has up until now been perceived as an ecological problem.

Acknowledgements

We thank Timm Hoffman, Kobus Els, Peter Novellie, Lucius Moolman, and Bev Geach for valuable discussion on thicket ecology and economics.

References

Acocks, J.P.H. 1975. Veld types of South Africa. *Memoirs of the Botanical Survey of South Africa* **40**: 1–128.

African National Congress. 1994. *The reconstruction and development programme*, Umanyano Publications, Johannesburg.

Antrobus, G.G., G.C.G. Fraser, M. Levin and H.R. Lloyd. 1994. *An overview of the agricultural economy of Region D: Final report*, Unit for Statistical Analysis, Port Elizabeth.

Aucamp, A.J. 1976. The role of the browser in the bushveld of the Eastern Cape. *Proceedings of the Grassland Society of Southern Africa* **11**: 135–138.

Birch, T. 1991. A farmer's perspective of Fish River Valley Bushveld. In *Proceedings of the First Valley Bushveld/Subtropical Thicket Symposium*, Grassland Society of Southern Africa, Howick, pp. 27–29.

Brown, G. and W. Henry. 1993. The viewing value of elephants, in E.B. Barbier (Ed.), *Economics and ecology: New frontiers and sustainable development*, Chapman and Hall, London, pp 141–155.

Bruton, M.N. and F.W. Gess. 1988. Introduction, in M.N. Bruton and F.W. Gess (Eds.), *Towards an environmental plan for the Eastern Cape*, Rhodes University, Grahamstown, p. iii.

Castley, J.G. 1992. Role of mammals as seed dispersers in the Alexandria Dunefield. Unpublished MSc thesis, University of Port Elizabeth.

Cowling, R.M. 1983. Diversity relations in Cape shrublands and other vegetation in the south-eastern Cape, South Africa. *Vegetatio* **54**: 103–127.

Cowling, R.M. 1984. A syntaxonomic and synecological study in the Humansdorp region of the fynbos biome. *Bothalia* **15**: 175–227.

Cowling, R.M. and P.M. Holmes. 1991. Subtropical thicket in the south Eastern Cape: A biogeographical perspective. In *Proceedings of the First Valley Bushveld/Subtropical Thicket Symposium.* Grassland Society of Southern Africa, Howick, pp. 3–4.

Dean, W.R.J. and I.A.W. MacDonald. 1993. Historical changes in stocking rates of domestic livestock as a measure of semi-arid and arid rangeland degradation in the Cape Province, South Africa. *Journal of Arid Environments* **26**: 281–298.

Eastern Province Herald. 1919. The Addo elephants, Rand view of extermination; suggestion for preservation; can they be trained? Port Elizabeth: *Eastern Province Herald*, 26 July 1919.

Els, J.F. 1994. The performance of the boergoat-kudu production system at the J.C. Steyn Prison Farm. (In Afrikaans). In *Information day: Genl. J.C. Stevn Prison Farm.* Kirkwood, 30 August 1994. East Cape Region, Department of Agriculture, Stutterheim, pp. 2–5.

Everard, D.A. 1987. A classification of the subtropical transitional thicket in the Eastern Cape, based on syntaxonomic and structural attributes. *South African Journal of Botany* **53**: 329–338.

Everard, D.A. 1991. Is subtropical thicket the "missing biome" of South Africa? In *Proceedings of the First Valley Bushveld/Subtropical Thicket Symposium*, Grassland Society of Southern Africa, Howick, pp. 1–3.

Fairall, N. 1994. Game research on the Genl. J.C. Steyn Prison Farm. (In Afrikaans). In *Information day: Genl. J.C. Stevn Prison Farm.* Kirkwood. 30 August 1994. East Cape Region, Department of Agriculture, Stutterheim, pp. 10–14.

Hall-Martin, A.J. 1980. Elephant survivors. *Oryx* **15**: 355–362.

Hall-Martin, A.J. and D.G. Barratt. In press. The effect of elephants *Loxodonta africana* on thicket vegetation in the Addo Elephant National Park. *African Journal of Ecology.*

Hall-Martin, A.J. and M.H. Knight. 1994. Conservation and management of black rhinoceros in South African national parks, in B.L Penzhorn and N.P.J. Kriek (Eds.), *Rhinos as game ranch animals*, University of Pretoria, Onderstepoort, pp. 11–19.

Hall-Martin, A.J. and B.L. Penzhorn. 1977. Behaviour and recruitment of translocated black rhinoceros *Diceros bicornis*. *Koedoe* **20**: 147–162.

Hanekom, D and L. Liebenberg. 1994. Utilization of national parks with special reference to the costs and benefits to communities. *Bulletin of the Grassland Society of Southern Africa* **5**: 25–36.

Haschick, S.L. 1994. Foraging and digestion by bushbuck (*Tragelaphus scriptus*) and boergoats (*Capra hircus*) in captivity. Unpublished MSc thesis, University of Port Elizabeth.

Hilton-Taylor, C. and A. Le Roux. 1989. Conservation status of the fynbos and karoo biomes, in B.J. Huntley (Ed.), *Biotic diversity in southern Africa: Concepts and conservation*, Oxford University Press, Cape Town, pp. 202–223.

Hobson, F.O., G.C. Stuart-Hill and M.L. Swart. 1994. Research results and local consensus data as basis for adaptive management of succulent bushveld. In *Information day: Genl. J.C. Steyn Prison Farm*. Kirkwood, 30 August 1994. East Cape Region, Department of Agriculture, Stutterheim, pp. 23–42.

Hoffman, M.T. 1989. A preliminary investigation of the phenology of subtropical thicket and karroid shrubland in the lower Sundays River Valley, SE Cape. *South African Journal of Botany* **55**: 586–597.

Hoffman, M.T. 1993. Major P.J. Pretorius and the decimation of the Addo elephant herd in 1919–1920: Important reassessments. *Koedoe* **36**: 23–44.

Hoffman, M.T. and R.M. Cowling. 1990a. Vegetation change in the semiarid eastern Karoo over the last 200 years: An expanding Karoo – fact or fiction? *South African Journal of Science* **86**: 286–294.

Hoffman, M.T. and R.M. Cowling. 1990b. Desertification in the lower Sundays River Valley, South Africa. *Journal of Arid Environments* **19**: 105–117.

Hoffman, M.T. and R.M. Cowling. 1991. Phytochorology and endemism along aridity and grazing gradients in the lower Sundays River Valley, South Africa: Implications for vegetation history. *Journal of Biogeography* **18**: 189–201.

Hoffman, M.T. and D.A. Everard. 1987. Neglected and abused – The Eastern Cape subtropical thickets. *Veld & Flora* **73**: 43–45.

Kopke, D. 1988. The climate of the Eastern Cape, in M.N. Bruton and F.W. Gess (Eds.), *Towards an environmental plan for the Eastern Cape*, Rhodes University, Grahamstown, pp. 44–52.

La Cock, G. 1992. *The conservation status of subtropical transitional thicket and regeneration through seeding of shrubs in the xeric succulent thicket of the Eastern Cape*, East Cape Scientific Services, Cape Nature Conservation, Port Elizabeth.

La Cock, G., A.R. Palmer and D.A. Everard. 1990. Re-assessment of the area and conservation status of subtropical transitional thicket (valley bushveld) in the Eastern Cape, South Africa. *South African Journal of Photogrammetry, Remote Sensing and Cartography* **15**: 231–235.

Lubke, R.A., D.A. Everard and S. Jackson. 1986. The biomes of the eastern Cape with emphasis on their conservation. *Bothalia* **16**: 251–261.

Manry, D.E and R.S. Knight. 1986. Lightning density and burning frequency in South African vegetation. *Vegetatio* **66**: 67–76.

Meissner, H.H. 1982. Substitution values of various classes of farm and game animals in terms of a biologically defined large stock unit. Department of Agriculture, C3, Pretoria, p. 4.

Midgley, J.J. 1991. Valley bushveld dynamics and tree euphorbias. In *Proceedings of the First Valley Bushveld/Subtropical Thicket Symposium*, Grassland Society of Southern Africa, Howick, pp. 8–9.

Midgley, J.J. and R.M. Cowling. 1993. Regeneration patterns in Cape subtropical transitional thicket: Where are all the seedlings? *South African Journal of Botany* **59**: 496–499.

Midgley, J.J. and D. Joubert. 1991. Mistletoes, their host plants and the effects of browsing by large mammals in Addo Elephant National Park. *Koedoe* **34**: 149–152.

Moolman H.J. and Cowling, R.M. 1994. The impact of elephant and goat grazing on the endemic flora of South African succulent thicket. *Biological Conservation* **68**: 53–61.

Nicol, I.G. 1988. The geomorphology of the Eastern Cape, in M.N. Bruton and F.W. Gess (Eds.), *Towards an environmental plan for the Eastern Cape*, Rhodes University, Grahamstown, pp. 24–36.

Novellie, P. 1988. The impact of large herbivores on the grassveld in the Addo Elephant National Park. *South African Journal of Wildlife Research* **18**: 6–10.

Palmer, A.R. 1990. A qualitative model of vegetation history in the Eastern Cape midlands, South Africa. *Journal of Biogeography* **17**: 35–46.

Palmer, A.R., B.J.S. Cook and R.A. Lubke. 1988. Aspects of the vegetation and soil relationships in the Andries Vosloo Kudu Reserve, Cape Province. *South African Journal of Botany* **54**: 309–314.

Palmer, A.R and G.D. La Cock. 1991. A contingency table model of the orders of subtropical transitional thicket based on median annual rainfall and altitude. In *Proceedings of the First Valley Bushveld/Subtropical Thicket Symposium*, Grassland Society of Southern Africa, Howick, pp. 6–7.

Penzhorn, B.L., P.J. Robbertse and M.C. Olivier. 1974. The influence of the African elephant on the vegetation of the Addo Elephant National Park. *Koedoe* **17**: 137–158.

Pienaar, U. de Villiers, P. van Wyk and N. Fairall. 1966. An aerial census of elephant and buffalo in the Kruger National Park and the implications thereof on intended management schemes. *Koedoe* **9**: 40–107.

Roux, P.W. and G.K. Theron. 1987. Vegetation change in the Karoo biome, in R.M. Cowling and P.W. Roux (Eds.), *The Karoo biome: A preliminary synthesis*, part 2 – Vegetation and history, *South African National Scientific Programmes Report* **142**: 70–95.

Rutherford, M.C. and R.H. Westfall. 1986. Biomes of southern Africa – An objective categorization. *Memoirs of the Botanical Survey of South Africa* **54**: 1–98.

Seydack, A.H.W. and R.C. Bigalke. 1992. Nutritional ecology and life history tactics in the bushpig (*Potamochoerus porcus*): Development of an interactive model. *Oecologia* **90**: 102–112.

Skead, C.J. 1987. *Historical mammal incidence in the Cape Province*, vol. 2: The eastern half of the Cape Province, including the Ciskei Transkei and East Griqualand, Chief Directorate of Nature and Environmental Conservation, Cape Town.

Smithers, R.H.N. 1983. *The mammals of the southern African subregion*, University of Pretoria, Pretoria.

Stuart-Hill, G.C. 1989. Adaptive management: The only practicable method of veld management, in J.E. Danckwerts and W.R. Teague (Eds.), *Veld management in the Eastern Cape*, East Cape Region, Department of Agriculture, Stutterheim, pp. 4–7.

Stuart-Hill, G.C. 1992. Effects of elephants and goats on the Kaffrarian succulent thicket of the Eastern Cape, South Africa. *Journal of Applied Ecology* **29**: 699–710.

Stuart-Hill, G.C. and A.J. Aucamp. 1993. Carrying capacity of the succulent valley bushveld of the Eastern Cape. *African Journal of Range and Forage Science* **10**: 1–10.

Stuart-Hill, G.C. and J.E. Dankwerts. 1988. Influence of domestic and wild animals on the future of Succulent Valley Bushveld. *Pelea* **7**: 45–56.

United Nations. 1977. *United Nations Conference on Desertification: Round-up, plan of action and resolutions, 29 August-9 September 1977*, United Nations, New York.

Western, D. 1987. Africa's elephants and rhinos: Flagships in crisis. *Trends in Ecology and Evolution* **2**: 343–346.

White, F. (editor). 1983. *Vegetation of Africa*, Unesco, Paris.

Whiting, B.H. 1991. Aspects of the germination of *Portulacaria afra* (L.) Jacq. In *Proceedings of the First Valley Bushveld/Subtropical Thicket Symposium*, Grassland Society of Southern Africa, Howick, pp. 32–35.

DESERTIFICATION IN AUSTRALIA: AN EYE TO GRASS ROOTS AND LANDSCAPES

JOHN A. LUDWIG and DAVID J. TONGWAY
Commonwealth Scientific and Industrial Research Organisation (CSIRO), Division of Wildlife and Ecology, PO Box 84, Lyneham, Canberra 2602, Australian Capital Territory

Abstract. Desertification in some form is estimated to have occurred over about 42% of the 5 million km^2 of arid and semiarid lands in Australia. The most common form of desertification is loss of perennial grasses from grasslands, savannas, and open woodlands, often with a replacement by inedible shrubs. Desertification continues to be a problem, especially during droughts when grazing pressures reduce ground cover, laying bare landscapes to wind and water erosion. But two national programs, Drought Alert and Landcare, are giving new hope in controlling land degradation. Both use a grassroots approach by promoting action through local pastoralist and farmer groups and by encouraging the use of effective techniques for rehabilitating landscapes. A strategic application of ponding banks and contour traps with an eye to the landscape has proven successful in stopping and reversing desertification processes.

1. How Big a Problem Is Desertification in Australia?

Like all developed countries with extensive arid and semiarid lands, Australia has a desertification problem. In the "land down-under" this problem is national in scope. More than 70% of the continent, some 5 million km^2, is arid or semiarid (Figure 1) (Parkinson, 1986). About 42% of this area, some 2 million km^2, is estimated to be either in a severe (26%) or very severe (16%) state of desertification (Christie, 1986). We believe that to some degree more than 50% of arid and semiarid lands of Australia that have been traditionally used for pastoralism have been desertified, although problems with very severe soil erosion tend to be localized, for example, around watering points.

The increased density and encroachment of shrubs into arid and semiarid grasslands, savannas, and woodlands is perhaps the most obvious and commonly cited form of desertification in arid and semiarid regions used for pastoralism (Mabbutt, 1978; reviews in Harrington *et al.*, 1984). But this form of desertification is usually preceded by a more serious change, a decline in the density and cover of long-lived, perennial grasses. In more-arid regions this decrease in grasses often results in extensive bare areas without shrubs, with only a cover of ephemeral forbs and grasses following rainy periods. What other forms of desertification are evident?

Loss of perennial grass cover leads to increased rates and extent of soil erosion by water and wind, clearly observable as rills, gullies, scalds, blowouts, and unconsolidated dunes (Mabbutt, 1978). However, our view is that these acute symptoms

Environmental Monitoring and Assessment **37**: 231–237, 1995.

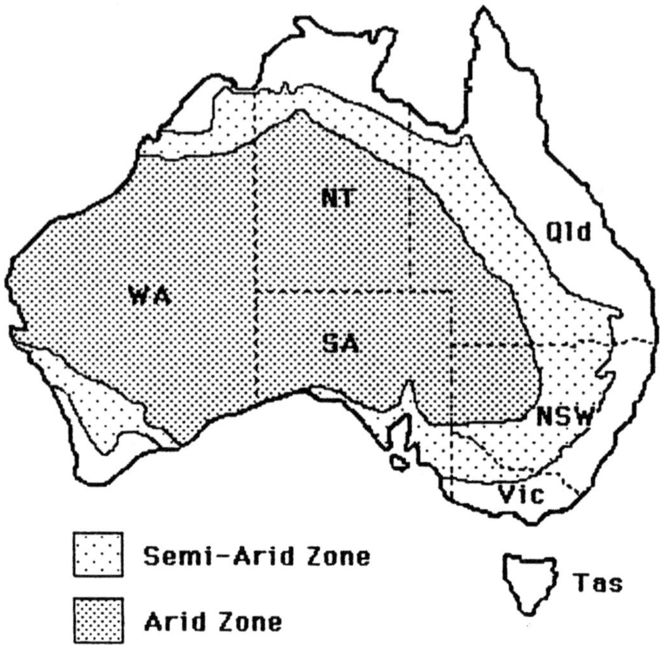

Fig. 1. The arid and semiarid climatic zones of Australia (after Christie, 1986).

of erosion are relatively rare compared to a more general and chronic condition of the land, the loss of landscape function (Tongway, 1991; Tongway and Ludwig, 1993).

Natural, undegraded semiarid landscapes tend to be patchy, heterogeneous systems (Tongway and Ludwig, 1990, 1994). Our results from the semiarid woodlands of eastern Australia show that this patchiness functions to conserve and use limited water and nutrients within the landscape, that is, to prevent the loss of and efficiently cycle these vital resources (Ludwig and Tongway, 1995). Simulations of these systems using historic rainfall regimes suggest that an optimal level of landscape patchiness conserves resources (Ludwig *et al.*, 1994b). Further, landscapes subject to desertification will lose this ability to conserve resources to a far greater extent than landscapes subject to global climate change (Ludwig *et al.*, 1994a).

2. Why Is Desertification Happening?

Desertification in Australia has happened and continues to happen because of naive land use, that is use of landscapes without an appreciation and understanding of the nature and dynamics of vegetation and soils and the vagaries of Australia's arid and semiarid climates (Mabbutt, 1978; Noble and Tongway, 1986). Europeans who settled arid and semiarid Australia in the late 1800s brought with them livestock grazing and land management practices that have proven in retrospect not to suit these landscapes (Harrington *et al.*, 1984). Stocking pressures often exceeded carrying capacities, and fencing constrained stock. These new pastoralists tended to manage the country for short-term production goals, not for the long-term sustainability of the land.

These new pastoralists also suppressed fire without realizing that small, patchy fires were a natural process within many of these landscapes, whether ignited by lightning storms or Aborigines (Hodgkinson *et al.*, 1984). These changes in fire management practices, in combination with other land management practices, have caused the signs of desertification that are so evident today. But has desertification stopped?

Recent assessments suggest that desertification processes continue to degrade Australian landscapes and continue to frustrate land managers attempting to halt and reverse these processes (Payne *et al.*, 1987). The rate of desertification is accelerated during drought. One had only to watch the television coverage of the 1994 drought in western New South Wales and Queensland to see landscapes depleted of vegetation by holding livestock on grazing lands during the drought.

Case studies have clearly documented the severe impacts of drought on the desertification of landscapes, for example the Western Australia Gascoyne Basin case study (Williams *et al.*, 1980). The town of Carnarvon was subjected to the impacts of massive flooding and erosion due to cyclonic rains falling on landscapes denuded by sheep grazing during the droughts preceding the cyclone.

3. Why Can't We Control Desertification?

Basically Australia lacks a strong national approach to controlling desertification. The political realities are that Australia was constituted as a Federation of States only in 1901, and each state has retained the right to regulate the management of its own land as it had as an independent crown colony. This situation has militated against a unified national approach to desertification management. Some states have well-developed desertification monitoring and assessment programs aimed at improving land management practices; other states have less developed programs.

Each state or territory uses different techniques to monitor its lands. Some states are using remote sensing technologies to assess and monitor desertification at a broad regional scale. But such monitoring is currently inadequate for detecting small desertification changes. On the other hand, more-sensitive ground-based assessments at the landscape and soil level are often fragmentary and insufficient, and a standard and objective technique has only recently been available (Tongway, 1994).

Further, poor control of desertification is also related to the extreme social isolation of pastoralists in out-back stations or ranches (Rickson *et al.*, 1987). Isolation of pastoralists away from the coastal population centers of Australia causes severe social, cultural, and economic difficulties. The great distance to neighbors, towns, schools, and markets works against the socioeconomic viability of these stations in the bush. Yet until recently the national economy has traditionally depended on the export earnings of the rural and mining industries, which face all the vagaries of world market fluctuations without Federal Government subsidies.

4. Is There Hope for Controlling Desertification?

We feel that there is hope for controlling desertification, largely because of new national programs aimed at improving land management. The National Drought Alert Program uses National Oceanic and Atmospheric Administration (NOAA) satellite imagery to detect trends in "greenness" that suggest pending drought conditions within important pastoral and farming regions (Brook *et al.*, 1992). Media alerts can then be issued to encourage land managers within these regions to take steps to avoid the economic and ecological impacts of drought. Such alerts are coordinated with educational campaigns on drought management.

A computer-based package called "Rainman" can be used as part of this drought alert program (Clarkson and Owens, 1991). Rainman provides forecasts on the likelihood of rainfall in the next few months given the current strength of the Southern Oscillation, which is a see-saw of air pressure across the equatorial Pacific Ocean (Partridge, 1991). The strength of the Southern Oscillation is measured as an index (SOI) of the air-pressure difference between Tahiti and Darwin. Normally the SOI is positive or occasionally only slightly negative. When the SOI is strongly negative (below -5), the equatorial current flowing toward Australia is slowed or reversed due to weak trade winds. Moreover, seawater temperatures are cool off the Australian coast and warm off the Peruvian coast – the reverse of normal conditions – and air pressures are high at Darwin and low at Tahiti. Strongly negative SOIs are correlated with drought conditions in vital pastoral and farming regions of Australia – so-called El Niño events.

The Landcare Australia program is another national campaign proving to be a successful grassroots approach to stopping and reversing desertification (Campbell, 1992). Federal funds are being used to encourage pastoralists and farmers to form local Landcare Groups and to use proven techniques for restoring desired vegetation on degraded areas. In farming communities this program has worked in conjunction with a Greening Australia campaign to plant farm trees as windbreaks and wildlife habitat. In pastoral communities Landcare Groups work to foster the use of treatments to repair degraded lands.

5. What Treatments Control Desertification?

As part of the mission of our CSIRO Division of Wildlife and Ecology and in collaboration with the Soil Conservation Service of New South Wales (now NSW

Conservation and Land Management), our National Rangelands Program has been working with Landcare Groups to control desertification by promoting the use of treatments that will restore landscape function and rebuild habitats on degraded areas. These treatments aim to prevent the loss of water and nutrient resources from landscapes by building ponding banks and contour furrows (Green, 1989) and by constructing contour traps (Ludwig and Tongway, in review; Tongway and Ludwig, in review). On steeper country (slopes $> 1\%$) contour banks and furrows have proven effective in retaining runoff water and hence preventing soil erosion and loss of nutrients from such landscapes (Green, 1989; Hacker, 1989).

On flatter but still sloping country (slopes $< 1\%$) contour traps – built from piles of large tree branches and shrubs obtained locally and placed in patches along contours – have proven to effectively trap soil and litter moving in runoff water as surface sheetflow (Tongway and Ludwig, 1993). Such contour traps become rich in nutrients and effective in storing runoff water, developing infiltration rates 10 times that of adjacent nonpatch areas. On level or near level country waterponding banks have proven useful in reclaiming "scalded" areas (Cunningham, 1987; Green, 1989).

Improved microclimates and enhanced growth of plants behind contour banks and within contour traps create habitats favorable for such invertebrates as ants and termites (Whitford *et al.*, 1992) and other soil organisms (Greenslade and Smith, 1994). These organisms are important in processes that enhance nutrient cycling. The buildup of biological activities and processes is a positive indicator of landscape rehabilitation (Ludwig and Tongway, 1993).

Applying rehabilitation treatments to landscapes must be coupled with other land management practices. For such treatments to be effective in the long term, grazing pressures on the land must be relatively light (Pressland and Graham, 1989; Ludwig and Tongway, in review). Grazing pressure can be decreased by reducing livestock numbers and by lightening grazing pressure from feral grazing animals (e.g., by harvesting goats and rabbits). Native grazers, such as kangaroos, may also need to be controlled. The important management goal should be to reduce total grazing pressure from all these herbivores.

Rehabilitation can also be enhanced by reinstating a fire regime that creates patches across the landscape and aims to control high densities of shrubs (Leigh and Noble, 1981; Noble *et al.*, 1986). Such a fire regime also decreases the probability of wildfires that have swept across semiarid country creating widespread destruction (Noble, 1991).

6. Conclusions

Most of the vast arid and semiarid regions of Australia have undergone desertification in the form of conversion of grasslands and savannas to more desert-like barren lands and shrublands although in most areas the severe impacts of soil erosion have been localized. Desertification continues today, especially during periods of drought when grazing pressures have removed vegetation cover, making landscapes vulnerable to wind and water erosion. Two new national programs are helping Australian pastoralists turn the corner on reversing desertification. The National Drought Alert Program gives an early warning of impending drought to pastoralists and farmers, so that they can take steps to manage for the drought before it deepens. The National Landcare Program is a grassroots movement aimed to organized pastoralists and farmers into local groups collectively working to rehabilitate degraded landscapes.

References

Brook, K.D., J.O. Carter, T.J. Danaher, G.M. McKeon, N.R. Flood and A. Peacock. 1992. The use of spatial modelling and remote sensing to forecast drought-related land degradation events in Queensland. In *Proceedings, Sixth Australasian Remote Sensing Conference* **1**: 140–149. Wellington, New Zealand.

Campbell, A. 1992. *Landcare in Australia: Taking the long view in tough times.* Third Annual Report, National Soil Conservation Program, Canberra.

Christie, E.K. 1986. *Pastoralism and ecology in arid Australia*, Longman Cheshire Pty. Ltd, Melbourne, pp. 1–8.

Clarkson, N.M. and D.T. Owens. 1991. *Rainman ver. 1.1 – Rainfall information for better management*, Queensland Department of Primary Industries, Brisbane.

Cunningham, G.M. 1987. Reclamation of scalded land in western New South Wales – A review. *Journal Soil Conservation Service New South Wales* **43**: 52–61.

Green, D.R. 1989. Rangeland restoration projects in western New South Wales. *Australian Rangeland Journal* **11**: 110–116.

Greenslade, P. and D. Smith. 1994. Soil fauna responses to restoration by mulching of degraded semi-arid soils at Lake Mere, New South Wales, in C.E. Pankhurst (Ed.), CSIRO Publications, Melbourne, pp. 67–69.

Hacker, R.B. 1989. An evaluation of range regeneration programmes in Western Australia. *Australian Rangeland Journal* **11**: 89–100.

Harrington, G.N., A.D. Wilson and M.D. Young (editors). 1984. *Management of Australia's rangelands*, CSIRO Publications, Melbourne.

Hodgkinson, K.C., G.N. Harrington, G.F. Griffin, J.C. Noble and M.D. Young. 1984. Management of vegetation with fire, in G.N. Harrington, A.D. Wilson and M.D. Young (Eds.), *Management of Australia's rangelands*, CSIRO Publications, Melbourne, pp. 141–156.

Leigh, J.H. and J.C. Noble. 1981. The role of fire in the management of rangelands in Australia, in A.M. Gill, R.H. Groves and I.R. Noble (Eds.), *Fire and the Australian biota*, Australian Academy of Sciences, Canberra.

Ludwig, J.A., S.G. Marsden and D.J. Tongway 1994a. A simulation of resource losses from semi-arid landscapes subject to degradation and climate change. In *Proceedings, Ecological Society of Australia 1994 Open Forum and Symposium Conference*, Alice Springs, p. 60.

Ludwig, J.A. and D.J. Tongway. 1993. Monitoring the condition of Australian arid lands: Linked plant soil indicators, in D.H. McKenzie, D.E. Hyatt and V.J. McDonald (Eds.), *Ecological indicators, vol. 1*, Elsevier Scientific, Essex, pp. 763–772.

Ludwig, J.A. and D.J. Tongway. 1995. Spatial organization of landscapes and its function in semi-arid woodlands, Australia. *Landscape Ecology* **10:**: 51-63.

Ludwig, J.A. and D.J. Tongway. In review. Rehabilitation of semi-arid landscapes in Australia. II. Restoring vegetation patches. *Restoration Ecology.*

Ludwig, J.A., D.J. Tongway and S.G. Marsden. 1994b. A flow-filter model for simulating the conservation of limited resources in spatially heterogeneous, semi-arid landscapes. *Pacific Conservation Biology* **1**: 209–213.

Mabbutt, J.A. 1978. *Desertification in Australia*, Water Research Foundation of Australia, Kingsford, Report No. 54, pp. 3–11.

Noble, J.C. 1991. Behaviour of a very fast grassland wildfire on the Riverine Plain of southeastern Australia. *International Journal of Wildland Fire* **1**: 189–196.

Noble, J.C., G.N. Harrington and K.C. Hodgkinson. 1986. The ecological significance of irregular fire in Australian rangelands, in P.J. Joss, P.W. Lynch and O.B. Williams (Eds.), *Rangelands: A resource under siege*, Australian Academy of Science, Canberra, pp. 577–580.

Noble, J.C. and D.J. Tongway. 1986. Pastoral settlement in arid and semi-arid rangelands, in J.S. Russell and R.F. Isbell (Eds.), *Australian Soils: The human impacts*, Queensland University Press, St. Lucia, pp. 217–242.

Parkinson, G. (editor). 1986. *Climate. Atlas of Australian resources*, third series, vol. 4, Division of National Mapping, Commonwealth Government Printers, Canberra, pp. 6–20.

Partridge, I. (editor). 1991. *Will it rain? El Niño and the Southern Oscillation*, Queensland Department of Primary Industries, Brisbane.

Payne, A.L., P.J. Curry and G.F. Spencer. 1987. An inventory and condition survey of rangelands in the Carnarvon Basin. In *Technical Bulletin No. 76*. Western Australia Department of Agriculture, Perth.

Pressland, A.J. and T.W.G. Graham. 1989. Approaches to the restoration of rangelands – The Queensland experience. *Australian Rangeland Journal* **11**: 101–109.

Rickson, R., P. Saffigna, F. Vanclay and G. McTainsh. 1987. Social bases of farmers' responses to land degradation. In A. Chisholm and R. Dumsday (Eds.), *Land degradation: Problems and policies*, Cambridge University Press, London, pp. 187–200.

Tongway, D.J. 1991. Functional analysis of degraded rangelands as a means of defining appropriate restoration techniques. In *Proceedings of the IVth International Rangeland Congress*, Association Française de Pastoralisme, Montpellier, France, pp. 166–168.

Tongway, D.J. 1994. *Rangeland soil condition assessment manual*, CSIRO Publication, Melbourne, pp. 1–16.

Tongway, D.J. and J.A. Ludwig. 1990. Vegetation and soil patterning in semi-arid mulga lands of eastern Australia. *Australian Journal of Ecology* **15**: 23–34.

Tongway, D.J. and J.A. Ludwig. 1993. Rehabilitation of minesite and pastoral land: The ecosystem function approach. In *Proceedings, Goldfields International Conference on Arid Landcare*, Milena di Russo, Inc, Hamilton Hill, pp. 51–57.

Tongway, D.J. and J.A. Ludwig. 1994. Small-scale resource heterogeneity in semi-arid landscapes. *Pacific Conservation Biology* **1**: 201–208.

Tongway, D.J. and J.A. Ludwig. In review. Rehabilitation of semi-arid landscapes in Australia. I. Recreating productive soil patches. *Restoration Ecology.*

Whitford, W.G., J.A. Ludwig and J.C. Noble. 1992. The importance of subterranean termites in semi-arid ecosystems of south-eastern Australia. *Journal of Arid Environments* **22**: 87–91.

Williams, O.B., H. Suijdendorp and D.G. Wilcox. 1980. The Gascoyne Basin, in M.R. Biswas and A.K. Biswas (Eds.), *Desertification: Associated case studies prepared for the United Nations Conference on Desertification*, Pergamon Press, Paris, pp. 3–106.

REGIONAL LANDSCAPE-ECOLOGICAL PLANNING AND DESERTIFICATION CONTROL IN ARID REGIONS OF THE COMMONWEALTH OF INDEPENDENT STATES

E. V. MILANOVA
Faculty of Geography, Moscow State University, Moscow 119899, Russia

Abstract. Scale-dependent present-day landscape mapping and assessment were used to study the relationship among physical environment, land use, and degree of landscape modification in the Aral Sea region and the Karakum Desert, areas prone to desertification in the Commonwealth of Independent States (CIS). Applying geographic information system (GIS) techniques at global (1:15,000,000), regional (1:1,000,000), and local (1:300,000) scales, researchers found that large-scale landscape assessment and mapping allow them to recognize landscape changes under desertification processes and assess the type and intensity of these processes. Remote sensing has been widely used to evaluate data reliability, to fill information gaps, and to reveal the dynamics of land use types resulting from landscape changes.

1. Landscape Concept and Methodology

Desertification is among the major environmental problems in arid, semiarid, and dry subhumid regions, seriously threatening the sustainable development of these regions. To most effectively study global environmental problems (including desertification) and their solutions, scientists need the best quality information on the present status and dynamics of environmental changes. Landscape assessment provides part of the basis for understanding the world as a system of interrelated components, each with a different set of environmental conditions. In response to this problem, a scale-dependent hierarchical landscape classification scheme is proposed for global and regional mapping.

The term "landscape" is commonly understood by the geographic and landscape-ecological research community as a general regional unit that integrates parts of the geobiosphere and socioeconomic sphere. A landscape system of any spatial dimension represents an ecological-economic system where two subsystems – natural and anthropogenic – coexist and interact within the boundaries of their comparatively stable natural basis. The present-day uniqueness of any landscape system is expressed in the character of the anthropogenic transformation of its natural pattern.

The notion of "land cover" accepted by the International Geosphere-Biosphere Program, Human Dimensions Program (IGBP-HDP) Land Use/Cover Project as an object of studies relating land use and land cover change (Turner *et al.*, 1993) has proved to be a close synonym of the recently defined hierarchical concept of present-day landscapes (or PDL) (Milanova, 1990).

Environmental Monitoring and Assessment **37:** 239–244, 1995.

Within the framework of PDL mapping it is feasible to investigate how regulatory tools at various political or administrative levels addressing different parts of the systems will influence the environment. Likewise it is recognized that different aspects of the landscape relate to such human driving forces as regulatory instruments (local, regional, national, or international). The complexity of these systems needs further investigation to determine if simpler analytical approaches would suffice. A likely outcome is that the systems are complex because social, economic, technical, and ecological processes operate on a wide range of spatial and temporal scales.

A classical methodology in landscape study of any scale begins by analyzing the systems and components and by observing the similarities and differences among systems. The only feasible method of preliminarily comparing landscape units is to view them synoptically from an aerial perspective. As such, the land cover and PDL concepts could be created only after remote sensing techniques were developed.

The medium-resolution synoptic coverage of the Landsat-TM and Cosmos-1939 sensing systems are ideally suited for landscape analysis. The high temporal but coarse spatial resolution images from NOAA polar-orbiting satellites are extremely useful for intra-annual (phenological) classification of cover types. A variety of conventional data sources describing natural and socioeconomic components of landscapes are used, as well as limited field observations for ground verification. Landscape functioning parameterization, essential for regional- and local-scale characterization, needs the use of landscape dynamics simulation and systems analysis techniques.

2. Landscape Mapping and Assessment at Global Scales

A group of scientists at the Faculty of Geography at Moscow State University (MSU) implemented global landscape mapping as a continuation of the MSU Global Thematic Map Wall-Chart 1:15,000,000 Series. The integrative World Map of Present-Day Landscapes was printed in 1993 under the sponsorship of the United Nations Environment Programme (UNEP). Offering a colorful presentation of the territorial distribution of present-day landscapes and the degree of their transformation under human impact, the map presents landscape units as complex systems whose natural and anthropogenic parts coexist and interact within geographically defined boundaries.

All PDLs have been classified into two groups by the degree of transformation: modal (primary or essentially unaffected) and natural-anthropogenic. Natural-anthropogenic landscapes have been further subdivided into derivative (or secondary) landscapes, landscape anthropogenic modifications, and technogeneous complexes.

The main criteria used for determining the degree of landscape transformation are the more sensible landscape components, such as vegetation, as well as the extent and intensity of the current human impact. An easily detectable yet dynamic part of a landscape, vegetation is widely used as an indicator of the spatial distribution of certain properties of a region, such as climatic and soil differences and groundwater levels, movement, and chemical composition. Commonly the vegetation cover is the first aspect of the landscape to be altered by human impacts.

This feature was used to develop a simple and reasonable approach to defining four major degrees of landscape transformation. In this way, the mapping procedure for this study implies four degrees of vegetation cover transformation. The first degree represents landscapes with practically no anthropogenic transformation of vegetative components. The second is associated with the emergence of secondary biotic successions under the influence of human action. The third reflects a vegetation cover that has been fundamentally altered so that human-induced vegetation dominates. And the fourth transformation includes areas where technogeneous structures have almost totally replaced the natural vegetation cover.

Specific goals of the global mapping program of present-day landscapes are as follows:

- to provide a quick reference and easily understandable presentation of the general geographical distribution of different stages of landscape evolution under human impact – from landscapes that are to all intents and purposes unaffected or "natural" to landscapes affected to varying degrees by human action – as an aid to understanding how far the status and patterns of present-day landscapes are from their potential natural conditions;
- to show the current trends in landscape evolution in each geographical zone or region (e.g. development of secondary savanna and shrubland in place of tropical rain forests resulting from clear-cutting, agriculturally induced secondary steppe advancement in continental temperate forests, anthropogenic deserts in semiarid and dry subhumid areas);
- to help rapidly identify landscape areas with similar or analogous present status in order to locate as precisely as possible sample territories with a better or more advanced environmental management experience that can be applied in similar circumstances elsewhere;
- to reveal the territorial extent of areas with the most heavily transformed landscapes that may need prompt rehabilitation.

3. Global Landscape Gis and Database Methodology

The digital version of the World Map of Present-Day Landscapes is a follow-up to the UNEP/MSU map, but it appears as a separate product, the "Global Landscapes GIS and Database," with interactive data display and query capabilities. The project has been organized as a collaborative venture between Moscow State University and the GIS Research Center of the Institute of Geography of the Russian Academy of Sciences. Map units have been digitized as polygons at the source scale of 1:15,000,000. All landscape polygons are characterized by three independent variables – LAND COVER, RELIEF, and MODIFICATION – that respectively represent the machine codes of individual natural landscape zones, landscape orographic classes, and landscape transformation trends. The identifiers of these variables are linked to the landscape attribute database that contains a set of characteristic attribute values and value ranges for every zone (average climatic and soil-vegetation patterns), class (elevation ranges, erosion potentials, vertical zonality types) and transformation trend (dominant and subdominant land use patterns, population densities, and growth rates).

A specially developed DOS- and Windows-compatible software application allows for interactive menu-driven data visualization, user-defined category selection and queries, and customized report generation, as well as hard copy, output map projection selection, and data export in Arc/Info and other GIS formats.

4. Regional and Local Levels of Arid Landscape Mapping and Assessment

The approach described earlier has also been tested in a detailed inventory and assessment of the present-day landscapes at several regional-scale (1:1,000,000) pilot areas in the CIS and other countries (Tunisia, India) (United Nations Environment Programme, 1993).

The Aral Sea basin was studied as a pilot test using prior field research and remote sensing analysis. The Aral Sea is fed by waters of the two largest Middle Asian rivers – the Amu-Darya and Syr-Darya. The deltas of these rivers have provided the focus for several ancient civilizations, most notably that of the Khorezm. Today the Amu-Darya delta is a densely inhabited agricultural region producing cotton, rice, grain, and vegetables. But the drive to increase irrigated agricultural production in this area, dating from the 1950s, has helped give the Aral Sea region the largest ecological crisis (or desertification problem) in the arid part of the CIS.

The Aral Sea has shrunk dramatically. In about 30 years its water level has fallen by more than 14 m, its area has declined by more than 40%, and its volume has decreased by more than 60%. The economic, social, and environmental effects of the decline in the Aral Sea have been catastrophic. Most fish species have become extinct, and other food reserves have disappeared in the increasingly saline waters. The exposed seabed of the Aral has become the source of large-scale dust storms. The heavy use of toxic chemical pesticides in irrigated areas has contaminated local drinking water supplies, which have also deteriorated through inadequate purification and sewage treatment plants.

The map of present-day landscapes showing desertification processes and their influence on the environment was compiled from field observations and remote sensing data at a scale of 1:1,000,000. The satellite imagery used for this project consists of composite images from the Salyut orbital station (70 m resolution) and the Meteor resource satellite (10–30 m resolution).

Four major groups of desertification processes have been found: drying of the marine coast, drying of the adjacent delta-alluvial plains, pasture degradation, and salinization and waterlogging of irrigated lands. All landscape components (including relief, vegetation and soil cover, surface water flow, and groundwater characteristics) have changed under human impact.

Analysis of the map allows one to identify landscape transformations and trace landscape evolution influenced by desertification processes. Widespread desiccation of the delta lakes and reed beds has been one result of this impact. Secondary salinization and waterlogging plague the irrigated areas, while saline water flowing back into the Amu-Darya from irrigated fields causes serious problems as water quality has declined.

The described approach has also been tested in a local-scale detailed inventory and analysis of the present-day landscapes of the Chardjou region of the Karakum Desert in the Central Asia region of the CIS (Turkmenistan), where intensive human activities and use of natural resources often alter the ecological equilibrium, which can trigger the onset of desertification. The scale of the landscape mapping (1:300,000) can show not only quantitative but also qualitative characteristics of processes of landscape degradation.

The PDL, considered as pasture and ecological-economic systems, was initially ranked according to the dominant type of grazing lands, and their productivity was measured by annual feed stock. Later the landscapes were ranked according to a grazing pressure coefficient, which was calculated as the ratio between the norm (i.e. standard carrying capacity) and the real grazing area per livestock unit.

Systems analysis and quantitative modeling were used to study the PDLs and their functioning. Six scenarios of landscape dynamics were selected for computer simulation, each with a different value of the main variable – livestock numbers: (1) conserving the existing land use type in the future, (2) establishing a reserve regime, (3) normal (official standard) type of grazing, (4) decreasing grazing pressure, (5) increasing grazing pressure, and (6) strongly increasing grazing pressure. The model includes two data set inputs: primary and secondary productivity. Computer simulation showed that the optimal land use for the Chardjou region is to decrease grazing pressure and improve pastures. Assuming constant market conditions, capital expenditures for pasture improvement will be compensated by economic benefits in less than 2 years.

The resulting map of landscapes shows the spatial differentiation of present-day complexes. These complexes have different natural features, and their economic potential is therefore also different. The "land unit" as an expression of landscape as a system helps to address topologic and chorologic landscape ecology relationships through the simultaneous use of characteristics of the most obvious (mappable) land attributes: landform, soil, and vegetation (including human alteration of these three). The land unit is the basis of the map legend but may be expressed by these three land attributes. The more dynamic land attributes are less suitable as diagnostic criteria but often link units through information/energy fluxes. The landscape maps have been created according to the landscape concept, which states that one can delineate the elementary units of landscapes, after which one can easily generate any thematic map (e.g. soil or vegetation map as well as land use maps) by aggregating polygons of these units according to a corresponding classification. This principle is used in most geographic information systems.

The most critical step in this process is creating the classification system with which the computer system can assign the proper physical and social characteristics (such as soil class, vegetation, land use) to each of the elementary units. It is convenient for the process to have this classification as a matrix that relates numerical identifiers of elementary units to identifiers of polygons on thematic maps.

5. Conclusions

1. The landscape concept can enhance land use/cover assessments and help delineate the world's landscapes as a system of regions with different patterns of land use/cover relationships.
2. The mapping of landscape units allows the comparison of regions prone to desertification. The World Map of Present-Day Landscapes and its digital version emphasize modeling of the trends of landscape evolution.
3. A key level of landscape assessment offers possibilities to identify types of landscapes according to qualitative and quantitative parameters. The map analogies and contrasts of existing environmental management practices and their results may give strong impetus to the elaboration of scientifically grounded landscape-ecological planning and regional economic policies in arid regions.

References

Milanova E.V. 1990. Geographical approach to landscape zonation. In *Proceedings of the Ad Hoc Consultative Meeting on the Assessment of Desertification*, United Nations Environment Programme-DC/PAC, Nairobi, pp. 281–287.

Milanova, E.V., A.V. Kushlin and N.J. Middleton (editors). 1993. *World map of present-day landscapes*, Soyuzkarta, Moscow.

Turner, B.L., R.H. Moss and D.L. Skole (editors). 1993. *Relating land use and global land-cover change*. IGBP Report No. 24. International Geosphere-Biosphere Program, Stockholm.

United Nations Environment Programme. 1993. *World atlas of desertification*, Edward Arnold, London.

SOUTH AFRICA'S ARID AND SEMIARID RANGELANDS: WHY ARE THEY CHANGING AND CAN THEY BE RESTORED?

SUZANNE J. MILTON and W. RICHARD J. DEAN
FitzPatrick Institute, University of Cape Town, Rondebosch 7700, South Africa

Abstract. Since the mid-19th century settled livestock ranching has been the major form of land use in South Africa, occupying 68% of the land surface. Decreases in livestock densities and ranch numbers during the past century imply that carrying capacities for domestic herbivores are falling. Differences in carbon isotope signals with soil depth and abrupt shifts in dominant plant species across ranch boundaries reveal that southern African rangelands are changing. Case studies suggest ways to control altered grassland composition, bush encroachment in arid savanna, and dominance by toxic and halophytic shrubs in arid shrublands. But climatic and biological factors constrain rates of passive recovery, and guidelines for active restoration are poor and techniques costly. Moreover, conservation of remaining good rangeland is seldom enforced, and economic considerations usually outweigh the land user's desire to sustain diversity and productivity.

1. Introduction

Ranching occupies 68% of South Africa's land surface (Directorate Agricultural Information, 1991). Although it generates only 2% of the gross domestic product and supports little more than 1 million people, ranching supplies most of the local demand for meat and wool. Desertification, or a functionally irreversible decline in rangeland productivity, therefore has serious social and economic implications for this country. As the theme of this conference is desertification in developed countries, we will not discuss the problems of the underdeveloped 15% of South Africa (Boonzaier *et al.*, 1990; Hatch and Grossman, 1993), where 30% of the nation's population subsists on communal grazing land (Directorate Agricultural Information, 1991). Most of South Africa's 706,000 km^2 of rangeland is owned by commercial ranchers. The present review focuses on this sector, which is "developed" in its access to education, transport, finance, and markets.

To understand desertification in the developed parts of South Africa we review changes in land use and the evidence for declining productivity on commercial rangeland at various spatial scales. We then explore the biological processes behind decreases in carrying capacity for livestock and discuss constraints on rangeland rehabilitation. Finally, we discuss the social and economic factors that motivate landowners to overexploit their rangelands.

Environmental Monitoring and Assessment **37:** 245–264, 1995.

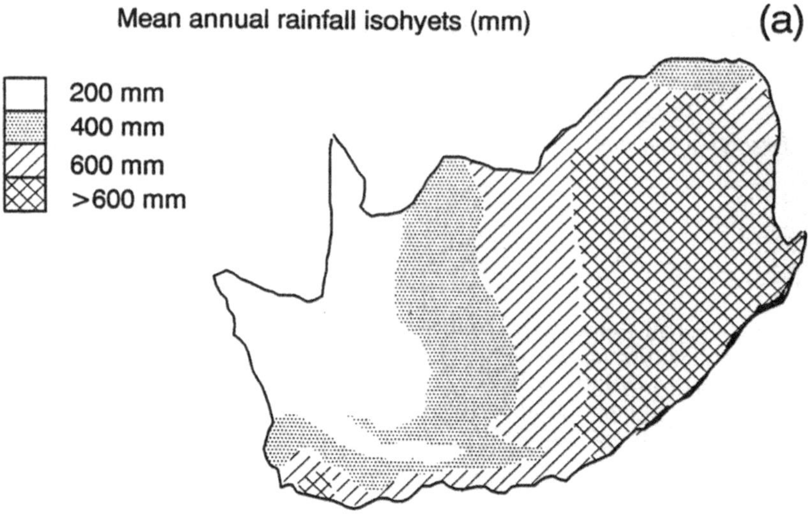

Fig. 1a.

Figs. 1(a)–(c). Rainfall isohyets (a), vegetation (b) and distribution of livestock (c) in South Africa.

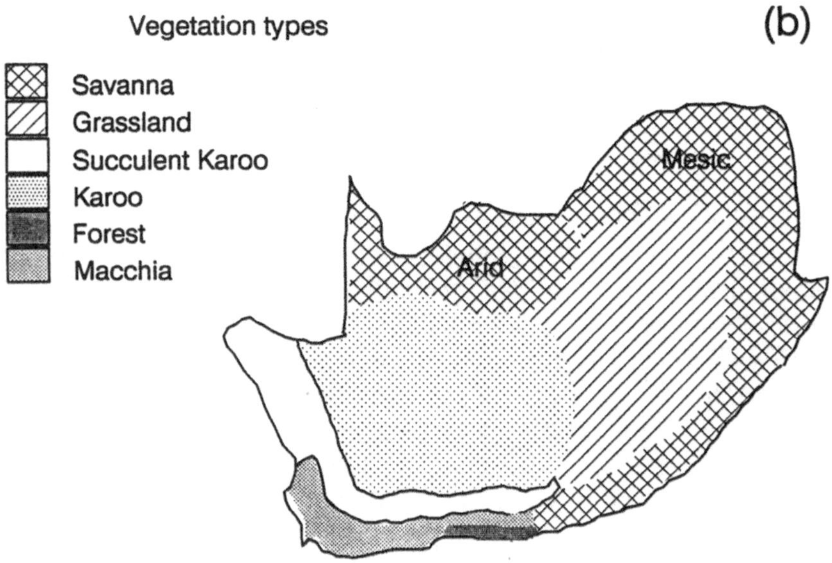

Fig. 1b.

2. Changes in Land Use

The indigenous grazing and browsing ungulates of South Africa's semiarid Karoo and savanna regions (Figures 1a and 1b) were nomadic. Eland (*Taurotragus oryx*), gemsbok (*Oryx gazella*), hartebeest (*Alcelaphus buselaphus*) and springbok (*Anti-

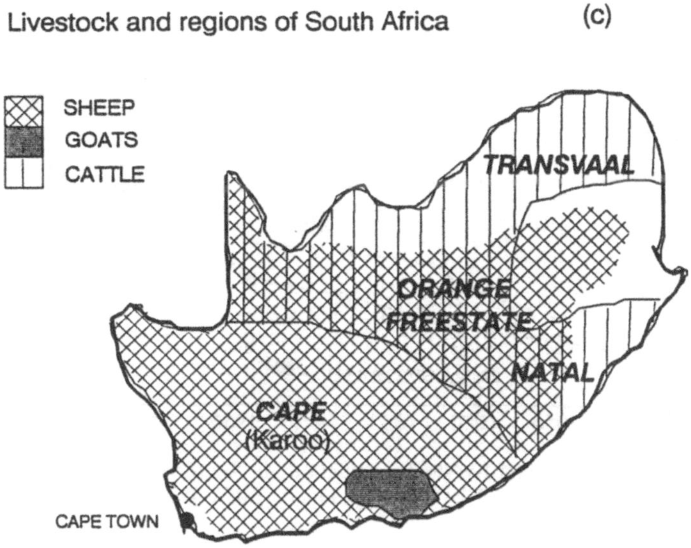

Fig. 1c.

dorcas marsupialis) moved seasonally or opportunistically to greener pastures (Skead, 1980). The native hunters and pastoralists of the region followed the game or moved with their herds of sheep (Smith, 1986). Seasonal shortages of surface water necessitated a similar lifestyle among the European ranchers (*trekboers*) who colonized the Cape Colony of South Africa in the 18th century (Talbot, 1961; Penn, 1986). Trekking between summer and winter pastures continued after formal alienation of land in the 1820s. But the Land Settlement Act of 1912, together with the state-assisted introduction of wind-driven waterpumps and fencing, ended the era of nomadic ranching, even in the most arid regions (Talbot, 1961). By this time cattle, sheep, and goats (Figure 1c) had largely replaced the indigenous ungulate populations, which had been reduced by efficient hunting and fencing (Skead, 1980).

3. Evidence for Declining Productivity

Evidence for declining secondary productivity and changes in vegetation composition has been accumulating since the early 1900s. We discuss this evidence at national, regional, ranch, and patch scales. This distinction is necessary because desertification may be a patchy rather than a homogeneous process (Dean *et al.*, 1995). Therefore, changes at the ranch scale cannot be inferred from productivity trends at the national scale.

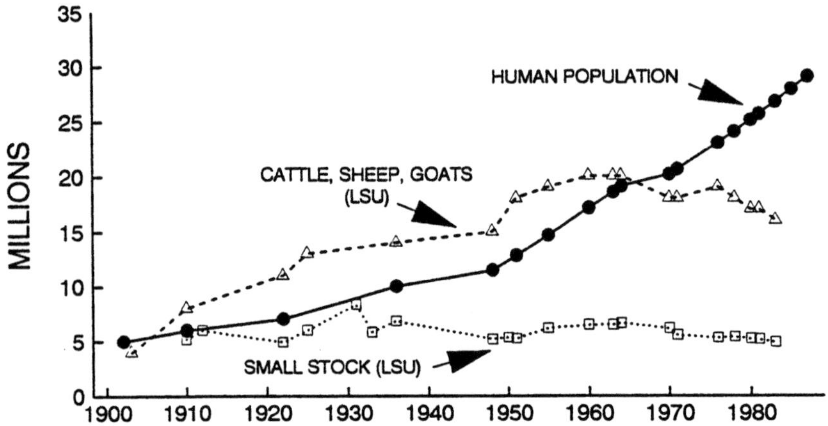

Fig. 2. Trends in human and livestock populations in South Africa. Livestock populations are expressed as large stock units (LSU). (Data from Fuggle, 1983; Directorate Agricultural Information, 1991.)

3.1. National trends

Government enquiries into drought and rangeland desertification began in the early 1900s (Union of South Africa, 1914, 1923) and were followed a generation later by other investigations (Kokot, 1948; Union of South Africa, 1951). All concluded that annual rainfall showed no negative trend and that the observed drying up of rivers and rangelands was the result of overgrazing and soil erosion.

Generating a lucrative market for wool, the industrial revolution in Europe led to a rapid increase in sheep ranching in South Africa between 1840 and 1850. Small stock (sheep and goat) populations reached a peak of 50 million in 1930 shortly after the fencing and water provision on ranches, but fell to 35 million in 1933 and appears to have stabilized at this level (Downing 1978; Macdonald, 1989). The human population in South Africa has grown by 2–3% annually since 1950 (Fuggle, 1983), whereas total livestock populations (large stock units – LSUs – equivalent to one bovid weighing 420 kg (Dean and Macdonald, 1994) have remained stable or decreased during this period (Figure 2).

Ranch numbers peaked in 1950 as a result of subdivision of deceased estates*, but many of the units were subeconomic (Union of South Africa, 1951). We infer from the decrease in ranch numbers and increase in average ranch size (Figure 3a) that rangeland productivity declined on a national scale thereafter. The change in ranch sizes was greater in the arid grasslands and shrublands of the Cape and

* A problem peculiar to South Africa was caused by subdividing ranches when an estate was left jointly to a rancher's sons. Each section of the original ranch was then independently ranched by the inheritor, often with disastrous results for the owner and the land. The process of fragmentation continued when the inheritor died and the subsection was again subdivided.

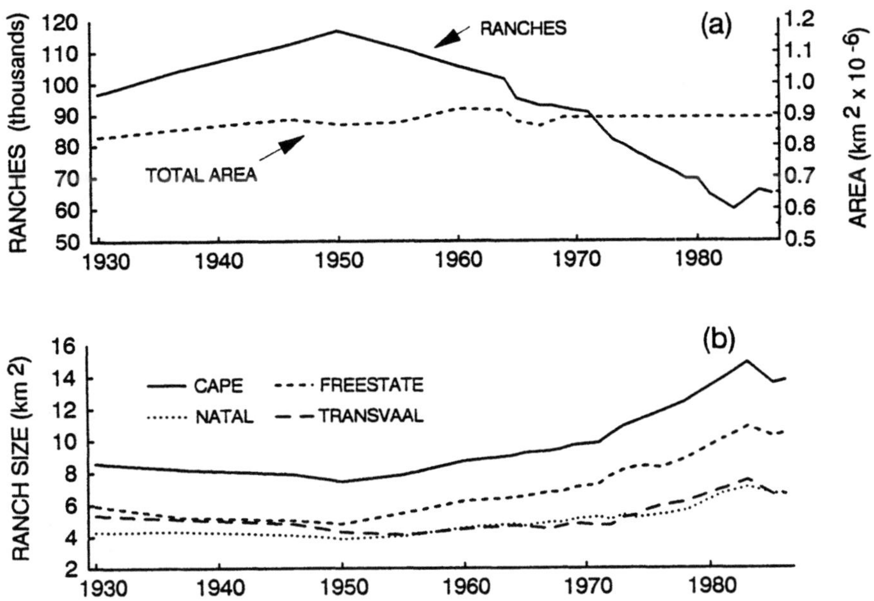

Fig. 3. Ranch numbers, total ranch area (a), and average ranch sizes in four regions of South Africa (b). (Data from Directorate Agricultural Information, 1991.)

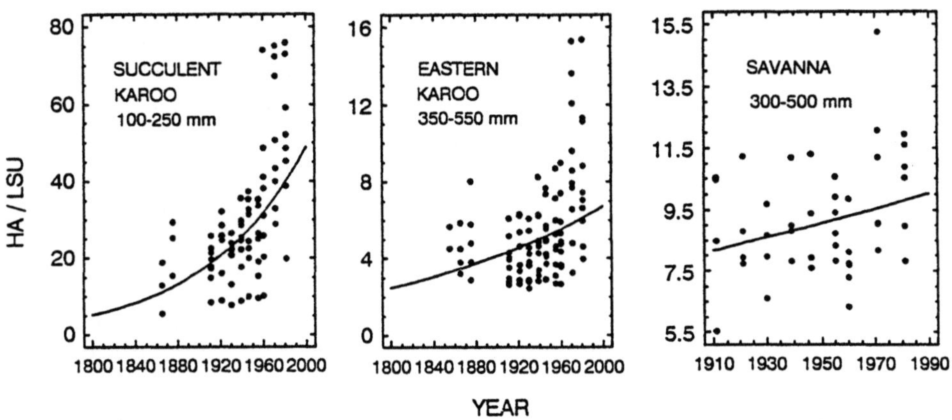

Fig. 4. Changes in livestock densities (hectares per LSU) in three vegetation types. (Redrawn from Dean and Macdonald, 1994.)

Orange Free State provinces than in the relatively mesic savannas of the Transvaal and Natal (Figure 3b).

3.2. Regional trends

Livestock densities did not follow the same trends in all vegetation types. Dean and Macdonald (1994) reported a negative relationship between mean annual rainfall

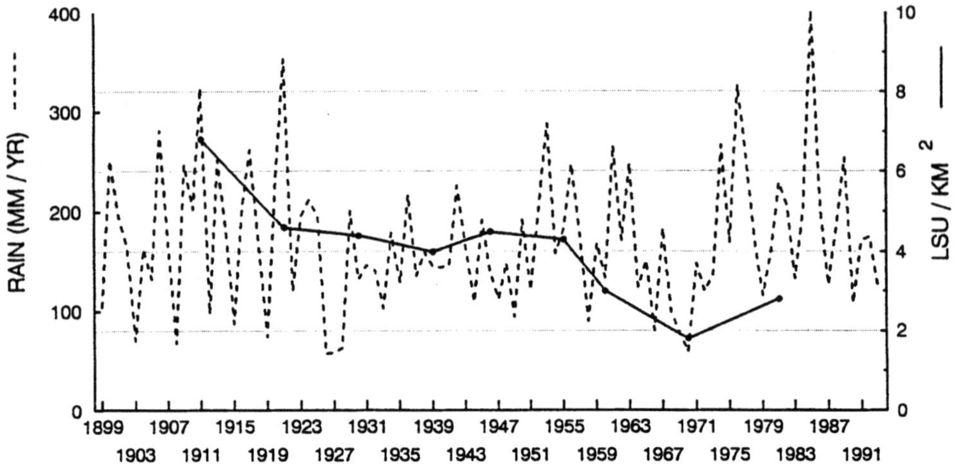

Fig. 5. Annual rainfall and stocking rates (LSU/km^2) in the Prince Albert District of the Western Cape, South Africa. (Unpublished data from stock censuses and records of the South African Weather Bureau.)

and percentage reduction in stocking rate. Decreases in the numbers of livestock in magisterial districts were greater in Karoo vegetation than in savanna (Figure 4). But history rather than aridity may explain why carrying capacity has decreased more in the southeastern than the northern Cape (Dean *et al.*, this volume). European rangeland acquisition spread inland from the southern and eastern Cape between 1806 and 1911 (Christopher, 1982), so that the south has been exposed to settled ranching for a century longer than the north.

Acocks (1953) suggested that ranching had led to a replacing of grassland by shrubland in parts of the Karoo. Differences in carbon isotope signals with soil depth (Bond *et al.*, 1994; Hoffman *et al.*, this volume) were greater in the Karoo-grassland ecotone of the eastern than in southern Karoo shrublands. These data (Bond *et al.*, 1994) suggest that ranching has led to a decrease in the abundance of C_4 grasses in ecotonal areas but that the winter-even rainfall southern Karoo was a nongrassy shrubland before European settlement.

Grass was probably a minor component of the vegetation of the arid (<200 mm rainfall) Prince Albert District in the southern Succulent Karoo even before European settlement (Bond *et al.*, 1994). Rainfall in this district shows no long-term trend, but livestock densities declined by 60% between 1910 and 1983 (Figure 5), indicating a decrease in forage availability.

3.3. RANCH-SCALE CHANGES

The composition of the herbaceous layer of nine sites in eastern Transvaal grassy savanna differed with ranch management history (O'Connor and Pickett, 1992). Perennial, large-seeded grasses (*Themeda*, *Heteropogon*) were cover dominants

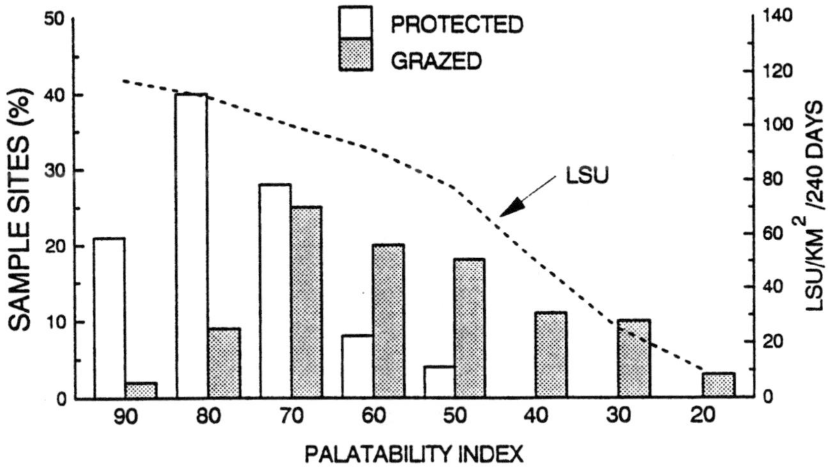

Fig. 6. Palatability of protected and grazed vegetation and carrying capacity for cattle (LSU/km^2/240 days) in Transvaal grasslands. (Data from Barnes, 1990.)

and abundant in the seed bank on protected and lightly grazed sites, existed in lower densities under moderate grazing, and were absent under heavy grazing. Cover of relatively short-lived xerophilic grasses (*Aristida*, *Urochloa*) showed the opposite trend although these grasses maintained seed banks at all sites. The palatable forage component was lower in the grazed half of 112 paired fenceline contrasts in the southeastern Transvaal (Figure 6), decreasing the estimated carrying capacity of the rangeland for cattle by 50 to 80% (Barnes, 1990).

On many ranches in arid and mesic savanna the density of woody plants (*Acacia*, *Rhigozum*) has increased (Moore, 1989; Teague and Smit, 1992), a phenomenon known as bush encroachment. In heavily grazed western Transvaal grassland, tufted grasses were replaced by thorn trees (*Acacia karroo*), arborescent succulents (*Aloe*), and rhizomatous grasses (Friedel, 1987). These replacement species are less accessible to livestock and consequently support fewer livestock than perennial grasslands.

Goat browsing has reduced or eliminated the once-dominant arborescent succulent *Portulacaria afra* on many ranches in eastern Cape subtropical thicket (Figure 7), resulting in an increase in deciduous spinescent shrubs (*Lycium*, American *Opuntia* spp.), annual grasses, and forbs (Hoffman and Cowling, 1990a; Stuart-Hill, 1992). This replacement of tropical thicket species by weedy and desert-affiliated species (Hoffman and Cowling, 1991) entailed a shift from a perennial to an ephemeral resource base that supplied less forage during droughts.

Photographic and survey data show that the relative abundance of grasses, succulents, and shrubs in Karoo shrublands varies with rainfall. A significant improvement in rangeland condition has been reported from some ranches (Palmer *at al.*, 1990; Hoffman and Cowling, 1990b) and a deterioration on others. Data

Fig. 7a.

Figs. 7(a)–(b). Fenceline contrasts. (a) Willowmore District: loss of arborescent succulent *Portulacaria afra* from an overstocked ranch to right of fence and (b) Prince Albert District: reduction of vegetation cover by ostriches to right of fence.

Fig. 7b.

from 80 pairs of fenceline contrast surveys in arid succulent Karoo shrublands show that overutilization reduces cover and favors toxic shrubs over palatable shrubs, halophytes over nonhalophytes, and ephemerals over perennials (Milton and Dean, 1992; Stokes, 1994). The structure and composition of rangeland with a history of heavy grazing resembled the vegetation of disturbed sites in more arid areas (Milton, 1990).

Ranch-scale changes have been measured in productivity and profit. An evaluation of the economic viability of commercial ranching in the eastern Cape savanna found that only 30% of the ranches in this district were profitable. Rangelands in poor condition need to be four to five times larger than comparable rangelands in good condition to generate a modest income (Moore and Odendaal, 1987; Danckwerts and Marais, 1989). Where grazing has led to bush encroachment in

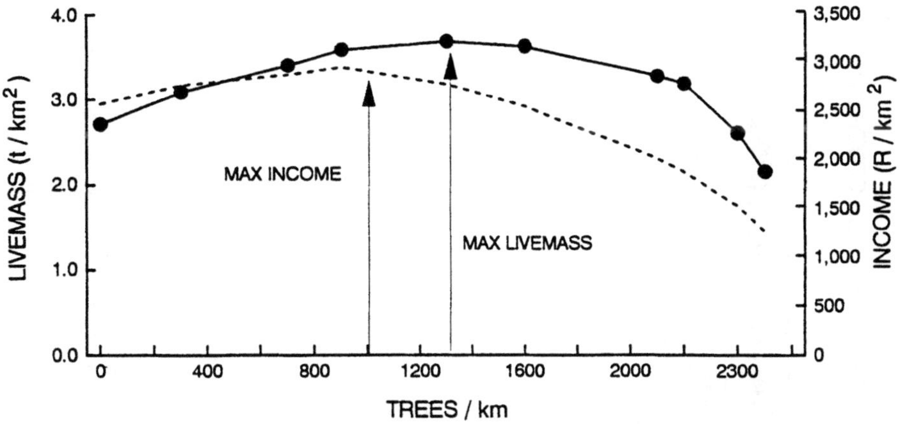

Fig. 8. The influence of bush encroachment on livestock (cattle plus goats) production and income. (Data from Stuart-Hill, 1987.)

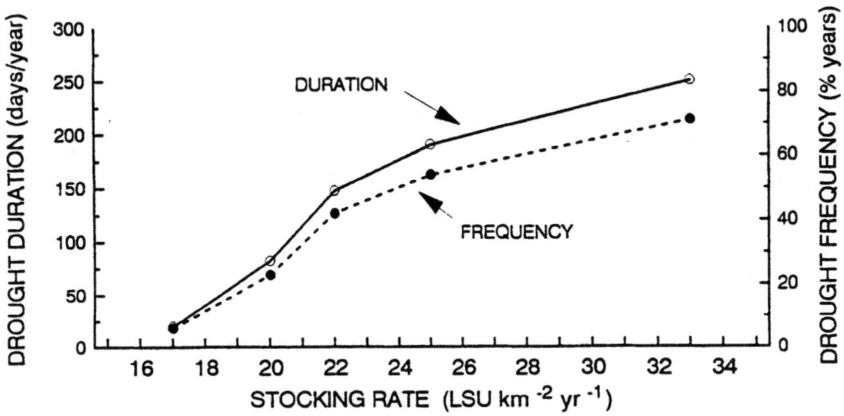

Fig. 9. Stocking rate and simulated duration and frequency of drought. (Data from Fouché *et al.*, 1985.)

grassy savanna, livemass production and profitability decrease once tree densities exceed 1200 trees/ha (Figure 8), even where cattle and goats are ranched together (Stuart-Hill, 1987). Forage shortages are equivalent to drought. Drought in semiarid grassland, defined as <100 kg/ha forage dry matter (Fouché *et al.*, 1985), increased in duration and frequency with an increase in stocking rate (Figure 9). Since forage shortages require the purchase of supplementary feed, maximum profits were expected on moderately stocked ranches.

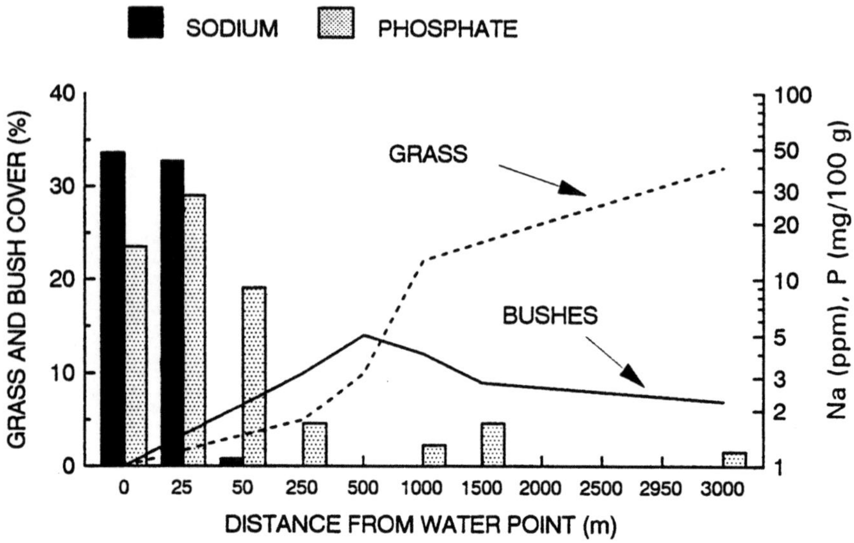

Fig. 10. Changes in vegetation cover and soil sodium and phosphate concentrations with distance from a cattle watering point. (Data from Martens, 1971; Perkins and Thomas, 1993a.)

3.4. Patch-scale changes

Both wild and domestic herbivores forage more intensively in some patches than in others (Hatch and Tainton, 1990; Novellie, 1990; Fuls, 1992). Nutrient-rich soils of bottomlands and termitaria (Macdonald, 1978; Milton *et al.*, 1992; Stokes, 1994) and areas within 2 km of stock watering points are particularly heavily used (Bosch and Gauch, 1991; Perkins and Thomas, 1993a, b; Stokes, 1994). Intensive trampling, dunging, and grazing of patches changes vegetation composition, reduces cover, compacts the soil, and increases alkalinity and salinity (Figure 10). The net effect of these changes is accelerated erosion (Roux and Opperman, 1986) and localized aridification (Stokes, 1994) or salinization (Wentzel *et al.*, 1994), which endures for decades.

Although some authors suggest that ranch-scale desertification results from the coalescence of aridified patches (Kellner and Bosch, 1992; Dean *et al.*, 1995), others have argued that desertified patches are spatially confined and have limited impact on the overall grazing potential (Perkins and Thomas, 1993a; Scoones, 1992).

4. Processes Underlying Rangeland Change

Here we review the ways in which rainfall and herbivory influence processes at the organism scale, ultimately changing the composition and productivity of rangelands.

4.1. Competition

Plants in southern African rangelands appear to compete primarily for water. Experimental investigation of tree-grass interactions in savanna at Nylsvley in the Transvaal (reviewed by Scholes and Walker, 1993) found that woody plants are consistently stronger competitors for water. Stunted tree seedlings persist in the grass layer and eventually outgrow and out compete the layer. Grassy savanna can therefore be maintained only by browsing, fire, wood cutting, and other processes that kill trees or suppress sapling growth. Reducing the grass layer, particularly in the absence of browsing, therefore increases tree seedling recruitment, leading to bush encroachment (reviewed by Scholes and Walker, 1993).

This conceptual model of effects of grazers on competition between grasses and woody plants, although generated in mesic savanna, is also suitable for semiarid *Acacia* savanna in the eastern Cape (du Toit, 1972a, b; Teague and Smit, 1992) and arid *Acacia* and *Rhigozum* savanna in the northern Cape, Botswana, and Namibia (Walter, 1954; Moore and Odendaal, 1987; Moore, 1989; Skarpe, 1990).

In the ecotone between winter and summer rainfall zones, rainfall seasonality influences the relative abundance of summer and winter growing plants. Long-term grazing trials (reviewed by O'Connor, 1985) show that summer rainfall has a greater effect on grass abundance than does grazing but that grazing during the growing season maintains grass basal cover below that of shrubs in most years. Shrub and succulent cover was more stable than grass but increased following above-average winter rainfall (Roux, 1966; Hoffman *et al.*, 1990). Shrub canopy cover in the succulent Karoo was correlated with rainfall of the previous winter growing season, and the response was more marked among palatable plants than defended species (Stokes, 1994). Frequent defoliation depletes carbohydrate reserves of palatable species, reducing the competitive advantage they generally gain through rapid regrowth (Van der Heyden, 1992).

In semiarid Karoo shrubland, composition usually changes during exceptionally wet or dry periods (O'Connor, 1985). Similarly, herbivory and other forms of disturbance that release resources can lead to mass seedling recruitment. Removing established plants (succulents and dwarf shrubs) prolonged soil moisture availability at 150 mm below the soil surface, improved seedling survival, and increased flowering in plants adjacent to cleared areas (Milton, 1995). By reducing competition for water, overgrazing probably accelerates the rate of change in these shrublands and directs the course of change.

4.2. Reproduction

Reducing the reproductive potential of forage species has recently been named the primary cause of composition change in savanna grassland, Karoo shrubland, and subtropical thicket on South African rangeland.

A simple model of seedbank dynamics of savanna grassland explained the changes in composition associated with overgrazing (O'Connor, 1991). Seeds of perennial savanna grasses have a limited survival in the seed bank, so that the seed bank size is determined primarily by the abundance of species in the vegetation. Species that produce few seeds (e.g. *Themeda triandra*) are easily eliminated by sustained cattle grazing and replaced by less-competitive species (e.g. *Aristida* spp.) that seed abundantly (O'Connor and Pickett, 1992).

Similarly, sheep in succulent and mixed Karoo shrublands reduce and in some cases prevent seed production of preferred forage plants by consuming flowerheads (Milton, 1994; Stokes, 1994). Palatable shrubs are generally longlived and do not maintain seedbanks (Esler *et al.*, submitted) so that replacement of natural mortality depends on regular seed production. Shrub species that are less acceptable as forage for sheep continue to produce seed on rangeland (Milton and Dean, 1990), and the frequency of these species in the plant community increases. In this way some rangelands in species-rich vegetation types become dominated by almost monospecific stands of toxic or spinescent shrubs (Milton and Hoffman, 1994).

The arborescent succulent *Portulacaria afra*, highly palatable to browsers, including elephant and domesticated goats, reproduces mainly by rooting where its drooping branches touch the soil. Elephant browsing in the canopy does not interfere with vegetative reproduction, but goat browsing from below prevents cloning and eventually snaps the trunks of these top-heavy "trees" (Stuart-Hill, 1992). Precluding asexual reproduction together with aridification of the hilly habitat by overstocking has caused local extinctions of this species on many eastern Cape ranches. Sheep are known to have a similar effect on the vegetatively reproducing dwarf shrub *Pentzia incana* when they browse it so short that its pendent branches cannot reach the soil surface and take root.

Drought favors species that have seed banks or survive and reproduce shortly after a drought is broken (Henrici, 1940; Milton and Dean, 1995). Grazing that prevents sexual or asexual reproduction of surviving palatable perennials results in a rapid transition to dominance of regenerating vegetation by annuals and unpalatable perennials (Roux and Theron, 1987; Danckwerts and Stuart-Hill, 1988).

4.3. Establishment

Grazing reduces vegetation cover, thereby doubling rainfall losses to runoff, increasing evapotranspiration, and reducing plant biomass production by 75% (Snyman and Fouché, 1991). The much greater temperature fluctuations on open soil than beneath canopies of shrubs and succulents (Beukman, 1991) precludes establishing such radiation and frost-sensitive species as *Portulacaria afra* in exposed microsites (Stuart-Hill, 1992). The increased frequency of arid and exposed microsites in overgrazed vegetation explains why plant assemblages on overgrazed rangelands resemble those on more arid sites (Stokes, 1994).

Small-seeded plants, particularly those of annual grasses and succulents, are dispersed to and establish on smooth surfaces. Many Karoo shrubs have wind-dispersed seeds that roll across open ground until trapped by litter or shrubs (Esler and Cowling, submitted) so that undisturbed vegetation normally consists of a mosaic of mixed-species shrub clumps (Yeaton and Esler, 1990). Denuded rangelands lose this structure because they offer more establishment sites for annuals and succulents than for shrubs.

5. Restoration

Attempts to restore productivity to rangelands by passive management (withdrawal of livestock) and active intervention (browsing, burning, clearing, reseeding, and cultivating) have generally been more successful in mesic savanna than other vegetation types.

5.1. LIVESTOCK WITHDRAWAL

In savanna bushes encroach rapidly, but because of the long lifespans of woody plants, this encroachment is slow to reverse itself through self-thinning (Scholes and Walker, 1993). Rates of restoration of palatable grasses to overgrazed grasslands are limited by seed availability and short seed-dispersal distances. Recovery times may exceed 70 years (O'Connor, 1991). Karroid shrubland in poor condition appears to be in a fairly stable state under continued grazing and did not further deteriorate during 20 years of monitoring (Stokes, 1994). No measurable improvement in species composition could be detected 15 and 50 years after livestock removal although the cover of unpalatable shrubs increased (Fairall and le Roux, 1991; Stokes, 1994).

5.2. BROWSING, BURNING, AND BUSH CLEARING

Running a combination of cattle and goats is more productive than running cattle alone in savanna rangeland with a high density of woody plants (Stuart-Hill, 1987). Moreover, continuous browsing, combined with burning, prevents further increases in tree density (du Toit, 1972a, b). Grass biomass increases following mechanical and chemical eradication of woody plants, but Scholes and Walker (1993) point out that this increase is often short lived. On nutrient-poor soils, grass cover decreases again within a few years after the nutrients associated with islands of fertility around dead trees have been depleted. Although a 75% reduction in *Rhigozum trichotomum* density on sandy soils in the arid northern Cape doubled grass production, relatively unpalatable annuals accounted for >95% of the grass biomass because perennial grass species were absent from the seed bank (Moore, 1989). Clearing of Karoo shrubland resulted in abundant seedling recruitment, but the seedling composition was once again determined by localized seed availability (Milton, 1995).

5.3. Reseeding

Large-scale reseeding trials have rarely been well monitored. In some cases, competition from established plants precluded establishment of new seedlings, whereas in others, changes in soil and microclimate appeared to limit recruitment. Without intervention, less palatable grasses can remain dominant in grassland for >20 years. Addition of seeds of perennials results in limited improvement, but rapid restoration usually requires plowing, fertilizing, and resowing (Edwards, 1981). Reseeding into undisturbed Karoo vegetation results in little or no recruitment of shrubs. When existing vegetation was cleared or where the soil was cultivated to a depth of 30 mm, seedling survival improved (Joubert and van Breda, 1976; Milton, 1994). Mud-pelleting of Karoo grass and shrub seed reportedly improved its germination in the field relative to uncoated seeds (Van Breda, 1939). Ripping, bushcutting, and reseeding on Karoo ranches with carrying capacities <10 LSU/km^2 is only economically feasible at the patch scale (Roberts, 1981).

5.4. Cultivating

Large areas (1–15 km^2) of denuded, saline, and impenetrable soil surfaces develop in bottomlands where overgrazing has accelerated drainage (Walters, 1951). Revegetation techniques for such areas include building low barrages to spread water and reduce the rate of flow, using mechanical cultivation, furrowing or ponding the topsoil to trap seed and water (Coetzee, 1992), and applying gypsum and organic mulches (Walters, 1951). Some landowners then sow locally collected seed, forage grasses, or Australian salt bush (*Atriplex* spp.) in the treated area, whereas others rely on dispersal of seed from surrounding rangelands to initiate the revegetation process. These costly measures, supported by government subsidies ($1000/$km^2$), are reported to double (2 to 4 LSU/km^2) carrying capacity for livestock (Coetzee, 1992) and have been applied to vast areas of the central Karoo.

6. Motivation for Overexploitation

6.1. Optimism

European colonists understandably had little understanding of the aridity and variability of the South African climate or of the carrying capacity of semidesert rangelands for livestock. An optimistic overestimation of carrying capacity continued into the 20th century. Department of Agriculture estimates of carrying capacity for small stock (sheep and goats) in the Karoo region are growing more conservative as the knowledge base improves (Figure 11). The difference between the

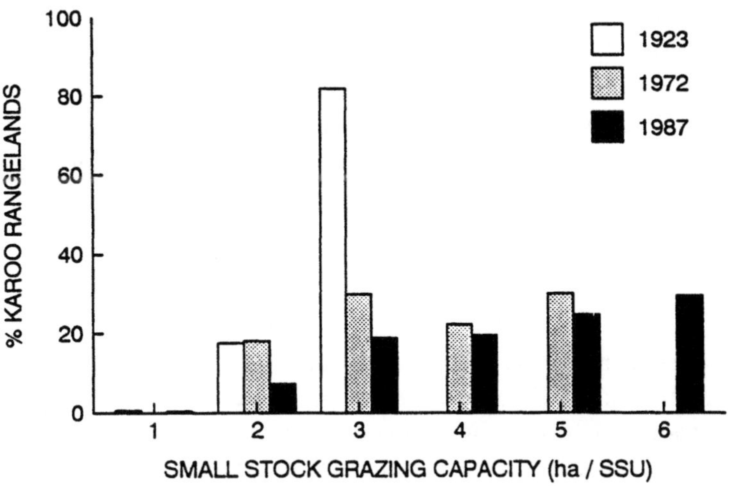

Fig. 11. Grazing capacity estimates for small stock in the Karoo. (Data from Scotney, 1988.)

1923 and 1972 estimates probably shows rangeland deterioration (Scotney, 1988), but readjustments made in 1987 reflect changes in perception and assessment techniques.

Actual stocking rates generally exceed recommended capacities by 5–20% (King and Bembridge, 1988; Dean and Macdonald, 1994). Ranchers generally consider their rangelands to be in better condition (96% of ranches rated fair-good) than do extension officers who placed only 61% of these ranches into this category (Düval and Scholtz, 1992). Few of the ranchers interviewed in this study thought that stock reduction and prolonged resting of camps were economically feasible.

6.2. Markets

The types of livestock kept on southern African rangelands are dictated by market forces as well as by environmental conditions. Variations in the relative abundance of draught animals, ostriches, mohair goats, and wool and pelt sheep reflect global changes in lifestyle and fashion (Scotney, 1988). A lucrative market for wool led to a wool boom (and subsequent rangeland degradation) in the 1930s. High prices now paid for ostrich skins on the world market tempt ranchers to stock these animals on rangeland at 200 times their natural density (Milton *et al.*, 1994), thereby trading sustainability for short-term profit.

6.3. Buffering

Although many South African ranches are too small to sustain ranching and are unprofitable (Danckwerts and Marais, 1989), their existence has been perpetuated by past State buffering in the form of interest-free loans for land acquisition and subsidies for supplementary feeds, boreholes, fencing, labor, range rehabilitation, and stock reduction. Drought assistance provided by the State has continually maintained livestock on land from which game and nomadic herders would have emigrated. Subsidies and loans that facilitate ranching in marginal areas have contributed to desertification and national debt. The debt burden of South African farmers to the State and to commercial banks and suppliers doubled between 1980 and 1987 (Directorate Agricultural Information, 1991).

6.4. Land Values and Alternatives

Inflated land prices encourage the ranching of subeconomic units in marginal areas. Danckwerts and King (1984) showed that on average the annual increase in the value of property in the Smaldeel area of the eastern Karoo was three times the property's productive value. Government planners have recently suggested that a land tax be introduced to compensate for environmental damage by ranching. Such a tax would force marginal ranchers off the land but may also encourage overstocking.

Ranching supplies only 2% of South Africa's gross domestic product, employs only 1 million of 8 million economically active people, is reducing rangeland diversity and productivity, and appears to have reached carrying capacity. Alternatives to ranching must be found to support the rapidly growing population. Ecotourism, hunting, and game plus stock mixes may provide more jobs and income. The other option is for South Africa to move from a natural-resource-based economy to an urban-centered manufacturing economy (Archer, 1994).

7. Conclusions

From the evidence presented in this review we conclude that rangelands on many South African ranches have been changed to a less productive state. Although productivity and vegetation composition continue to fluctuate with rainfall, the transformed state is stable in that livestock production and vegetation composition do not appear to be deteriorating further. Rangelands are also stable in the sense that improved production cannot be achieved by passive management. Research on sustainable use of rangeland, rangeland rehabilitation, and alternatives to ranching are current research priorities in South Africa.

Acknowledgements

This paper is a contribution to the Desertification Programme of the FitzPatrick Institute, University of Cape Town and was funded by the Foundation for Research Development, the Department of Environmental Affairs, and the Southern African Nature Foundation. The Bureau of Land Management, U.S. Department of the Interior sponsored our participation in the International Symposium and Workshop on Desertification in Developed Countries. We are grateful to Beaumont McClure and Walter Whitford, who facilitated our attendance at the symposium.

References

Acocks, J.P.H. 1953. Veld types of South Africa. *Memoirs of the Botanical Survey of South Africa* **28**: 1–192.

Archer, S.F. 1994. Economic perspectives on Karoo economy. *Arid Zone Ecology Forum abstracts*, Foundation for Research Development, Pretoria, pp. 39–40.

Barnes, D.L. 1990. A survey of grazed and ungrazed grassland in the south-eastern Transvaal highveld. I. Palatability composition and grazing capacity. *Journal of the Grassland Society of Southern Africa* **7**(4): 217–222.

Beukman, R.P. 1991. The role of nurse plants in the vegetation dynamics of the succulent Karoo. MS thesis, University of Cape Town, South Africa.

Bond, W.J., W.D. Stock and M.T. Hoffman. 1994. Has the Karoo spread? A test for desertification using carbon isotopes from soils. *South African Journal of Science* **90**(7): 391–397.

Boonzair, E.A., M.T. Hoffman, F.M. Archer and A.B. Smith. 1990. Communal land use and the tragedy of the commons: Some problems and development perspectives with specific reference to semi-arid regions of southern Africa. *Journal of the Grassland Society of Southern Africa* **7**(2): 77–80.

Bosch, O.J.H. and H.G. Gauch. 1991. The use of degradation gradients for the assessment and ecological interpretation of range condition. *Journal of the Grassland Society of Southern Africa* **8**(4): 138–146.

Christopher, A.J. 1982. Towards a definition of the nineteenth century South African frontier. *South African Geographical Journal* **64**(2): 97–113.

Coetzee, J. 1992. Norme vir herwinning van kaal kolle. *Landbouweekblad* **754**: 48–49.

Danckwerts, J.E. and King, P.G. 1984. Conservative stocking or maximum profit: A grazing management dilemma. *Journal of the Grassland Society of Southern Africa* **1**(4): 25–28.

Danckwerts, J.E. and J.B. Marais. 1989. An evaluation of the economic viability of commercial pastoralism in the Smaldeel area of the eastern Cape. *Journal of the Grassland Society of Southern Africa* **6**(1): 1–7.

Danckwerts, J.E. and G. C. Stuart-Hill. 1988. The effect of severe drought and management after drought on mortality and recovery of semi-arid grassland. *Journal of the Grassland Society of Southern Africa* **5**(4): 218–222.

Dean, W.R.J., M.T. Hoffman, M.E., Meadows and S.J. Milton. 1995. Desertification in the semi-arid Karoo, South Africa: Review and reassessment. *Journal of Arid Environments* **30**: 247–264.

Dean, W.R.J. and I.A.W. Macdonald. 1994. Historical changes in stocking rates of domestic livestock as a measure of semi-arid and arid rangeland degradation in the Cape Province, South Africa. *Journal of Arid Environments* **26**(3): 281–198.

Dean, W.R.J., S.J. Milton and M.A. du Plessis. This volume. Where, why and to what extent have rangelands in the Karoo, South Africa, desertified. *Environmental Monitoring and Assessment*.

Directorate of Agricultural Information, 1991. *Abstract of agricultural statistics*, Directorate of Agricultural Information, Pretoria, pp. 1–110.

Downing, B.H. 1978. Environmental consequences of agricultural expansion in South Africa since 1850. *South African Journal of Science* **74**: 420–422.

Du Toit, P.F. 1972a. *Acacia karroo* intrusion: The effect of burning and sparing. *Proceedings of the Grassland Society of Southern Africa* **7**: 23–27.

Du Toit, P.F. 1972b. The goat in a bush-grass community. *Proceedings of the Grassland Society of Southern Africa* **7**: 44–50.

Düval, G.H. and H.P.J. Scholtz. 1992. The incompatibility of controlled selective grazing systems with farmers' needs. *Journal of the Grassland Society of Southern Africa* **9**(1): 24–29.

Edwards, P.J. 1981. Sour grassveld, in N.M. Tainton (Ed.), *Veld and pasture management in South Africa*, Shuter and Shooter, Pietermaritzburg, pp. 395–400.

Esler, K.J. and R.M. Cowling. Submitted. Patterns of seedling establishment in semi-arid, succulent Karoo vegetation: The role of facilitation and dispersal. *Journal of Vegetation Science*.

Esler, K.J., R.M. Cowling and J.M. Juritz. Submitted. Dynamics of soil-stored seed banks in semi-arid, succulent Karoo vegetation. *Journal of Vegetation Science*.

Fairall, N. and A. le Roux. 1991. Game management in arid areas: The non-equilibrium alternative, in L.A. Renecker and R.J. Hudson (Eds.), *Wildlife production: Conservation and sustainable development*, University of Alaska, Fairbanks, pp. 251–254.

Fouché, H.J., J.M. de Jager and D.P.J. Opperman. 1985. A mathematical model for assessing the influence of stocking rate on the incidence of droughts and for estimating the optimal stocking rates. *Journal of the Grassland Society of Southern Africa* **2**(3): 4–6.

Friedel, M.H. 1987. A preliminary investigation of woody plant increase in the western Transvaal and implications for veld assessment. *Journal of the Grassland Society of Southern Africa* **4**(1): 25–30.

Fuggle, R.F. 1983. Population growth and resource demands, in R.F. Fuggle and M.A. Rabie (Eds.), *Environmental concerns in South Africa*, Juta, Cape Town, pp. 23–29.

Fuls, E.R. 1992. Ecosystem modification created by patch-overgrazing in semi-arid grassland. *Journal of Arid Environments* **23**: 59–69.

Hatch, G.P. and D. Grossman. 1993. The role of the grassland scientist in agriculture in developing areas, in G.P. Hatch and P.J.K Zacharias (Eds.), *Proceedings of the First Developing Areas Symposium*, Grassland Society of Southern Africa, Scottsville, pp. 28–30.

Hatch, G.P. and N.M. Tainton. 1990. A preliminary investigation of area-selective grazing in the southern tall grassveld of Natal. *Journal of the Grassland Society of Southern Africa* **7**(4): 238–242.

Henrici, M. 1940. Fodder plants of the broken veld, part 2, Fauresmith District. *Department of Agriculture Science Bulletin* **213**: 1–42. Government Printer, Pretoria.

Hoffman, M.T., G.D. Barr and R.M. Cowling. 1990. Vegetation dynamics in the semi-arid Karoo, South Africa: The effect of seasonal rainfall and competition on grass and shrub basal cover. *South African Journal of Science* **86**: 462–463.

Hoffman, M.T., W.J. Bond and W.D. Stock. This volume. Desertification of the eastern Karoo, South Africa: Conflicting paleontological, historical and soil isotopic evidence. *Environmental Monitoring and Assessment*.

Hoffman, M.T. and R.M. Cowling. 1990a. Desertification in the lower Sundays River Valley, South Africa. *Journal of Arid Environments* **19**: 105–117.

Hoffman, M.T. and R.M. Cowling. 1990b. Vegetation change in the semi-arid eastern Karoo over the last 200 years: An expanding Karoo – fact or fiction? *South African Journal of Science* **86**: 286–294.

Hoffman, M.T. and R.M. Cowling. 1991. Phytochorology and endemism along aridity and grazing gradients in the lower Sundays River Valley, South Africa: Implications for vegetation history. *Journal of Biogeography* **18**: 189–201.

Joubert, J.G.V. and P.A.B. van Breda. 1976. Methods of establishing *Osteospermum sinuatum* in Little Karoo veld. *Proceedings of the Grassland Society of Southern Africa* **11**: 123–124.

Kellner, K. and O.J.H. Bosch. 1992. Influence of patch formation in determining the stocking rate for southern African grasslands. *Journal of Arid Environments* **22**: 99–105.

King, P.G. and T.J. Bembridge. 1988. An extension approach for the National Grazing Strategy based on farmer characteristics in the eastern Cape. *Journal of the Grassland Society of Southern Africa* **5**(1): 8–14.

Kokot, D.F. 1948. An investigation into the evidence bearing on recent climatic changes over southern Africa. In *Irrigation Department Memoir*, Government Printer, Pretoria, p. 160.

Macdonald, I.A.W. 1978. Pattern and process in a semi-arid grassland in Rhodesia. *Proceedings of the Grassland Society of Southern Africa* **13**: 103–109.

Macdonald, I.A.W. 1989. Man's role in changing the face of southern Africa, in B.J. Huntley (Ed.), *Biotic diversity in southern Africa*, Oxford University Press, Cape Town, pp. 51–77.

Martens, E. 1971. The effect of tribal grazing patterns on the habitat in the Kalahari. *Botswana Notes and Records* (special edition) **1**: 234–241. The Botswana Society, Gaborone.

Milton, S.J. 1990. Life styles of plants in four habitats in an arid Karoo shrubland. *South African Journal of Ecology* **1**: 63–72.

Milton, S.J. 1994. Small-scale reseeding trials in arid rangeland: Effects of rainfall, clearing and grazing on seedling survival. *African Journal of Range and Forage Science* **11**(2): 54–58.

Milton, S.J. 1995. Spatial and temporal patterns in the emergence and survival of seedlings in arid Karoo shrubland. *Journal of Applied Ecology* **32**: 145–156.

Milton, S.J. and W.R.J. Dean. 1990. Seed production in rangelands of the southern Karoo. *South African Journal of Science* **86**: 231–233.

Milton, S.J. and W.R.J. Dean. 1992. An underground index of rangeland degradation: Cicadas in arid Karoo shrublands, southern Africa. *Oecologia* **91**(2): 288–291.

Milton, S.J. and W.R.J. Dean. 1995. Effects of the 1990/91 drought on rangeland in the Steytlerville Karoo. *South African Journal of Science* **91**: 78–94.

Milton, S.J., W.R.J. Dean and C.P. Marincowitz. 1992. Preferential utilization of pans by springbok (*Antidorcas marsupialis*). *Journal of the Grassland Society of Southern Africa* **9**(3): 114–118.

Milton, S.J., W.R.J. Dean and W.R. Siegfried. 1994. Food selection by ostrich in southern Africa. *Journal of Wildlife Management* **58**(2): 234–248.

Milton, S.J. and M.T. Hoffman. 1994. The application of state-and-transition models to rangeland research and management in arid succulent and semi-arid grassy Karoo, South Africa. *Journal of the Grassland Society of Southern Africa* **11**(1): 18–26.

Moore, A. 1989. Die ekologie en ekofisiologie van *Rhigozum trichotomum*. PhD thesis, University of Port Elizabeth, South Africa, p. 163.

Moore, A. and A. Odendaal. 1987. Die ekonomiese implikasies van bosverdigtiging en bosbeheer soos van toepassing op 'n speenkalfproduksiestelsel in die doringbosveld van die Molopo-gebied. *Journal of the Grassland Society of Southern Africa* **4**(4): 139–142.

Novellie, P. 1990. Habitat use by indigenous grazing ungulates in relation to sward structure and veld condition. *Journal of the Grassland Society of Southern Africa* **7**(1): 16–23.

O'Connor, T.G.O. 1985. A synthesis of field experiments concerning the grass layer in the savanna regions of southern Africa. *South African National Scientific Programmes Report* **114**: 1–119.

O'Connor, T.G.O. 1991. Local extinction in perennial grasslands: A life-history approach. *The American Naturalist* **137**(6): 753–773.

O'Connor, T.G.O. and G.A. Pickett. 1992. The influence of grazing on seed production and seed banks of some African grasslands. *Journal of Applied Ecology* **29**: 247–260.

Palmer, A.R., C.G. Hobson and M.T. Hoffman. 1990. Vegetation change in a semi-arid dwarf shrubland in the eastern Cape, South Africa. *South African Journal of Science* **86**: 392–395.

Penn, N.G. 1986. Pastoralists and pastoralism in the northern Cape frontier zone during the eighteenth century, in M. Hall and A.B. Smith (Eds.), *Prehistoric pastoralism in southern Africa*, The South African Archaeological Society Goodwin Series **5**: 62–68.

Perkins, J.S. and D.S.G. Thomas. 1993a. Environmental responses and sensitivity to permanent cattle ranching, semi-arid western central Botswana, in D.S.G. Thomas and R.J. Allison. (Eds.), *Landscape sensitivity*, Wiley and Sons, New York, pp. 273–286.

Perkins, J.S. and D.S.G. Thomas. 1993b. Spreading deserts or spatially confined environmental impacts? Land degradation and cattle ranching in the Kalahari Desert of Botswana. *Land Degradation and Rehabilitation* **4**: 179–194.

Roberts, B.R. 1981. Karoo, in N.M. Tainton (Ed.), *Veld and pasture management in South Africa*, Shuter and Shooter, Pietermaritzburg, pp. 415–425.

Roux, P.W. 1966. Die uitwerking van seisonale reenval en beweiding op gemengde karooveld. *Proceedings of the Grassland Society of Southern Africa* **1**: 103–110.

Roux, P.W. and D.P.J. Opperman. 1986. Soil erosion, in R.M. Cowling, P.W. Roux and A.J.H. Pieterse (Eds.), *The Karoo biome, a preliminary synthesis*, part 1 – Physical environment, *South African National Scientific Programmes Report* **124**: 92–111, Foundation for Research Development, Pretoria.

Roux, P.W. and G.K. Theron. 1987. Vegetation change in the Karoo biome, in R.M. Cowling and P.W. Roux (Eds.), *The Karoo biome, a preliminary synthesis*, part 2 – Vegetation and history, *South African National Scientific Programmes Report* **142**: 50–69, Foundation for Research Development, Pretoria.

Scholes, R.J. and B.H. Walker. 1993. *An African savanna – Synthesis of the Nylsvley study*, Cambridge University Press, Cambridge, pp. 213–229.

Scoones, I. 1992. Land degradation and livestock production in Zimbabwe's communal areas. *Land Degradation and Rehabilitation* **3**: 99–113.

Scotney, D.M. 1988. The agricultural areas of southern Africa, in I.A.W. Macdonald and R.J.M. Crawford (Eds.), *Long-term data series relating to southern Africa's renewable natural resources, South African National Scientific Programmes Report* **157**: 316–333, Foundation for Research Development, Pretoria.

Skarpe, C. 1990. Structure of the woody vegetation in disturbed and undisturbed arid savanna, Botswana. *Vegetatio* **87**: 11–18.

Skead, C.J. 1980. *Historical mammal incidence in the Cape Province*, vol. 1. Cape Nature Conservation, Cape Town, pp. 494–597.

Smith, A.B. 1986. Competition, conflict and clientship: Khoi and San relationships in the western Cape, in M. Hall and A.B. Smith (Eds.), *Prehistoric pastoralism in southern Africa*, The South African Archaeological Society Goodwin Series **5**: 35–41.

Snyman, H.A. and H.J. Fouché. 1991. Production and water use efficiency of semi-arid grasslands of South Africa as affected by veld condition and rainfall. *Water SA* **17**(4): 263–268.

Stokes, C.J. 1994. Degradation and dynamics of succulent Karoo vegetation. MS thesis, University of Natal, Pietermartizburg, South Africa.

Stuart-Hill, G.C. 1987. Refinement of a model describing forage production, animal production and profitability as a function of bush density in the false thornveld of the eastern Cape. *Journal of the Grassland Society of Southern Africa* **4**(1): 18–24.

Stuart-Hill, G.C. 1992. Effects of elephants and goats on the Kaffrarian succulent thicket of the eastern Cape, South Africa. *Journal of Applied Ecology* **29**: 699–710.

Talbot, W.J. 1961. Land utilization in the arid regions of southern Africa, part 1: South Africa, in L.D. Stamp (Ed.), *A history of land use in arid regions, arid zone research 17*, Unesco, Paris, pp. 299–338.

Teague, W.R. and G.N. Smit. 1992. Relations between woody and herbaceous components and the effects of bush clearing in southern African savanna. *Journal of the Grassland Society of Southern Africa* **9**(2): 60–71.

Union of South Africa. 1914. *Droughts, rainfall and soil erosion* (Senate Select Committee Report No. 2). Government Printer, Pretoria, pp. 55.

Union of South Africa. 1923. *Final report of the Drought Investigation Commission* (UG49). Government Printer, Cape Town, pp. 210.

Union of South Africa. 1951. *Report of the Desert Encroachment Committee* (UG59). Government Printer, Pretoria, pp. 27.

Van Breda, N.G. 1939. An improved method of sowing grass and Karoo shrub seed. *South African Journal of Science* **36**: 328–335.

Van der Heyden, F. 1992. Effects of defoliation on regrowth and carbon budgets of three semi-arid Karoo shrubs. PhD thesis, University of Cape Town, South Africa.

Walter, H. 1954. *Grundlagen der weidewirtschaft in Südwesafrika*, Eugen Ulmer, Stuttgart, pp. 99–107.

Walters, M.M. 1951. Kaal kolle in die oostelike gemengde Karoo. *Boerdery in Suid-Afrika* **26**: 125–132.

Wentzel, H.E., M.W. van Rooyen, G.K. Theron and A.J. de Villiers. 1994. *Mesembryanthemum guerichianum*: Dominance in old kraals. *Arid Zone Ecology Forum abstracts*. Foundation for Research Development, Pretoria, p. 45.

Yeaton, R and K.J. Esler. 1990. The dynamics of a succulent Karoo vegetation: A study of species association and recruitment. *Vegetatio* **88**: 103–113.

IMPACT OF DESERTIFICATION ON TEMPERATURE TRENDS IN THE MIDDLE EAST

HASSAN A. NASRALLAH* and ROBERT C. BALLING, JR.**
* *College of Health Sciences, Department of Environmental Sciences, Public Authority for Applied Education and Training, Kuwait City, Kuwait;* ** *Office of Climatology, Arizona State University, Tempe, Arizona, USA*

Abstract. The intense interest in desertification and climate change has stimulated detailed studies of temperature records in many areas of the world. In this investigation, the temperature records from the Middle East region are analyzed over the period 1950–1990. Results reveal a linear, statistically significant temperature increase of 0.07 °C/decade over the 41-year period. An analysis of spatial controls on these temperature changes reveals a warming effect associated with both overgrazing and the degree of human-induced desertification. The results of this study are consistent with theoretical and empirical studies predicting and demonstrating a warming signal associated with these land surface changes in the world's dryland areas.

1. Introduction

Within the past few years intense popular and scientific interest in desertification and climate change has stimulated detailed studies of temperature records in many areas of the world. While many scientists are focusing their investigations on temperature responses to the continued buildup of greenhouse gases (Houghton *et al.*, 1990, 1992), others have attempted to link observed warming signals to land surface processes and in particular to desertification and/or overgrazing (e.g. Balling, 1991; Nasrallah and Balling, 1994). This investigation examined temperature records from the Middle East and measured potential linkages to desertification and overgrazing.

2. Background

For several decades the climatic impact of desertification processes has been a focus of numerical modeling experiments and empirical research. A landmark article by Charney (1975) reported a numerical modeling study of a biogeophysical feedback that linked desertification to surface cooling and a reduction in local precipitation. But once surface hydrological considerations were included in the model to perturb the surface energy balance, the direct link of desertification to surface cooling became less clear. In fact, desertification was linked to surface warming in many modeling experiments (e.g. Charney *et al.*, 1975; Druyan, 1989; Novak, 1990; Franchito and Rao, 1992; Sud *et al.*, 1993; Bounova and Krishnamurti, 1993a, b; Yang *et al.*, 1994).

Environmental Monitoring and Assessment **37:** 265–271, 1995.

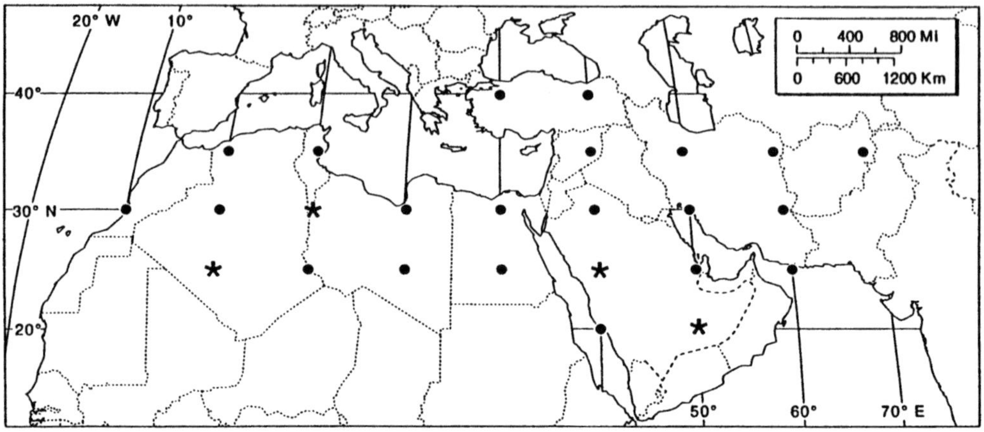

Fig. 1. Map of Middle Eastern temperature grid points. Grid points labeled "*" have inadequate temperature data and are not used in this study.

Recognizing the results from these modeling studies, many scientists have conducted field experiments to measure the actual temperature response to desertification. For example, Bryant *et al.* (1990) examined a large-scale vegetation discontinuity along the United States-Mexico border and found that the overgrazed Mexican landscape was nearly 4 °C warmer than adjacent lands in the United States during summer afternoon periods. Similarly, many other scientists (e.g. Jackson and Idso, 1975; Ripley, 1976a, b; Idso, 1977; Idso and Deardorff, 1978; Wendler and Eaton, 1983) have reported measurements linking desertification to surface warming. These researchers argued that removal of vegetation in drylands lowers soil moisture levels, reduces radiant energy used to evaporate and transpire water, and ultimately increases surface and near-surface temperatures.

Attempting to move from local and regional scales to even larger geographical scales, Balling (1991) showed that many areas of the world that have undergone desertification have shown a differential warming of approximately 0.5 °C per century when compared to adjacent areas that have not been affected by desertification. Similarly, Nasrallah and Balling (1994) demonstrated that in land areas of the Northern Hemisphere between 10° N and 50° N, overgrazed areas were warming significantly faster than non-overgrazed lands. The analysis in this study of the Middle East should provide another empirical test on how desertification affects temperature trends in this dryland region.

3. Data Requirements

The Middle East is defined in this study as an area from Morocco on the west to Afghanistan on the east and from Turkey on the north to the Arabian Sea on the

TABLE I
Selected data for grid points used in the analyses

Latitude (° N)	Longitude (° E)	Country	Temperature Change (°C/decade)	Desertification Class	Overgrazing Class
40	30	Turkey	–0.06	1	No
40	40	Turkey	–0.08	0	Yes
35	0	Algeria	–0.04	3	No
35	10	Tunisia	0.12	3	Yes
35	40	Syria	–0.02	3	Yes
35	50	Iran	0.21	3	Yes
35	60	Iran	0.19	3	Yes
35	70	Afghanistan	0.07	0	Yes
30	–10	Morocco	0.03	0	No
30	0	Algeria	0.10	1	No
30	20	Libya	0.05	2	Yes
30	30	Egypt	0.03	1	No
30	40	Saudi Arabia	–0.09	1	Yes
30	50	Iran	0.09	3	No
30	60	Iran	0.18	3	Yes
25	10	Algeria	0.18	2	No
25	20	Libya	0.01	1	No
25	30	Egypt	–0.08	1	No
25	50	Saudi Arabia	0.07	2	Yes
25	60	Iran	0.23	3	Yes
20	40	Saudi Arabia	0.02	2	Yes

south (Figure 1). To represent conditions in this region, three basic data sets are used:

(1) Temperature data were taken from an updated version of the gridded (5° latitude × 10° longitude) data set prepared by Jones *et al.* (1986). Although 25 of these grid points are within the Middle East, 4 had a large amount of missing temperature data and were eliminated from further consideration. The remaining 21 grid points have temperature records that are more than 90 percent complete during the 1950–1990 study period (Table I).

(2) A map prepared for the United Nations by Dregne (1977) was used to estimate the severity of human-induced desertification at each of the 21 Middle East grid points. Only 3 of the 21 grid points are in the "no" desertification class (0), 6 are in the "slight" desertification class (1), 4 are in the "moderate" class (2), and 8 are in the "severe" class (3) (Table I). Despite its shortcomings, we assume

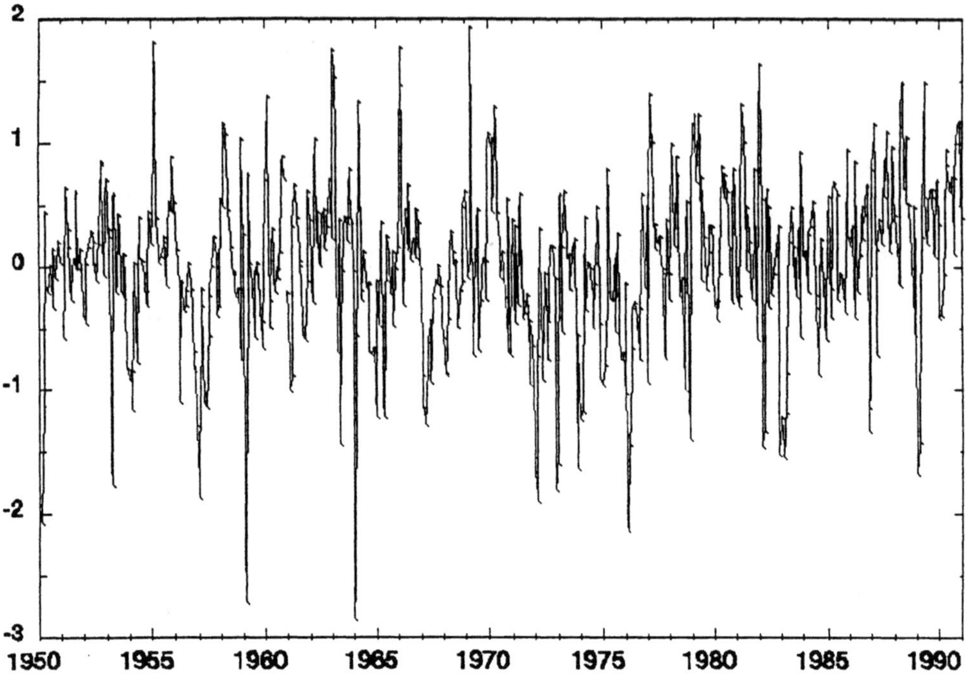

Fig. 2. Middle Eastern monthly temperature anomalies (1950–1990).

that the Dregne map provides a crude surrogate for trends in desertification that may have occurred during our 1950–1990 study period.

(3) The United Nations Environment Programme's (UNEP, 1992, p. 24) *World Atlas of Desertification* contains a map of "Areas Affected by Overgrazing." According to the UNEP description of the map, "overgrazing" is meant to include both the actual removal of vegetation and other effects of livestock such as trampling and resultant compaction of soils. Overgrazed areas, particularly those in arid and semiarid lands, typically have a decrease in vegetation cover, increased erosion, soil degradation, and encroachment of shrubs into grazing lands. As seen in Table I, 9 of the 21 Middle Eastern grid points were not classed as overgrazed, whereas UNEP classed the remaining 12 as overgrazed areas.

4. Results

A plot of the areally averaged mean monthly temperature anomalies (based on a 1950 to 1979 base period) for the Middle East region for the period 1950–1990 is presented in Figure 2. The plot shows high intermonthly variance with an underlying, statistically significant (α = 0.05) linear trend of +0.07 °C ($\pm$

0.05 °C)/decade. Although the plot reveals some cooling from about 1960 to the mid-1970s and warming from the mid-1970s to the present, second- and third-order polynomial curves did not significantly increase the variance explained by the linear fit. On a seasonal basis, winter months (DJF) warmed 0.01 °C/decade, spring months (MAM) warmed 0.13 °C/decade, summer months (JJA) warmed 0.08 °C/decade, and autumn months (SON) warmed 0.05 °C/decade. Summer and spring were warming at a statistically significant rate, whereas the warming rates in autumn and fall were not statistically significant.

The goal of this investigation was not only to examine the temporal trend in temperatures in the Middle East region but also to analyze the impact of desertification on any changes in temperature. To perform such an analysis, a new matrix was constructed. The matrix contained 21 rows, one for each grid point, and seven columns including latitude, latitude squared, longitude, longitude squared, the product of latitude and longitude, the Dregne desertification category, and the calculated change in temperature (Table I). A stepwise multiple regression analysis was conducted on an annual and seasonal basis with linear temperature change as the dependent variable. Independent variables could be added to the regression equations only if (a) the variable significantly improved the predictability of the dependent variable, (b) the variable was not significantly related to other predictors already in the equation, and (c) the variable's regression coefficient significantly differed from zero.

The resulting equation for the annual data took the following form:

$$\Delta T/\Delta \text{decade (Annual)} = -0.03 + 0.05\,(\pm\,0.03) \times \text{Desert} \tag{1}$$

where "Desert" represents the Dregne desertification class (0–3). The R value equals 0.55, implying that 30 percent of the spatial variance in the temperature trends can be explained by the desertification variable. The Spearman rank order correlation coefficient, which does not assume normality in the independent variable, equals 0.56. The Durbin-Watson statistic for the equation equals 1.98, showing no significant autocorrelation in the residual series. Interpretation of the resulting regression equation is relatively straightforward; the results show that the degree of human-induced desertification has a positive, statistically significant impact on the temperature trends in the Middle East. Very interestingly, the intercept suggests that in the absence of desertification (Desert = 0), the region would have cooled by 0.03 °C/decade over the past 41 years instead of the warming that has occurred.

On a seasonal basis, no statistically significant linkage could be determined between temperature trends and the independent variables in the winter and spring seasons. But for the summer months a regression equation was determined as follows:

$$\Delta T/\Delta \text{decade (JJA)} = -0.07 + 0.09\,(\pm\,0.08) \times \text{Desert} \tag{2}$$

where the R value equals 0.46 ($R^2 = 0.21$), and the Durbin-Watson statistic equals 2.48, thereby indicating no autocorrelation in the residuals. The Spearman rank order correlation coefficient is 0.45. In the autumn season, a regression equation was determined as follows:

$$\Delta T/\Delta \text{decade (SON)} = -0.09 + 0.08\ (\pm\ 0.06) \times \text{Desert} \quad (3)$$

where the R value equals 0.51 ($R^2 = 0.26$), and the Durbin-Watson statistic equals 1.98, again indicating no autocorrelation in the residuals. The Spearman rank order correlation coefficient for this final equation is 0.53. These seasonal analyses reinforce the results obtained for the annual data. In each case, statistically significant regression coefficients continue to show the warming effect of desertification in the region. Additionally, in the absence of desertification (Desert = 0), the summer and fall temperatures would have declined over the 41-year study period instead of producing the small amount of warming.

Finally, we stratified temperature trends according to the binary overgrazing categorization. On an annual basis, the mean temperature trend value for the overgrazed areas was 0.08 °C/decade compared to 0.03 °C/decade for the non-overgrazed grid points. Despite a nearly threefold difference in the temperature trend values, the difference between the means was not statistically significant ($t = 1.09$). And although the seasonal comparisons all showed differential warming associated with UNEP's (1992) overgrazing classification, none of the differences were statistically significant.

5. Conclusions

Our results show a linear warming of near 0.07 °C/decade across the Middle East for the period 1950–1990. The greatest warming is occurring in the spring, and the least warming is occurring in the winter. Analysis of spatial patterns reveals a statistically significant and positive relationship between temperature trends and the severity of human-induced desertification. The annual, summer, and fall regression equations show that in the absence of desertification, the region would have cooled over the study period. In addition, when the temperature trends are stratified by a binary overgrazing classification, Middle Eastern grid points affected by overgrazing are found to be warming faster than the grid points not affected by overgrazing, but the differences are not statistically significant.

The results from the drylands of the Middle East provide even more evidence of a local and regional warming signal associated with desertification and overgrazing. This signal must be considered in any attempt to link regional temperature trends to the atmospheric buildup of anthropo-generated greenhouse gases.

Acknowledgements

Funding for this research was provided by the Kuwait Foundation for the Advancement of Sciences, State of Kuwait (Project Number 92–05–15).

References

Balling, R.C., Jr. 1991. Impact of desertification on regional and global warming. *Bulletin of the American Meteorological Society* **72**: 232–234.

Bounova, L. and T.N. Krishnamurti. 1993a. Influence of soil moisture on the Sahelian climate prediction I. *Meteorology and Atmospheric Physics* **52**: 183–203.

Bounova, L. and T.N. Krishnamurti. 1993b. Influence of soil moisture on the Sahelian climate prediction II. *Meteorology and Atmospheric Physics* **52**: 205–224.

Bryant, N.A., L.F. Johnson, A.J. Brazel, R.C. Balling, Jr., C.F. Hutchinson and L.R. Beck. 1990. Measuring the effect of overgrazing in the Sonoran Desert. *Climatic Change* **17**: 243–264.

Charney, J.G. 1975. Dynamics of deserts and drought in the Sahel. *Quarterly Journal of the Royal Meteorological Society* **101**: 193–202.

Charney, J.G., P.H. Stone and W.J. Quirk. 1975. Drought in the Sahara: A biogeophysical feedback mechanism. *Science* **187**: 434–435.

Dregne, H.E. 1977. *Generalized map of the status of desertification of arid lands.* Prepared by FAO of the United Nations, UNESCO and WMO for the 1977 United Nations Conference on Desertification.

Druyan, L.M. 1989. Advances in the study of sub-Saharan drought. *International Journal of Climatology* **9**: 77–90.

Franchito, S.H. and V.B. Rao. 1992. Climatic change due to land surface alterations. *Climatic Change* **22**: 1–34.

Houghton, J.T., B.A. Callander and S.K. Varney (editors). 1992. *Climate change 1992: The supplementary report to the IPCC scientific assessment.* Cambridge University Press, Cambridge.

Houghton, J.T., G.J. Jenkins and J.J. Ephraums (editors). 1990. *Climate change: The IPCC scientific assessment*, Cambridge University Press, Cambridge.

Idso, S.B. 1977. A note on some recently proposed mechanisms of genesis of deserts. *Quarterly Journal of the Royal Meteorological Society* **103**: 369–370.

Idso, S.B. and J.W. Deardorff. 1978. Comments on "The effect of variable surface albedo on atmospheric circulation in desert regions." *Journal of Applied Meteorology* **7**: 560.

Jackson, R.D. and S.B. Idso. 1975. Surface albedo and desertification. *Science* **189**: 1012–1013.

Jones, P.D., T.M.L. Wigley and P.B. Wright. 1986. Global temperature variations between 1861 and 1984. *Nature* **322**: 430–434.

Nasrallah, H.A. and R.C. Balling, Jr. 1993. Spatial and temporal analysis of Middle Eastern temperature changes. *Climatic Change* **25**: 153–161.

Nasrallah, H.A. and R.C. Balling, Jr. 1994. The effect of overgrazing on historical temperature trends. *Agricultural and Forest Meteorology* **71**: 425–430.

Novak, M.D. 1990. Micrometeorological changes associated with vegetation removal and influencing desert formation. *Theoretical and Applied Climatology* **42**: 19–25.

Ripley, E.A. 1976a. Comments on the paper 'Dynamics of deserts and droughts in the Sahel' by J.G. Charney. *Quarterly Journal of the Royal Meteorological Society* **102**: 466–467.

Ripley, E.A. 1976b. Drought in the Sahara: Insufficient biogeophysical feedback? *Science* **191**: 100.

Sud, Y.C., W.C. Chao and G.K. Walker. 1993. Dependence of rainfall on vegetation: Theoretical considerations, simulation experiments, observations and influences from simulated atmospheric soundings. *Journal of Arid Environments* **25**: 5–18.

United Nations Environment Programme (UNEP). 1992. *World atlas of desertification*, Edward Arnold, London.

Wendler, G. and F. Eaton. 1983. On the desertification of the Sahel zone. *Climatic Change* **5**: 365–380.

Yang, R., J. Shukla and P.J. Sellers. 1994. The influence of changes in vegetation type on the surface energy budget. *Advances in Atmospheric Sciences* **11**: 139–161.

SATELLITE MONITORING OF DESERT PLANT COMMUNITY RESPONSE TO MOISTURE AVAILABILITY

A. J. PETERS and M. D. EVE
Department of Geography, New Mexico State University, Las Cruces, NM 88003, USA

Abstract. Our study demonstrates the utility of coarse spatial-resolution satellite spectra for analysis of vegetation phenophases and response to moisture availability in an arid ecosystem. We show the feasibility of deriving information on vegetation parameters such as stress and growth patterns in arid regions through the use of satellite-derived vegetation indices, despite the usual problems associated with a high ratio of soil to vegetation cover. Vegetation in our study area consists of Chihuahuan Desert grassland and scrub, including extensive zones of mixed desert scrub and grassland. Historic vegetation change has been well documented and is exemplified by decreasing grass cover and increasing shrub cover, a general trend of desertification. Our analysis suggests that satellite-based inputs can be used to improve our understanding of the spatial dynamics of climatic impacts on natural vegetation and to help us distinguish these processes from human-caused desertification.

1. Introduction

In arid and semiarid lands drought and anthropogenic stressors are believed to cause serious range degradation (Buffington and Herbel, 1965; Dobyns, 1981). Even though there is disagreement over the causes of deterioration in condition of these lands, vegetation changes must be accurately monitored if management is to be successful. Dregne (1977) estimated that 906,000 km^2 (approximately 10%) of the United States land area has undergone severe or very severe desertification. Warren and Hutchinson (1984) recognized that "the difference in timing of 'green-up' in shrubs and grasses may be important for distinguishing between shrub cover and grass cover with remotely sensed data." They also acknowledged that if the "time dimension" were included in remote sensing studies of rangelands, rangelands might be successfully monitored. Peters *et al.*, (in review) used a time series of coarse-resolution satellite spectra from a drought year to distinguish between native desert vegetation of C_4 grasses and C_3 shrubs in virtually the same study area as Warren and Hutchinson's (1984).

Successful classification of desert vegetation with multi-date satellite imagery is based on differences in the physiologies and therefore the phenologies of native plant species (Eidenshink and Hass, 1992; Peters *et al.*, in review; Warren and Hutchinson, 1984). A goal of our research is to demonstrate a technique for identifying unique vegetation communities in an arid region from greenness peaks and growth patterns (phenophases) resulting from variable moisture regimes and thereby to demonstrate the utility of coarse-resolution satellite spectra as a regional monitoring tool.

Environmental Monitoring and Assessment **37:** 273–287, 1995.

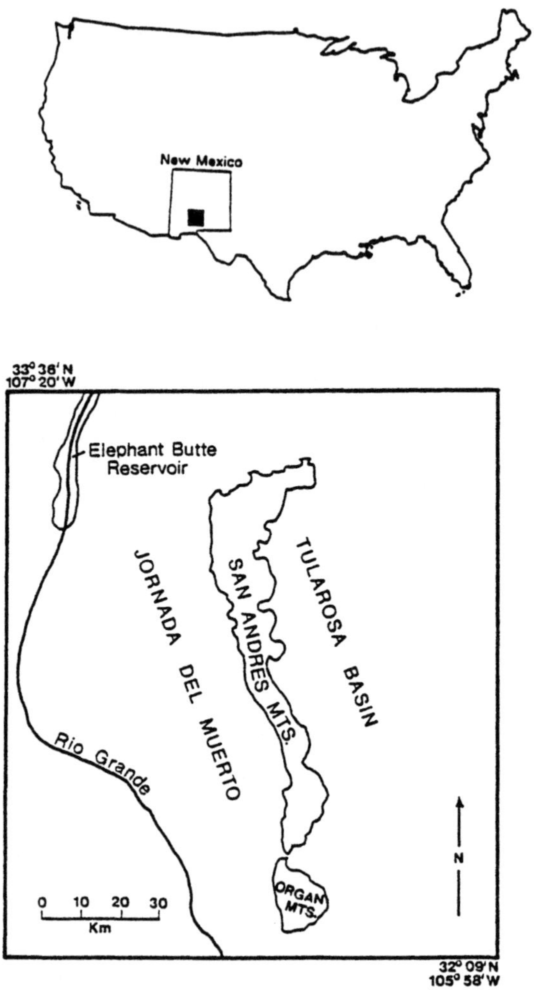

Fig. 1. Location and detail of the study area within the Chihuahuan Desert of southern New Mexico, USA.

2. Study Area

The study area includes part of the northern Chihuahuan Desert of southern New Mexico, USA (Figure 1). This region contains the gently sloping Tularosa and Jornada basins and is bisected by the San Andres and Organ mountains. Elevations in the intermountain basins range from 1180 to 1375 m. Precipitation in these desert basins is extremely variable, with an annual average of 230 mm and less in some areas. Most winter moisture comes from low-intensity cyclonic storms from the Pacific. Highly variable summer rain usually falls in localized convectional thunderstorms during the monsoonal period between mid-July and mid-September. These rains provide more than half the annual precipitation. In most years August has the highest monthly rainfall. High temperatures and low humidity cause large water losses to evaporation, especially during late spring and early summer. Annual temperatures average 24.5 °C, and June has the maximum average monthly high temperature (36 °C) (Paulsen and Ares, 1962).

In the Chihuahuan Desert of southern New Mexico vegetation consists of a grass-shrub complex, with extensive zones of mixed desert scrub and grassland (Paulsen and Ares, 1962; Warren and Hutchinson, 1984). Dominant shrubs of the region are C_3 species, which include creosotebush (*Larrea tridentata*), mesquite (*Prosopis glandulosa*), and tarbush (*Flourensia cernua*). Creosotebush is an evergreen shrub that attains maximum growth in the cool months of April-May and October-November (Fisher *et al.* 1988). Mesquite is a winter deciduous shrub that initiates leaf and stem growth between late April and late May, depending upon such landscape attributes as slope and aspect. Tarbush is a winter deciduous shrub whose spring leaf production depends upon winter/spring precipitation. Dominant perennial grasses are C_4 species that require relatively high night temperatures to produce new growth. Timing of green-up and maximum growth of desert grasses is mainly a function of water availability and temperature (Stephens and Whitford, 1993). These grasses include tobosa grass (*Hilaria mutica*), several species of grama grass (*Bouteloua* spp.), three-awns (*Aristida* spp.), dropseeds (*Sporobolus* spp.), burrograss (*Schleropogon brevifolius*), and various species of muhly (*Muhlenbergia* spp.). Data acquired during early land surveys in New Mexico have documented vegetation changes in our study area (Buffington and Herbel, 1965; Dick-Peddie, 1993; Gross and Dick-Peddie, 1979; York and Dick-Peddie, 1969). These studies show that former grama grasslands have been substantially replaced by shrubs, particularly mesquite and creosotebush (York and Dick-Peddie, 1969; Schlesinger *et al.*, 1990).

3. Method

Our data were derived from the National Oceanic and Atmospheric Administration's Advanced Very High Resolution Radiometer (NOAA-AVHRR). These operational weather satellites were intended mainly for observing cloud and sea-surface parameters. But their ability to monitor changes in land characteristics over large areas makes them invaluable for land-based studies (NOAA, 1991b; Tucker *et al.*, 1991). We obtained multispectral data acquired by the AVHRR from NOAA-10 for the 1989, 1990, and 1991 growing seasons because these years had precipitation conditions ranging from periods of drought to above-normal growing season precipitation. Data from the High Resolution Picture Transmission (HRPT) mode of the NOAA-10 have a spatial resolution of 1.1 km at satellite ground track (nadir). Other characteristics include high radiometric resolution (10-bit or 1024 gray levels), 2400 km scanning view, and a 07:30 equatorial overpass time (NOAA, 1991b). All data acquired have satellite nadir in or very near the study area.

3.1. Image preprocessing

We implemented a one-step algorithm for combining geometric and radiometric calibration and solar zenith angle corrections (Di and Rundquist, 1994). An image processing step accounts for per-date sensor scan-angle distortion by georeferencing each pixel to a latitude and longitude coordinate system. We conducted later image processing and analyses using the Earth Resource Data Analysis System (ERDAS) software on a personal computer, interactively coregistering all georeferenced images to within one pixel locational tolerance and retaining a 10-bit resolution throughout.

We minimized cloud obstruction in the imagery through image masking, using the thermal channel of the AVHRR sensor (10.3–11.7 μm) to locate cloud pixels, which are generally cooler than land pixels. We used the red image band to mask pixels lying in cloud shadows. We then produced a binary cloud mask for each date of imagery by designating clouds and cloud shadows as zero-value pixels and non-cloud pixels as one. By later multiplying the cloud mask and reflectance images, we eliminated clouds by converting to a value of zero. We standardized atmospheric attenuation of all red and near-infrared images using histogram minimization (Jensen, 1986) and used the signal from reflectance at the center of Elephant Butte Reservoir as the base value. And to ensure that reflectance over Elephant Butte Reservoir was the same for all imagery, we shifted histograms downward for each date of imagery, thereby normalizing atmospheric path radiance throughout the data.

3.2. Vegetation indices

Successful vegetation discrimination from satellite data depends upon the contrast in spectral radiance between vegetation and the surrounding soil (Tucker, 1979). A mathematical quantity referred to as the Normalized Difference Vegetation Index (NDVI) is routinely calculated from AVHRR data because of its sensitivity to the presence and condition of green vegetation and its ability to normalize atmosphere and background attenuation (Huete and Jackson, 1987; Huete and Tucker, 1991; Tucker *et al.*, 1991). Formulation for NDVI is as follows: (NIR − RED)/(NIR + RED), where NIR equals Near-Infrared reflected energy (0.725–1.10 μm) and RED equals Red-reflected energy (0.58–0.68 μm). Calculation of the NDVI results in pixels with an index value theoretically between −1.0 and +1.0. Vegetation will generally yield high index values, water will yield negative values, and bare soil will yield values near zero due to the reflectance characteristics of these surface materials (Lillesand and Kiefer, 1994; Tucker, 1979).

We studied an area where the amount of cover in native plant communities, consisting of desert grassland and desert shrub, is often less than 40% (Peters *et al.*, 1993). At such low cover, present satellite-derived vegetation indices are often confounded by soil-background conditions (Elvidge and Lyon, 1985; Frank, 1985;

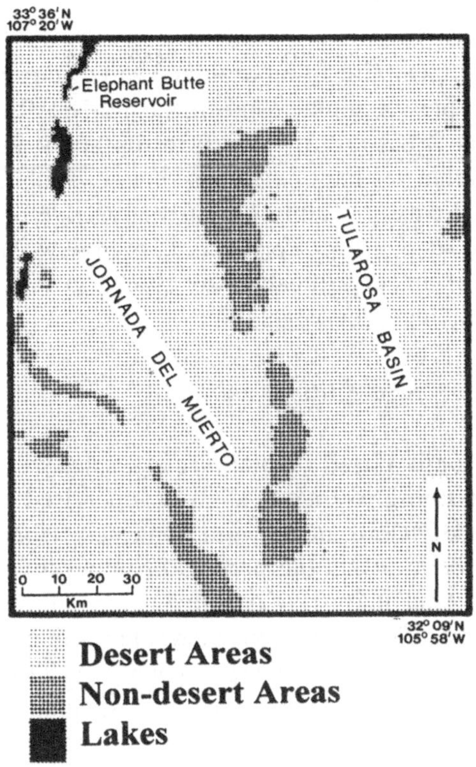

Fig. 2. Location of desert and nondesert vegetation based on image classification of eight cloud-free 1989 Normalized Difference Vegetation Index (NDVI) images.

Huete and Jackson, 1987; Huete and Tucker, 1991). For minimizing the influence of background reflectance, we suggest a methodology that can be applied under the worst conditions for vegetation analysis in an arid ecosystem. Our approach is based on extraction of a qualitative vegetation "signal" from the imagery in spite of the dominance of soil background. Soils at a given location do not change significantly during a growing season. In these desert ecosystems with their inherent low vegetation cover, the ratio of vegetation to soil background remains relatively constant (Huete and Tucker, 1991). Therefore a vegetation signal that would not be distinguishable on a single date of imagery provides meaningful information when analyses are conducted on a carefully controlled temporal sequence of imagery. Additionally, we have carefully selected our single-date HRPT imagery to minimize the effects of soil moisture and off-nadir atmospheric attenuation. Our concept of accounting for soil background is based on the use of image stratification to establish standard polygons that are used for annual and interannual monitoring of vegetation dynamics.

3.3. Initial Image Stratification

We conducted an initial stratification of image variance using unsupervised image classification techniques on eight cloud-free 1989 NDVI images. In this process we wished to isolate areas not of interest to our study, such as forested slopes, agriculture, and riparian vegetation. We based our post-classification sorting of the 10 resulting spectral classes mainly on our knowledge of the study area (Dick-Peddie, 1993; Peters *et al.*, 1993; Peters *et al.*, in review; Warren and Hutchinson, 1984). The results of this classification are shown in Figure 2. Later analysis of imagery focused only on desert vegetation.

3.4. Delineation of Desert Cover Classes

To derive polygons for the desert vegetation classes that we believe provide a useful delineation for monitoring, we input to an unsupervised classification algorithm eight cloud-free 1989 NDVI images with nondesert vegetation digitally masked (see Figure 2). We believe that the process of delineating polygons that remain consistent throughout our analysis of desert vegetation provides some standardization of soil background. These polygons therefore result in a consistent spatial unit for all later annual and interannual-temporal analyses. Temporal changes in scene reflectance can then be measured and compared qualitatively, even though we do not know the exact effect of soil background in each polygon (vegetation class).

The year 1989 in the study area experienced below-average spring and early-summer rainfall. Consequently C_4 grasses remained senescent in the early-growing season due to drought, and we used the distinct spring greenness peaks of C_3 shrubs and the late season greening of C_4 grasses to spectrally and temporally distinguish areas dominated by these plant communities.

Post-classification sorting using spectral-signature euclidian distances, image spatial relationships, existing maps, and personal knowledge resulted in three typical Chihuahuan Desert cover classes of grass, grass/shrub, and shrub-dominated communities. Figure 3 shows the geographic distribution of these classes, along with the nondesert classes derived during initial image stratification. This process was successful because the greenness peaks of C_3 shrubs in the early growing season (May-June) are clearly distinguishable from each community's C_4 grass component, which is evident during the late growing season of August and September (Figure 4). The magnitude of each curve is mainly related to species phenology and percent plant cover and indirectly to biomass. But since soil background influence in each pixel has only been nominally accounted for, these curves can only be qualitatively compared.

Our field transect data show that plant communities dominated by shrubs have a much lower cover (10–35%) than those consisting mostly of grasses (30–70%). Each of these classes is typical of the vegetation communities in the northern Chihuahuan Desert (Warren and Hutchinson, 1984). These curves are derived from

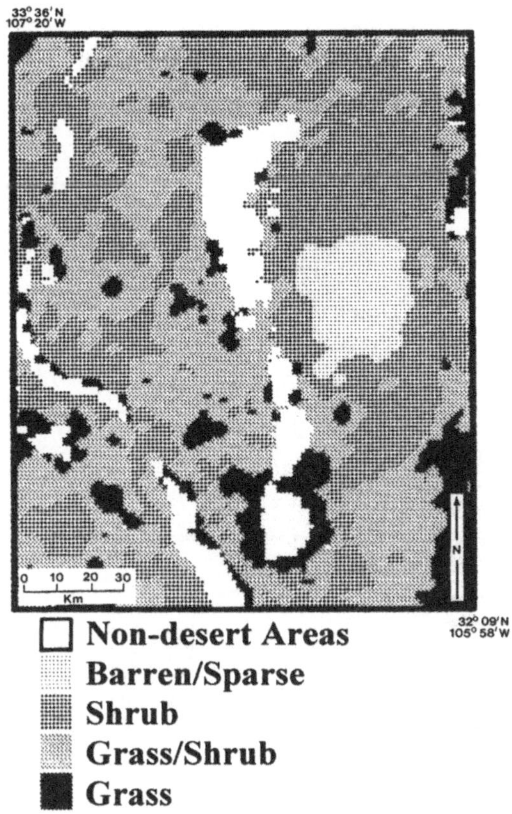

Fig. 3. Geographic distribution of desert grass, grass/shrub and shrub classes. Nondesert vegetation was digitally masked from the analysis.

our complete 1989 AVHRR data set, which includes the eight cloud-free images used in classification as well as six other images with clouds digitally masked. A barren/sparse vegetation class is shown for reference (Figure 4). This class consists mostly of the White Sands gypsum dune area. The occasional phenological spikes in the barren/sparse cover class are believed to be from photosynthesis of vegetation in low-lying interdunal areas.

4. Results and Discussion

The Southern Desert Climatic Division of New Mexico experienced a wide range of precipitation regimes during the 3 years selected for our study (Table I) (NOAA, 1989, 1990, 1991a). Below-normal annual precipitation was experienced in 1989, whereas 1990 and 1991 had 2 or more months below average during the early growing season (April through June). All years had below-normal precipitation during June, which is typically the warmest month in the region. During the late growing

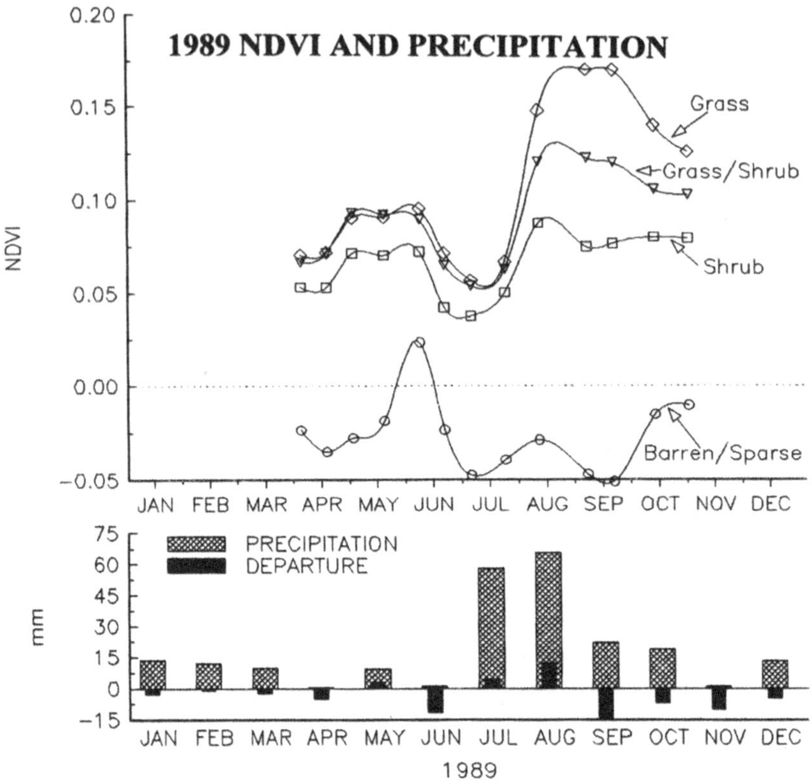

Fig. 4. Satellite-derived plant community growth characteristics (1989) from 14 dates of Normalized Difference Vegetation Index (NDVI) imagery for grass, grass/shrub, shrub and barren/sparse cover types. The bar graph shows monthly precipitation and departure from normal.

season (July-October), 1989 had below-normal precipitation, and 1990 and 1991 were well above normal (Table 1). Average total precipitation (263.66 mm) for the Southern Desert Climatic Division is higher than the 230 mm noted for these desert basins by Paulsen and Ares (1962) due to orographic precipitation received at some of the higher elevation weather stations in the foothill margins of the division.

4.1. Annual Plant Growth Characteristics

Figures 4–6 show annual satellite-derived NDVI curves depicting plant growth characteristics for each desert plant community during the 1989–1991 growing seasons. These graphs also show monthly rainfall (mm) departure from normal for the Southern Desert Climatic Division.

The best spectral separation of C_3 shrub and C_4 grass-dominated communities occurred during the 1989 growing season (Figure 4). Below-normal precipitation

TABLE I

Normal monthly and annual precipitation (1951–80), monthly and monthly departure from normal for 1989–1991 in the Southern Desert Climatic Division

	1951-80	1989		1990		1991	
	NORMAL	ACTUAL	DEPART.	ACTUAL	DEPART.	ACTUAL	DEPART.
JAN	16.26	13.21	-3.05	11.68	-4.58	18.29	+2.03
FEB	12.95	11.94	-1.01	11.43	-1.52	29.46	+16.51
MAR	11.94	9.65	-2.29	12.70	+0.76	21.34	+9.40
APR	5.59	0.25	-5.34	9.14	+3.55	0.25	-5.34
MAY	6.60	9.40	+2.80	10.92	+4.32	4.83	-1.77
JUN	12.70	1.02	-11.68	3.56	-9.14	11.68	-1.02
JUL	53.85	57.91	+4.06	79.76	+25.91	76.20	+22.35
AUG	53.34	65.28	+11.94	72.14	+18.80	83.31	+29.97
SEP	36.07	21.84	-14.23	78.49	+42.42	53.59	+17.52
OCT	25.65	18.80	-6.85	21.59	-4.06	11.43	-14.22
NOV	11.18	1.02	-10.16	35.81	+24.63	16.26	+5.08
DEC	17.53	12.95	-4.58	31.50	+13.97	117.09	+99.56
TOT	263.66	223.27	-40.39	378.72	+115.06	443.73	+180.07

during the early growing season (March-June) resulted in limited spectral differences between the vegetation classes. Throughout the 1989 growing season the shrub-dominated areas had the lowest NDVI response due to the relatively sparse vegetation cover (10–35%). Precipitation during the late growing season (July-October) was minimal, with only July and August experiencing above-normal rainfall (Figure 4). All vegetation classes developed a very distinct bimodal NDVI response during 1989. NDVI values for the shrub communities show similar early- and late-season growth patterns. Limited monsoonal moisture was insufficient for C_3 shrubs to develop very much new leaf material. But slightly above-average July-August precipitation was sufficient for C_4 grasses to develop, and their resulting NDVI peak is quite distinct.

Precipitation during the 1990 and 1991 growing seasons differed primarily during the March-June period (Figures 5 and 6). Although both years had an above-average monsoonal moisture regime, early-season moisture was quite limited in 1990, and plant growth was minimal as shown by the very flat NDVI response (Figure 5). From November 1990 through March 1991 above-average winter precipitation resulted in soil moisture buildup and a distinct early-season peak of C_3 and late-spring C_4 plants (Figure 6). The usual decline in photosynthesis during

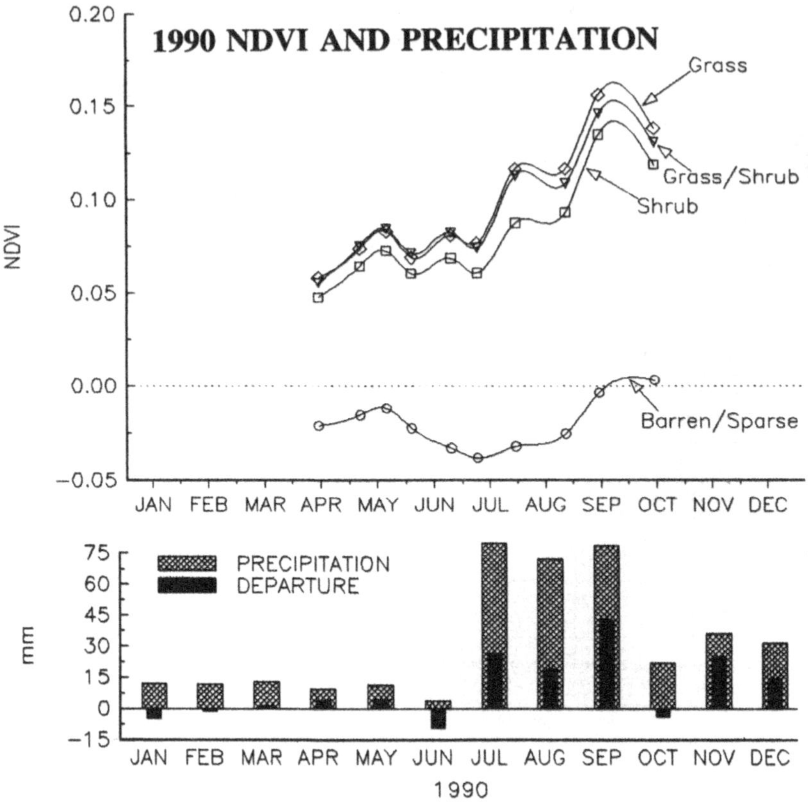

Fig. 5. Satellite-derived plant community growth characteristics (1990) from 14 dates of Normalized Difference Vegetation Index (NDVI) imagery for grass, grass/shrub, shrub and barren/sparse cover types. The bar graph shows monthly precipitation and departure from normal.

June and early July was scarcely detectable in 1990 (Figure 5) due to above-average precipitation during April and May. During the late season of 1990 and 1991 less distinct differences in our NDVI curves resulted from above-average precipitation. Field data recorded at the Jornada Long-Term Ecological Research Site show that the less distinct differences could have been due to summer annuals. Measurements at mesquite sites reveal that the percentage annual cover was significantly higher in 1990 and 1991 than in 1989. These data, however, are for only one small portion of our study area and may not be indicative of the whole area. Moreover, in this desert environment an annual flush commonly follows drought.

4.2. Interannual Plant Growth Characteristics

Figures 7–9 depict interannual comparisons for grass, grass/shrub, and the shrub plant communities. Seasonal variability resulting from differences in rainfall char-

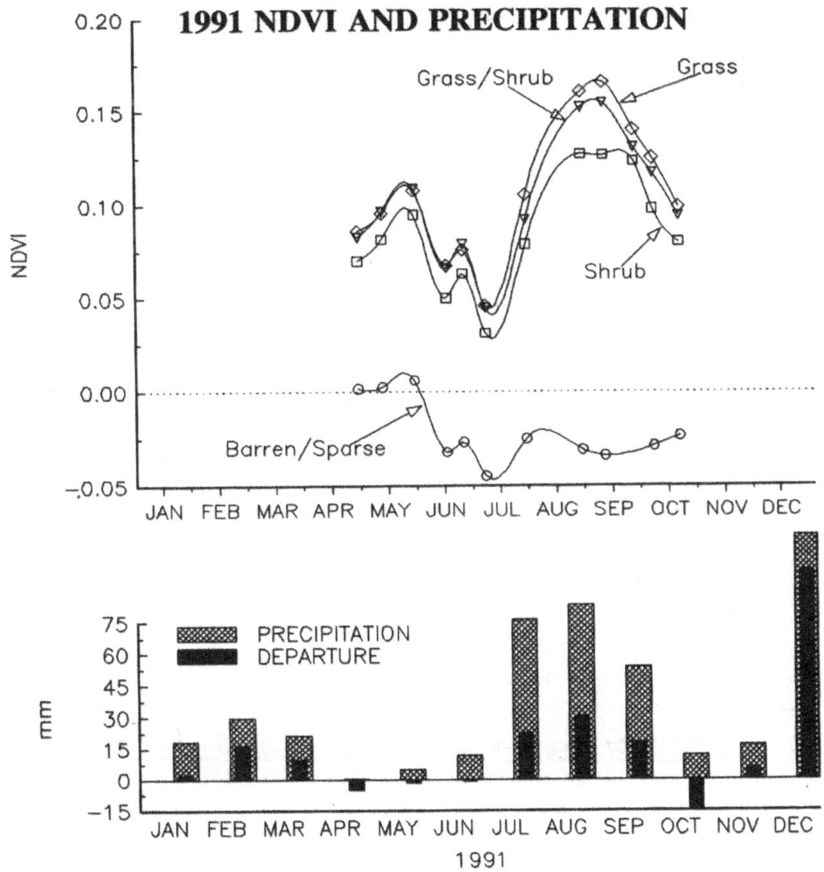

Fig. 6. Satellite-derived plant community growth characteristics (1991) from 14 dates of Normalized Difference Vegetation Index (NDVI) imagery for grass, grass/shrub, shrub and barren/sparse cover types. The bar graph shows monthly precipitation and departure from normal.

acteristics can be evaluated by comparing individual seasonal curves. Because of the minimal photosynthesis of C_4 grasses in the spring, we did not note a distinct spectral separation of grass and shrub communities. The typical June/July decline in photosynthesis due to high temperatures and low precipitation is apparent (Figure 7) as is the late-season peak of grassland productivity, when night temperatures are high and monsoonal moisture is present. NDVI response during the late season is relatively high, as cover in these communities ranges from 30 to 70 percent. Seasonal variability resulting from differences in rainfall characteristics can be evaluated by comparing individual seasonal curves. In mixed grass/shrub plant communities a bimodal character of photosynthesis is evident (Figure 8). The

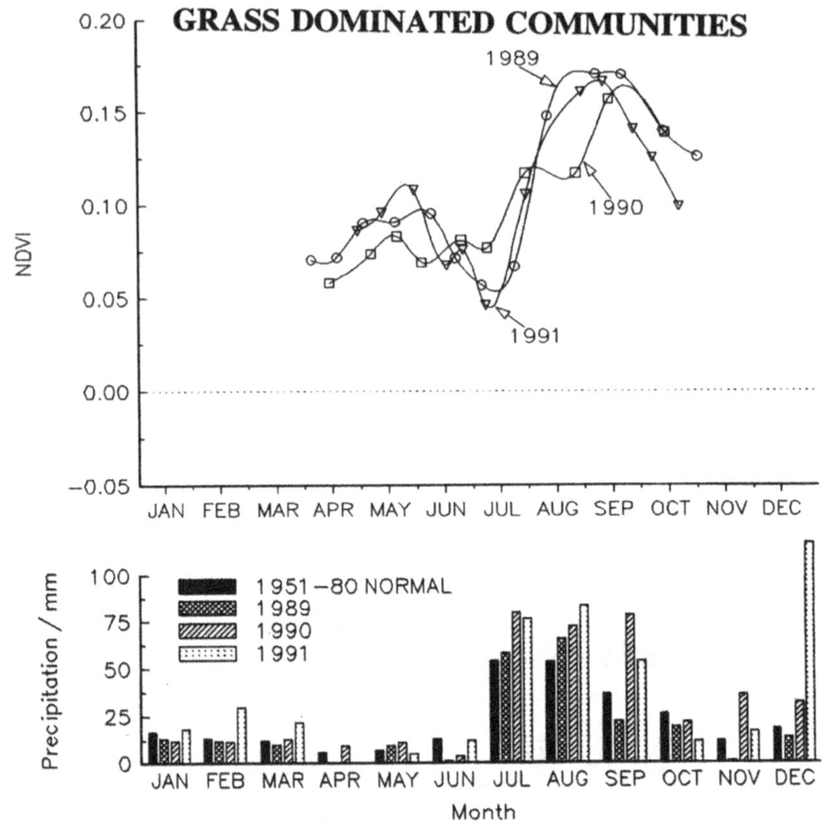

Fig. 7. Interannual (1989–1991) satellite-derived plant growth characteristics from 14 dates of Normalized Difference Vegetation Index (NDVI) imagery for the grass cover type. The bar graphs show monthly precipitation for 1989–1991 and 1951–80 normal.

late-season peak of photosynthesis is not as high as that of the grass-dominated communities. Interannual variability also characterizes plant growth in this cover class. Growth patterns of shrub communities vary considerably (Figure 9), showing lower NDVI values as a result of low plant cover, which ranges from 10–35%.

5. Conclusion

As a result of minimal photosynthesis of C_4 grasses in the spring, we noted limited spectral separability during this period. Spectral separability during the monsoon period was greater due to increased photosynthesis and differences in percent cover. Consequently, we believe that our method of desert plant community satellite moni-

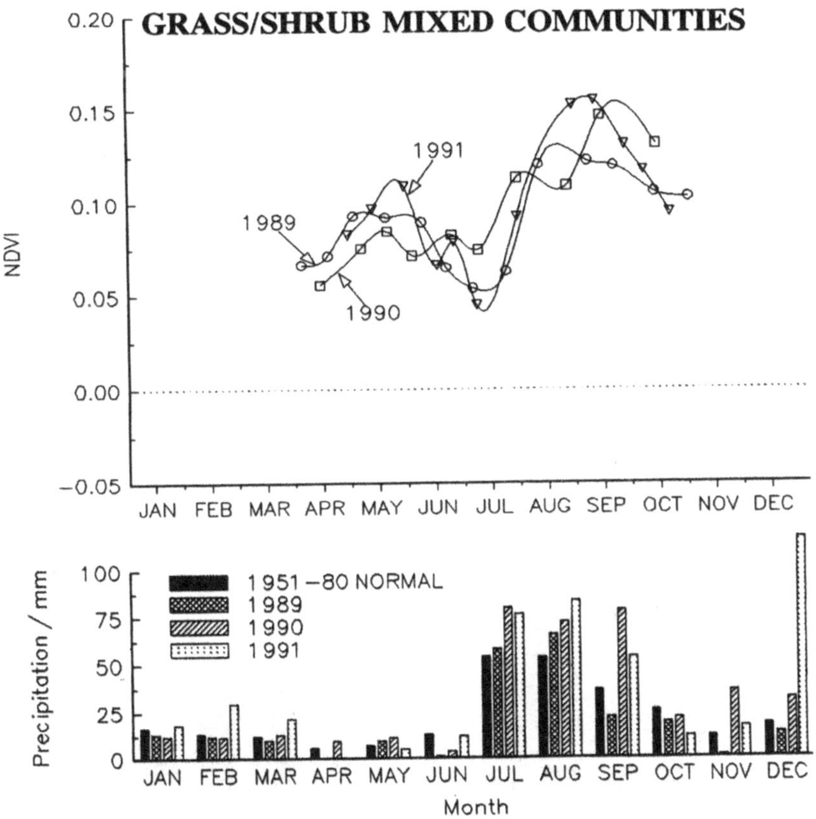

Fig. 8. Interannual (1989–1991) satellite-derived plant growth characteristics from 14 dates of Normalized Difference Vegetation Index (NDVI) imagery for the grass/shrub cover type. The bar graphs show monthly precipitation for 1989–1991 and 1951–80 normal.

toring could potentially be applied to larger regional-scale areas. Small areas would need to be evaluated in detail to describe parameters such as rainfall effectiveness and local precipitation patterns. We are now involved in research that could allow us to remove scene background reflectance on a per-pixel basis and quantitatively estimate such vegetation parameters as biomass and leaf-area index. This analysis also provides satellite-based inputs that we could use to improve our understanding of the spatial dynamics of climate on natural vegetation and that could help us distinguish these processes from anthropogenic causes of desertification.

We have found that we can use satellite-derived vegetation indices to derive information on arid-region vegetation parameters such as stress and growth patterns, despite the high ratio of soil to vegetation cover in the spectral response.

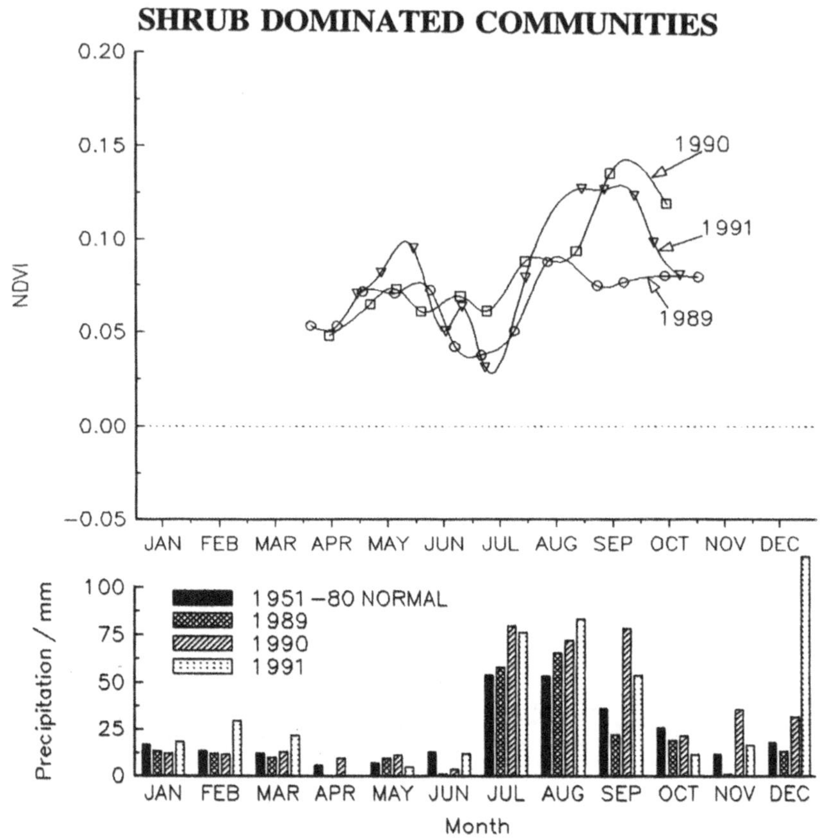

Fig. 9. Interannual (1989–1991) satellite-derived plant growth characteristics from 14 dates of Normalized Difference Vegetation Index (NDVI) imagery for the shrub cover type. The bar graphs show monthly precipitation for 1989–1991 and 1951–80 normal.

The result of our research is a spectral model of Chihuahuan Desert vegetation communities based on greenness peaks or growth patterns (phenophases), which in turn result from variable moisture regimes.

References

Buffington, L.C. and C.H. Herbel. 1965. Vegetational changes on a semidesert grassland range from 1858 to 1963. *Ecological Monographs* **35**: 139–164.

Di, L. and D.C. Rundquist. 1994. A one-step algorithm for correction and calibration of AVHRR level 1b data. *Photogrammetric Engineering and Remote Sensing* **60**: 165–171.

Dick-Peddie, W.A. 1993. *New Mexico vegetation: Past, present and future*, University of New Mexico Press, Albuquerque.

Dobyns, H.F. 1981. *From fire to flood: Historic human destruction of Sonoran Desert riverine oases*. Anthropological Papers No. 20, Ballena Press, Socorro, New Mexico.

Dregne, H.E. 1977. Desertification of arid lands. *Economic Geography* **3**: 329.

Eidenshink, J.C. and R.H. Hass. 1992. Analyzing vegetation dynamics of land systems with satellite data. *Geocarto International* **1**: 53–61.

Elvidge, C.D. and R.J.P. Lyon. 1985. Influence of rock-soil spectral variation on assessment of green biomass. *Remote Sensing of Environment* **17**: 265–279.

Fisher, F.M., J.C. Zak, G.L. Cunningham and W.G. Whitford. 1988. Water and nitrogen effects on growth and allocation patterns of creosotebush in the northern Chihuahuan Desert. *Journal of Range Management* **41**: 387–391.

Frank, T.D. 1985. Differentiating semiarid environments using Landsat reflectance indexes. *Professional Geographer* **37**: 36–46.

Gross, F.A. and W.A. Dick-Peddie, 1979. A map of primeval vegetation in New Mexico. *The Southwestern Naturalist* **24**: 115–122.

Huete, A.R. and R.D. Jackson. 1987. Suitability of spectral indices for evaluating vegetation characteristics on arid rangelands. *Remote Sensing of Environment* **23**: 213–232.

Huete, A.R. and C.J. Tucker. 1991. Investigation of soil influences in AVHRR red and near-infrared vegetation index imagery. *International Journal of Remote Sensing* **12**: 1223–1242.

Jensen, J.R. 1986. *Introductory digital image processing*, Prentice Hall, New York.

Lillesand, T.M. and R.W. Kiefer. 1994. *Remote sensing and image interpretation*, John Wiley, New York.

National Oceanic and Atmospheric Administration (NOAA). 1989. *Climatological data – New Mexico*. Vol. 93. NOAA National Climate Data Center, Asheville, North Carolina.

National Oceanic and Atmospheric Administration (NOAA). 1990. *Climatological data – New Mexico*. Vol. 94. NOAA National Climate Data Center, Asheville, North Carolina.

National Oceanic and Atmospheric Administration (NOAA). 1991a. *Climatological data – New Mexico*. Vol. 95. NOAA National Climate Data Center, Asheville, North Carolina.

National Oceanic and Atmospheric Administration (NOAA). 1991b. *NOAA Polar Orbiter data user's guide*, U.S. Department of Commerce, NOAA, NESDIS, NCDC and the Satellite Data Services Division, Washington.

Paulsen, H.A. and F.N. Ares. 1962. *Grazing values and management of black grama and tobosa grasslands and associated shrub ranges of the Southwest*. U.S. Department of Agriculture, Forest Service, Technical Bulletin No. 1270. Washington.

Peters, A.J., M.D. Eve and W.G. Whitford. In review. Analysis of desert plant community growth patterns with high temporal resolution satellite spectra. *Journal of Applied Ecology*.

Peters, A.J., B.C. Reed, M.D. Eve and K.M. Havstad. 1993. Satellite assessment of drought impact on native plant communities of southeastern New Mexico, USA. *Journal of Arid Environments* **24**: 305–319.

Schlesinger, W.H., J.F. Reynolds, G.L. Cunningham, L.F. Huenneke, W.M. Jarrell, R.A. Virginia and W.G. Whitford. 1990. Biological feedbacks in global desertification. *Science* **47**: 1043–1048.

Stephens, G. and W.G. Whitford. 1993. Responses of *Bouteloua eripoda* to irrigation and nitrogen fertilization in a Chihuahuan Desert grassland. *Journal of Arid Environments* **24**: 415–421.

Tucker, C.J. 1979. Red and photographic infrared linear combinations for monitoring vegetation. *Remote Sensing of Environment* **8**: 127–150.

Tucker, C.J., H.E. Dregne and W.W. Newcomb. 1991. Expansion and contraction of the Sahara Desert from 1980 to 1990. *Science* **253**: 299–301.

Warren, P.L. and C.F. Hutchinson. 1984. Indicators of rangeland change and their potential for remote sensing. *Journal of Arid Environments* **7**: 107–126.

York, J.C. and W.A. Dick-Peddie. 1969. Vegetation changes in southern New Mexico during the past hundred years, in W.G. McGinnies and B.J. Goldman (Eds.), *Arid lands in perspective*, University of Arizona Press, Tucson, pp. 157–166.

IMPLICATIONS OF THE DISINTEGRATION OF THE FORMER SOVIET UNION FOR DESERTIFICATION CONTROL

T. A. SAIKO
Department of Geographical Sciences, University of Plymouth, Drake Circus, Plymouth, Devon, PL4 8AA, United Kingdom

Abstract. Following the removal of censorship on environmental information in 1986 the magnitude of the Aral Sea disaster has been publicly acknowledged while the situation has continually worsened. Major efforts by the USSR Academy of Sciences as well as republic scientists since the 1970s have been supplemented by international expertise. The Soviet government adopted a special resolution on the Aral Sea in September 1988, but adequate financing was not available to solve this problem. With the disintegration of the USSR, the new independent states took full responsibility for their desertification control. In a corresponding tide of nationalism, Russia was solely accused of being responsible for the problem, and, not surprisingly, the controversial project of Siberian river diversion has been recently revived. There has been a transition from Russian to state language in all institutions, thus "squeezing out" the speaking of Russian. The Central Asian states have started to explore their own ways to deal with the catastrophe. But political, cultural, and ethnic rivalries between countries; growing nationalism and economic difficulties; and competition for water have not created the conditions to successfully solve desertification problems. Without change, the future of the Aral Sea appears to be bleak.

1. Introduction: Historical Background

As late as the early 1980s open discussion of the magnitude of desertification that had taken place in many parts of the former Soviet Union (FSU) could hardly be imagined. During the author's work in the Centre for International Projects (CIP) of the USSR Commission for the United Nations Environment Programme (UNEP) in Moscow, an introductory address from Soviet scientists to participants of international activities would normally include statements about the limited spread of desertification within the USSR and the possibilities for its control under the centralized planned economy. In publications on desertification control the following statements were popular: "The USSR's long-term experience in solving problems of arid land reclamation and the measures for the rational use of nature have made it possible to manage the national economy without essential violation of the ecological balance and desertification processes" (Babaev *et al.*, 1984).

In 1985 any mention of the Aral Sea disaster was deleted by the censor (*Glavlit*) from an international monograph on desertification (Glantz, 1986) prepared for publication by the author of this paper.

In 1986, with the advent of *glasnost* (publicity) in the Gorbachev's *perestroyka* (restructuring) period, environmental information was no longer censored. An avalanche of revelations followed. Estimates of the areas affected by desertification

Environmental Monitoring and Assessment **37:** 289–302, 1995.

TABLE I

Total desertification hazard of USSR arid lands in kilometers^2 and percent. (from Kharin, 1987)

Union Republic	Desertification types							Total desertified area	Badlands	Areas not affected by by deserti-fication	Total area
	Degradation of vegetation	Degradation around wells	Wind erosion	Salinization of irrigated lands	Technogenic desertification	Salinization due to sea level decline	Water erosion				
KAZAKH SSR	481,513	8,714	–	6,536	61,006	75,162	–	632,931	19,609	436,835	11,089,375
(%)	44.2	0.8	–	0.6	5.5	6.9	–	59.9	1.8	38.9	100
UZBEK SSR	47,127	51,256	11,024	3,993	41,884	7,993	–	163,277	13,507	98,841	275,625
(%)	17.1	18.6	4.0	1.4	15.2	2.9	–	59.2	4.8	36.0	100
TURKMEN SSR	129,305	42,168	48,012	20,875	12,942	11,690	12,228	277,220	25,468	114,812	417,500
(%)	31.2	10.1	11.5	5.0	3.1	2.8	2.8	66.5	6.1	27.4	100

in three of the five Central Asian republics made in 1987 (Table I) have revealed a situation far from harmonious: almost 633,000 km^2 (59.9%) of arid lands are subjected to desertification processes in Kazakhstan, 163,000 km^2 (59.2%) in Uzbekistan, and 277,000 km^2 (66.5%) in Turkmenistan (Kharin, 1987). Given the fact that only 4,000 km^2 were reportedly affected by salinization of irrigated lands in Uzbekistan while other experts (Craumer, 1992) guessed the amount as 13,400 km^2, these overall figures still seem to understate the problem.

But the most critical issue – the Aral Sea disaster – was already involving the whole basin. Environmental awareness has revealed what was already known by scientists: that the degradation processes in this region have passed the critical threshold after which their control can be achieved only through immediate concerted action provided with adequate financial support from the Soviet government. As Glantz (1994) noted, it was "a known crisis situation that crept up on policy makers over a period of thirty years."

The foundation for this environmental catastrophe was laid in the early 1950s. A large-scale irrigation campaign to attain "cotton independence" was started by a 1950 decree on the transition to a new mechanized irrigation system. One of campaign's major hydrotechnical giants, the hand-dug Karakum Canal, stretches for 1100 km through the deserts of Turkmenistan, and "no other single Soviet measure has found such a wide degree of support and appreciation among Turkmens" (Murat, 1975).

On the eve of the 40th anniversary of the start of the canal's construction, A. Ilamanov, the Turkmen Minister for Water Economy, emphasized the fundamental importance of the Karakum Canal for socioeconomic development in Turkmenistan. He stated that in the last 2 decades after the canal was completed cotton production from the newly irrigated 700,000 ha had increased six fold, the production of meat and milk had doubled, and the production of fruits and vegetables had increased by between two and seven times. Water supply to the cities of Turkmenistan had been improved, and many new settlements were created within the irrigation zone (Ilamanov, 1993).

This rapid economic development, however, was followed by immense desertification problems. As elsewhere in Central Asia, land has been irrigated by the least efficient method – surface furrow irrigation. At present three to five waterings per year are needed to desalinate the upper 1 m of soil (Glazovsky, 1990). Excessive norms result in waterlogging and soil salinization and increased water salinity. A total of 18% of 13.5 km^3 of the water annually diverted from the Amudarya River by the canal is lost to evaporation and infiltration through the unlined Karakum Canal bed.

By 1987 the irrigation system in Central Asia included 52,800 km of main canals, 390 reservoirs, and 230 dams and other hydrotechnical installations. From 1960 to 1990 about 1000 km^{3}of the Amudarya and Syrdarya waters did not reach the Aral Sea (Kuznetsov, 1993). This withdrawal ultimately led to the desiccation of the Aral Sea with a multitude of dramatic environmental and socioeconomic

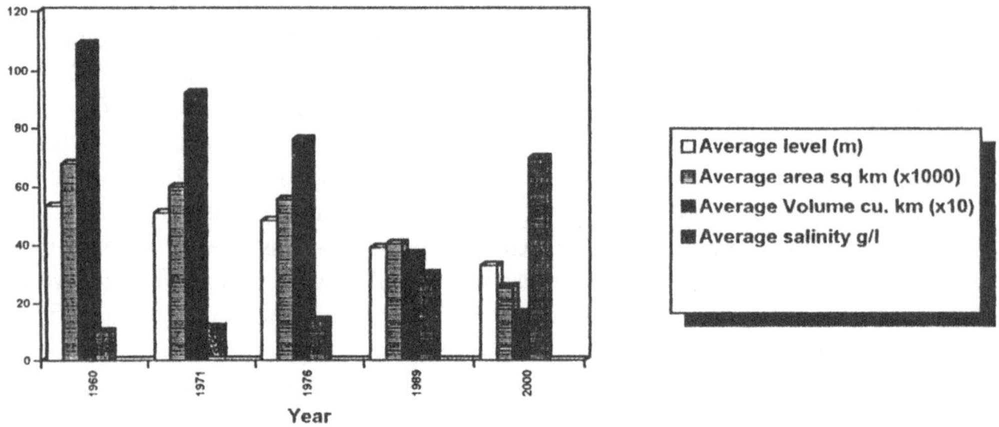

Fig. 1. Changes in the Aral Sea based on data from Micklin, 1992.

consequences widely discussed in recent years (Saiko and Zonn, 1994; Glantz *et al.*, 1993; Micklin, 1992; Sinott, 1992). In the 1980s the sea was divided into two parts: the Large Aral and the Small Aral. Basic changes in the average sea level, surface area, volume, and salinity since 1960 are diagrammatically shown in Figure 1.

These changes have affected the ecology of adjacent territories and the basin as a whole. Dust storms in the 1980s carried between 90 million and 140 million tons of salt and sand from the exposed seabed to Byelorussian farmlands 1,900 km to the northwest (Feshbach and Friendly, 1992). But the most dramatic effects of these complex desertification problems have apparently been on the local population of Kazakhstan and the Karakalpakstan Republic of Uzbekistan.

The health of at least 70% of these people has been affected, also due in part to the heavy pollution of water with chemicals. In the 1970s and 1980s the frequency of cancer, typhoid, hepatitis, and tuberculosis increased in comparison to other regions of the FSU and Central Asia. The infant mortality rate has reached 110 per 1,000 babies in the Bozatau district (United Nations Environment Programme, 1992). Food shortages and the absence of adequate medical aid and good-quality drinking water have aggravated this situation. In addition, mass unemployment due to the decline of the sea's fishing industry and the region's agriculture has created social tensions and outmigration from this "worst single instance of agricultural ecocide in the Soviet Union" (Feshbach and Friendly, 1992).

Even in the mid-1960s a group of scientists warned of the possible negative consequences (Glantz *et al.*, 1993; Micklin, 1992). But the expansion of irrigation was a policy of the Soviet government executed by the USSR Ministry for Water Management and Land Reclamation, which had envisaged the sacrifice of the sea in its initial plans. At the same time, some scientists predicted that the disappearance of the Aral would not influence the landscape of the region and that the drying out of the sea was much more advantageous than its preservation (Sinott, 1992).

Appeals from the suffering population have stimulated public "green" initiatives, and environmental opposition was organized in the summer of 1986 by the Uzbekistan Writers Union. Named The Committee for Saving the Aral, this group was particularly active in organizing expeditions in September 1988 to monitor and publicize the collected ecological data on the Aral Sea basin (Sinott, 1992). Public discussion were later arranged in the towns on the former sea banks.

Public activism and open scientific discussion of desertification problems in the press of Central Asian republics have apparently influenced the highest-level decisionmaking in Moscow. Particularly after the Chernobyl accident, environmental issues were given relatively high priority by Gorbachev and the Soviet government. A Commission on the Aral Sea, established by the government in April 1987, has developed a comprehensive program of actions.

Several other resolutions were adopted during the *perestroyka* period. The most important one was "On measures for a fundamental improvement of the ecological and sanitary conditions in the Aral Sea region and for effectiveness in the use, strengthening, and protection of water and land resources in this basin," passed on September 30, 1988.

Recognizing the need for all-union cooperation in solving the Aral crisis and the significant contribution of Russian scientists in studying this problem, the Aral Sea Scientific and Coordinating Center was established in 1988 within the Institute of Geography of the USSR Academy of Sciences.

A scientific ecological program for the integrated use and protection of water and land resources was elaborated and discussed in 1980s. By 1990 a competition for the possible solution of the Aral Sea crisis had resulted in 200 projects.

Nor did the catastrophe go unheeded by the world community. In 1990 the United Nations Environment Programme (UNEP)/USSR International Project "Assistance in the preparation of the Plan of Action for rehabilitation of the Aral Sea" was sponsored by UNEP and implemented by CIP and the "Aral" Centre. Russian and Central Asian scientists assessed the data in a diagnostic study document with further recommendations for action (United Nations Environment Programme, 1992).

In 1990 the Fund of Global Infrastructure of Japan was founded to help solve the problem (Zonn, 1993). The sense of urgency came through appeals of Central Asian leaders, who in June 1990 claimed that the growing deficit and pollution of water resources had placed the region on the brink of ecological disaster and that the problems were so urgent that "they could not be solved by regional efforts alone." In other words, without more water from outside, the problem could not be solved. Thus the "ghost" of the notorious Siberian water transfer project abandoned in 1986 reappeared. But because of a lack of financial resources and the half-hearted approach of Gorbachev and his policies, almost none of the needed measures were implemented before the Soviet Union dissolved.

The appearance of five newly independent states in the Central Asian region with the disintegration of the former Soviet Union in December 1991 had a multitude of political, economic, social, and infrastructural implications. The Aral Sea crisis has become an international (interstate) rather than a national problem. In this new political situation each of the states confronted its own ecological problems, being solely responsible for decisionmaking within its borders and for developing a new level of relations with regional partners and the outside world. Because of the high complexity of the problem, certain preconditions were needed to achieve any real progress in desertification control:

- Political stability within the region.
- Effective interregional cooperation.
- An agreed-upon regional desertification control policy.
- High priority assigned to desertification control within economic development plans of respective countries.
- Adequate financing of the agreed-upon programs.

This paper now analyzes processes in the political, economic, and social life of the Central Asian region and assesses the potential for solving desertification control problems in the Aral Sea basin.

2. Political Implications of Disintegration

First and foremost, the new geopolitical situation implied the end of colonial relationships between former Soviet republics. The concept of a colonial system has been widely discussed in the western literature describing the attitude of the Soviet Union toward the Central Asian republics (Clem, 1992). At the same time some experts did admit the positive side of such relations. As Liebowitz (1992) stated, "past development 'strategy' in Central Asia, condemned by many for being exploitive, has resulted in some great socioeconomic gains for the indigenous population and impressive growth of selected sectors of the region's economy." Craumer (1992) described changes in Soviet Central Asia in the recent decades as "the agricultural revolution" that "brought a major expansion of the area of irrigated land, increases in yields through use of fertilizers, biocides, and improved cultivation techniques, and a large improvement in levels of mechanisation for most types of crop and livestock production."

Indeed, irrigation development in Central Asia was accompanied by major investments in agriculture, particularly from 1971 to 1985. Total investments from 1955 to 1987, directed only to the agricultural development of Uzbekistan, reached 52.2 billion rubles (Glazovsky, 1990). By the end of the Soviet regime irrigated areas increased 1.6 times in Uzbekistan and Tadjikistan, 2.5 times in Kazakhstan, and 1.3 times in Kyrgyzstan (United Nations Environment Programme, 1992).

It is not surprising, therefore, that relations between Russia and each of the states of Central Asia have been in the forefront of the geopolitical situation. As Russia was the true successor state to the Soviet Union, all the faults of the Soviet Union were automatically assigned to Russia, particularly by the common people. Russia was successfully used by politicians making careers in the wake of increasing nationalism that immediately followed the declarations of independence. These attitudes varied, often depending on the share of Russian population in the country. But in one way or another the attitudes are still present in their current political lives.

Current dependence on Russia is based on the economic crisis experienced by most of the former Soviet republics, including Russia. This crisis has been caused by the disruption of strong economic, financial, legal, scientific, and infrastructural links that have existed for decades within a single state. Such a situation would be inevitable in the disintegration of this state with its overcentralized political power. Decisionmaking in the former Soviet Union did not favor balanced regional development as it pursued "national" interests of a group of people in power who could easily resettle whole nations or sacrifice the lives of millions of innocent people (mainly Russian nationals) to implement the decisions of the Communist Party.

Therefore, there are grounds for complaint and claims against former Soviet power in most of the former republics. Russia has probably suffered most from the policy of internationalism, lacking even such essential national attributes as its own academy of sciences.

But one should not underestimate the contribution of Russian science in the education of national cadres, particularly in Central Asia, where the population before the revolution was predominantly illiterate. In the 1950s all academics in Central Asian institutes were Russian or Armenian. Soviet power forcefully introduced the Russian language and neglected the need to develop national languages. At the same time Soviet power gave local people access to world literature and science through education in one of the official languages of the United Nations. Nationalism tends to subjective attitudes and actions that ignore the evident but disadvantageous aspects of reality.

As they used other ecological problems in the former republics, politicians used desertification problems to strengthen anti-Russian attitudes. By equating Russia with the former Soviet Union, politicians could easily reach the needed effect. Unofficially Russia is blamed for the ecological catastrophe of the Aral Sea (Glantz *et al.* 1993). Russia is also convicted of creating a monocultural agriculture and developing a distorted one-sided economy orientated toward the export of raw materials. It is not surprising that the idea of transferring water from Siberian rivers has recently reappeared. As Craumer (1992) mentioned, "Local outrage over cotton monoculture and its effects on the food situation has become an important issue of nationalism among the indigenous population."

One constructive element is present in nationalism, namely a stimulus for creating a national identity. This element is particularly important in the case of the Central Asian republics, which had never existed as sovereign states before they were incorporated into the Russian Empire. A complex of contradictory attitudes are expressed in dual policies in relation to Russia and the rest of the world. Thus all five Central Asian states belong to both the Commonwealth of Independent States (December 1991) and the Economic Cooperation Organization (July 1993), a Muslim trading bloc founded by Iran, Pakistan, and Turkey in 1979. These five states all have bilateral agreements with Russia, and at the same time all but Kyrgyzstan coolly received a recent initiative of the main "integrator" Nursultan Nazarbaev, President of Kazakhstan, to establish a Euroasian Union.

Unsurprisingly, in respect to Russian-Central Asian relations, political stability in the region is potentially explosive in only one region – North Kazakhstan. Culturally and historically part of Siberia, this region accommodates the majority of the Russian population in Kazakhstan, accounting for 36% of the total population or 6.2 million ethnic Russians (Terekhov, 1993). Kazakhs make up only 43% of the Kazakhstan population. With a mixture of 100 nationalities from several waves of voluntary and forced resettlement, Kazakhstan is the only Central Asian state whose constitution divides its population into indigenous and nonindigenous sectors.

In December 1992, 15,000 Russians in Ust-Kamenogorsk demonstrated for dual citizenship and state language (Sakwa, 1993). Kazakhstan is also one of the two relatively democratic states. (The second is Kyrgyzstan, with 26% of its population being ethnic Russian.) Kazakhstan strives to maintain its political stability by declaring national equality and attempting to balance interests between its Slavic population and Kazakh nationalists. At the same time, Kazakhstan encourages ethnic Kazakhs from other regions (e.g. Mongolia) to settle in its northwest areas to change the demographic balance. The rates of Russian outmigration (133,000 in 1993) suggest that Kazakhstan may reach its goal by the end of the century (Migranyan, 1994). Despite the lack of official Russian claims for northern areas, I. Solzhenitsyn has recently discussed such claims, and some influential people in the government support them. Outmigration of Russians from Central Asia reached 233,400 in 1993. Most of these emigrants were industrial and scientific workers (*Moscow News*, N.29, 17–27 July 1994).

Tadjikistan lost its political stability more than 2 years ago. After a short peaceful interval, a civil war representing a regional power struggle continued. This war has already distanced this country from decisions on environmental problems.

Nationalism is even stronger at the inter-ethnic level. A blend of nationalities has been normal in the former Turkestan republic since ancient times. Successive waves of invaders have contributed to intermixing of nations. In the 1920s Josef Stalin, guided by the principle of "divide and rule," decided to divide the Central Asian region and had his executives draw artificial borders based on nationality rather than regional criteria (Schwartz, 1992). This division created an area of maximum potential explosion – the fertile Fergana Valley shared by Uzbekistan,

Fig. 2. Potentially disputable areas in the Central Asian Region.

Kyrgyzstan, and Tadjikistan. The Fergana Valley was the site of acute ethnic clashes in 1989–1990. These boundaries have created many potentially disputable areas in most states (Figure 2).

Competition for deficient water supplies sometimes accompanies land disputes, as happened in Osh in June 1990 between Kyrgyz and Uzbek nationals. In 1991 several conflicts between Meskhetian Turks and Uzbeks led to the urgent evacuation of 30,000 Meskhetian Turks (World Bank, 1993). In 1990–1991 40,000 ethnic Russians left Andizhan after violent clashes. The most explosive situation is in Kyrgyzstan, the poorest country of the region, and Uzbekistan, where nationalism, supported by post-Soviet censorship and suppression methods, is the strongest.

Leaders of the Uzbek green opposition group "Birlik," who started the Aral Sea publicity campaign, have recently been sent to prison for 3 to 5 years (*The Economist*, 16 July 1994). In this way the absence of democracy limits the potential to solve environmental problems.

The Karakalpak Republic in Uzbekistan complains about its total dependence on Tashkent for any decisions. The whole nation is virtually on the verge of extinction, with almost no measures taken in the past years and an annual outmigration of 20,000 ecological refugees (Zhalilov, 1994). This is evidence that political issues are often closely connected with social ones.

In Turkmenistan the political situation concerning democratic traditions is probably the worst, since this state is actively pursuing the personality cult of its President Saparmurad Niyazov, the former Communist leader. The state seems to have forgotten its former ecological problems and now enjoys "Ten Years of Prosperity" arranged by the President (*The Economist*, 22 January 1994). At the same time Saparmurad Niyazov can boast the most stable political situation in the region. Russians (including those contributing to studies of desertification control) who cannot pass the Turkmen language test required for any relatively high post are being quietly squeezed out of the country. All five states have made their national language the official tongue, and only Kyrgyzstan has recently made a decision to support the use of both its language and Russian.

3. Economic and Social Implications of Disintegration

The dissolution of the USSR has left five states without a usual source of financial inflows from the central government to confront variations in accumulated economic wealth and potential natural resources. Thus any means to solve ecological problems, i.e. desertification control, can now be sought only within tight national budgets or from external donors.

Craumer (1992) believes that "the shift to republic self-financing or even economic sovereignty, however, will make generation of the capital needed to accomplish many of the agricultural reforms more difficult. In addition, the interest of the Russian Republic in allocating capital investment to the diversion of some of its water resources, will certainly be even less in a situation of sovereign republics." Cotton monoculture has distorted the economies of the region. Before disintegration, demands for grain and other foodstuffs were satisfied by exports from other fraternal republics. With independence, these exports have to be paid for by soft or hard currency needed for many other priorities. With the transition to national currencies in 1993, these problems have been aggravated.

Water deficiency in the region is relative. Per capita consumption of water has been more than twice the USSR average (Glazovsky, 1990), mainly because of wasteful technologies. With the disintegration of the Soviet Union, each former republic was free to determine its own balance of cotton versus food and fodder production. Each state can now make its own decisions on the proper use of water resources allocated not by quotas from Moscow but as agreed upon since 1992 by the Interstate Committee on Water Economy.

But only Uzbekistan has slightly decreased its areas under cotton (Saiko and Zonn, 1994). In 1993 Turmenistan decided to extend the Karakum Canal by another 300 km to irrigate the country's southwestern virgin lands, thus increasing the total irrigated areas to more than 1 million ha. Three new reservoirs will be built (Ilamanov, 1993), and another 6 km^3 of water will be diverted from the Amudarya River. This policy could be understood if cotton were the only source of needed currency. To the contrary, Turkmenistan boasts rich reserves of gas estimated at more than 8 trillion m^3 and oil estimated at 700 million tons (*The Economist*, 22

January 1994). There are almost no signs of economic reforms, and the political style of economic decisionmaking is strongly reminiscent of the Soviet period.

Kazakhstan and Kyrgyzstan have been particularly active in the transition to market economies, and both states have attained relatively good results despite the opposite initial levels. Kazakhstan is well endowed with natural resources, including immense oil reserves at Tengyz, developed jointly with the Chevron Company, and at the Kumkol oilfield in West Kazakhstan. A radical privatization campaign was launched in April 1994. Despite these successes, economic priorities are apparently much higher than ecological ones because in December 1993 President Nursultan Nazarbaev announced that all populist programs, including the Aral one, should be curtailed in favor of measures to strengthen the new national currency, the tenge.

Inflation in Kyrgyzstan is probably the lowest in Central Asia – 3.5% per month. The new currency, the som, has remained fairly stable, supported by the International Monetary Fund, the World Bank, and other foreign bankers (*The Economist*, 10 September 1994). But life is still hard for common people, and President Askar Akaev, a former professor of mathematics, has his inner political problems with an inefficient parliament, a recently resigned government, and nationalist opposition.

Uzbekistan, which is most affected by the Aral Sea crisis, experienced a particularly bad economic decline in 1992–1993. Its president, the former Communist Party Leader Ikram Karimov, is the most reluctant reformer of his populous country. Inflation has skyrocketed, unemployment has increased, and social unrest has compelled Karimov to declare the start of economic reforms in April 1994 (*The Economist*, 19 March 1994). With suppressed media and opposition, he will have limited chances to attract foreign investments.

Social tensions are closely related to the economic situation, being aggravated in most of the countries by rapid population growth exceeding 3% per year and the consequent concealed or unconcealed unemployment. The situation is particularly explosive in ecologically difficult regions neighboring the Aral Sea, both in Uzbekistan and Kazakhstan. Since the 1980s a deteriorating standard of living, combined with labor surpluses and massive unemployment, has been associated with inadequate medical services, sewage facilities, and drinking water (World Bank, 1993).

Economic difficulties and the "ecocide" of the local population make politicians explicitly state that "political priority centres on the improvement of regional water quality for human consumption" (Glantz *et al.*, 1993). One of the attributes of the Central Asian regions is its religion, Islam, an attribute on which forecasts for these states have been based. One of the experts in political economy has stated: "A marketised and rapidly growing... Central Asia could become a key part of a pan-Turkic revival led by Turkey, while a more feudal, less economically successful region is apt to feel the pull of Islamic fundamentalism..." (Ericson, 1992).

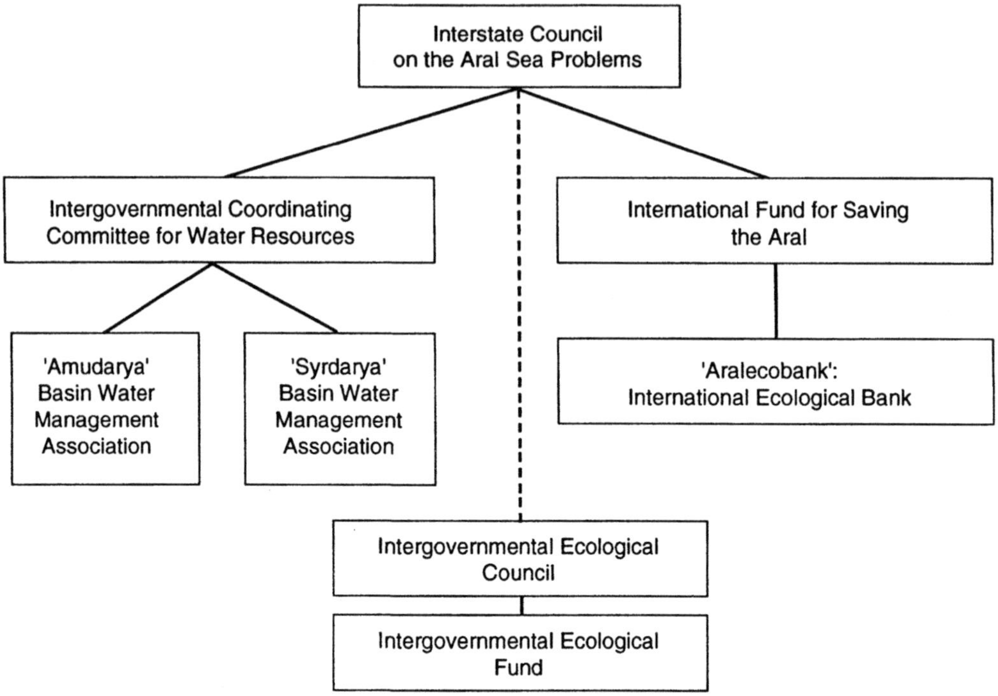

Fig. 3. Regional cooperation infrastructure for the Aral Sea problem (after Zonn, 1993).

The experience of the past 3 years has clearly shown that most countries have chosen the Turkic westernized way of development and that the danger of Islamic fundamentalism has been seriously overestimated, even in Tadjikistan (Migranyan, 1994). This choice can be explained by the fact that Islam for the Central Asian population was much less an ardent religion than a traditional way of life. Even today with Islam's revival in some countries (Islam is officially banned in Uzbekistan and Turkmenistan), its influence is probably felt only in Tadjikistan, where the Islamic Renaissance Party joined tribal opponents in opposing the ruling party. And Islam could even produce a positive effect through its teaching of the need for a careful and economical attitude toward water.

One should not underestimate the positive actions taken, including the creation of a regional cooperation infrastructure to solve the Aral Sea problem (Figure 3). Several regional meetings have been held (Zonn, 1993), the most important in Kzyl-Orda on 26 March 1993. At this meeting the states recognized the general tasks and agreed to establish a regional fund by allocating 1% of each country's gross national product. But this meeting did not reach a final agreement on the measures for improving the situation, nor were any real finances directed to this fund at the present time (*The Economist*, 15 October 1994).

Meanwhile the desiccation of the Aral Sea has gradually continued. By 1993 the sea's volume had decreased three fold. Its level had dropped by 16.5 m, and the average salinity has trebled, approaching that of open ocean waters (Glantz, *et al.*, 1993). A new Aralkum desert is being formed on the dried sea basin and adjacent territories.

4. Conclusion

Analysis of the political and socioeconomic implications of the disintegration of the Soviet Union has shown that after independence no successful solution to the Aral Sea crisis has been found with the exception of the infrastructure developed for regional cooperation. Political instability is evident at all levels of relationships. Serious nationalist tendencies have developed throughout the region, at times breaking out into open conflicts. No agreed-upon efficient desertification control policies have been adopted at the regional level. Because of the economic collapse, desertification control priorities are ranked very low in most states. For similar reasons, measures to improve the ecological situation in the Aral Sea basin have been inadequately financed. No external donors have offered financial assistance for practical measures. And if this situation does not drastically change before the end of this century, the Aral Sea faces the prospect of being doomed to turn into a series of saline lakes. Meanwhile, the adjacent population would be on the edge of extinction.

References

Babaev, A.G., I.S. Zonn and N.S. Orlovsky. 1984. *The USSR experience in desert reclamation and desertification control*, VINITI, Moscow.

Clem, R.S. 1992. The frontier and colonialism in Russian and Soviet Central Asia, in R.A. Lewis (Ed.), *Geographic perspectives on Soviet Central Asia*, Routledge, London and New York, pp. 19–36.

Craumer, P.R. 1992. Agricultural change, labour supply and rural outmigration in Soviet Central Asia, in R.A. Lewis (Ed.), *Geographic perspectives on Soviet Central Asia*, Routledge, London and New York, pp. 132–175.

Ericson, R.E. 1992. Economics, in T.J. Colton and R. Legvold (Eds.), *After the Soviet Union: From empire to nations*, The American Assembly, New York, pp. 49–83.

Feshbach, M. and A. Friendly. 1992. *Ecocide in the USSR*, Harper Collins, New York, pp. 73–88.

Glantz, M.H. (editor). 1986. *Arid land development and the combat against desertification: An integrated approach*, Centre for International Projects, Moscow, pp. 114–135.

Glantz, M.H. 1994. Creeping environmental phenomena in the Aral Sea basin. Paper presented at the NATO Advanced Research Workshop, Critical Scientific Issues of the Aral Sea Basin: State of Knowledge and Future Research Needs, 2–5 May 1994. pp. 1–28.

Glantz, M.H., A.Z. Rubinstein and I.S. Zonn. 1993. Tragedy in the Aral Sea basin. *Global Environmental Change* June 1993: 174–198.

Glazovsky, N.F. 1990. *Aralsky Krizis*, Nauka Publ., Moscow.

Ilamanov, A. 1993. *Garagumsky kanal: Proshloye i nastoyashchee*. Turkmenskaya iskra, 16.09, p. 6.

Kaiser, R.J. 1992. Nations and homelands in Soviet Central Asia, in R.A. Lewis (Ed.), *Geographic perspectives on Soviet Central Asia*, Routledge, London and New York, pp. 280–311.

Kharin, N.G. 1987. *Poyasnitelnaya zapiska k karte 'Antropogennoye opustynivaniye aridnyh territoriy SSSR'*. Ylym, Ashkhabad, pp. 3–32.

Kuznetsov, N. 1993. Luchshe byt pessimistom chem optimistom. *Zelyonyy Mir* **12**: 14.

Liebowitz, R.D. 1992. Soviet geographical imbalances in Soviet Central Asia, in R.A. Lewis (Ed.), *Geographic perspectives on Soviet Central Asia*, Routledge, London and New York, pp. 101–131.

Micklin, Ph.P. 1992. Water management in Soviet Central Asia: Problems and prospects, in J.M. Stewart (Ed.), *The Soviet environment: Problems, policies and politics*, Cambridge University Press, Cambridge, pp. 88–114.

Migranyan, A. 1994. Rossiya i blizhnee zarubezhye. *Nezavisimaya Gazeta*, 18 January, pp. 4–5.

Murat, A.B. 1975. Turkmenistan and the Turkmen, in Z. Katz (Ed.), *Handbook of major Soviet nationalities*, Macmillan, New York, pp. 262–282.

Saiko, T. and I. Zonn. 1994. Deserting a dying sea. *Geographical Magazine* **66**(7): 12–15.

Sakwa, R. 1993. *Russian politics and society*, Routledge, London and New York, pp. 366–391.

Schwartz, L. 1992. The political geography of Soviet Central Asia: Integrating the Central Asian frontier, in R.A. Lewis (Ed.), *Geographic perspectives on Soviet Central Asia*, Routledge, London and New York, pp. 37–73.

Sinnott, P. 1992. The physical geography of Soviet Central Asia and the Aral Sea problem, in R.A. Lewis (Ed.), *Geographic perspectives on Soviet Central Asia*, Routledge, London and New York, pp. 74–97.

Terekhov, V. 1993. Bezhentsy i emigranty: Kak predotvratity katastrophu. *Nezavisimaya Gazeta*, 12 January, p. 3.

United Nations Environment Programme (UNEP). 1992. *Diagnostic study for the preparation of an action plan for rehabilitation of the Aral Sea*. UNEP, Nairobi.

World Bank. 1993. *Humanitarian aid to the republics of Central Asia*. Report. World Bank, Washington.

Zhalilov, I. 1994. Aralskoye more umirayet. Ochered za karakalpakami. *Nezavisimaya Gazeta*, 10 August, p. 4.

Zonn, I.S. 1993. Problema Aralav svete novoi geopolitiki. *Problemy Osvoyeniya Pustyn* **3**: 9–17.

MONITORING SOIL PRODUCTIVE POTENTIAL

DAVID TONGWAY
Commonwealth Scientific and Industrial Research Organisation (CSIRO), National Rangelands Program, Division of Wildlife and Ecology, PO Box 84, Lyneham, Canberra 2602 ACT Australia

Abstract. Desertification involves the loss of soil productive potential, but a means of assessing and monitoring the progress of desertification on the soil has been elusive. Soil is so varied and complex that methods of assessing condition are too slow, tedious, and expensive for routine use. Moreover, differences in soil type can be confused with soil condition. This paper presents a structured method of assessing soil condition. This method is based on recognizing and classifying soil surface features and examining soil properties that reflect the status of the processes of erosion, infiltration, and nutrient cycling. Published in the form of a user manual, the method has the following three stages: (1) defining the geomorphic setting of the site, (2) recognizing patch/interpatch associations and the mode of erosion at the landscape scale, and (3) assessing soil surface condition ratings in quadrats sited within the landscape pattern patches. Stage 3 is achieved by observing each of 11 features in the field and classifying their status according to detailed fieldnotes and photographs. The method applies to a wide range of soil types and biogeographical regimes and has proven to be repeatable among observers and quickly transferred to new observers.

1. Introduction

The assessment of soil condition has traditionally used signs of erosion, or soil loss, as the criteria (Soil Conservation Service, 1976). But because of the wide range of soil types, a single unambiguous set of criteria was difficult to produce. In addition, researchers have noted marked changes in soil productive potential not accompanied by soil loss (Mott *et al.*, 1979; Bridge *et al.*, 1983). Many detailed measurements of the physical, chemical, and biological properties of soils, when considered together, may reveal the productive potential of a site. But these measurements are too slow, too tedious, and too expensive for routine or regular monitoring.

Tongway and Smith (1989) and Greene and Tongway (1989) studied desertified landscapes in the eastern Australian woodlands and found several physical and chemical soil properties that accounted for losses in pasture production (Table I). They also devised a set of morphological characteristics in the soil surface to reflect the differences in productive potential on different sites

Recent research in the Australian rangelands has been taking a more process-based line of inquiry into the structure and function of landscapes, with the objective of producing broadly based dynamic models of landscape function (Tongway and Ludwig, 1990; Tongway, 1991; Ludwig and Tongway, 1995; Tongway and Ludwig, 1994). This research confirmed the theoretical predictions of Noy-Meir (1973, 1981, 1985) that in landscapes where water was a limiting resource, production per

Environmental Monitoring and Assessment **37:** 303–318, 1995.

TABLE I

The relationship between four stages of desertification (1 is best, 4 is worst) and soil available nitrogen, organic carbon infiltration rate stability index and grass production. Differences in superscript indicate significant differences at $p < 0.05$ Tukey's HSD

Property	Desertification Stage			
	1	2	3	4
Available nitrogen (ppm)	81.90^a	64.40^b	35.40^c	23.60^d
Organic carbon (%)	1.90^a	1.45^a	0.95^b	0.71^b
Infiltration rate (mm/h)	49.20^a	21.00^b	18.00^b	7.80^b
Stability index	177.20^a	173.30^a	92.70^b	77.00^b
Grass production (g/m^2)	231.00^a	120.00^b	31.20^c	13.60^d

unit area would be maximized if the resources were distributed in patches rather than evenly. Tongway and Ludwig (1990) showed that this principle extended to all scarce resources in a landscape, such as nutrients and organic matter. Tongway (1990) proposed that the availability of scarce resources be coherent in time as well as space, i.e. the decoupling of nutrients and water was a hallmark of degraded lands. These studies also found that the spatial distribution of resource-rich patches was not random but organized into a pattern of repetitive zones of depletion linked to zones of accumulation. Other work found that this organization was multi-scaled, consisting of nested sets of resource-poor and resource-rich patches varying in dimension from a few centimeters (e.g. grass tussocks) to hundreds of meters (e.g. groves of mulga trees) (Tongway *et al.*, 1989; Ludwig and Tongway, 1995; Tongway and Ludwig, 1994).

At landscape scale, therefore, the productive potential of soil will be patchy, not uniform, even under the best of condition circumstances. Random sampling or sampling that does not consider this organization is unlikely to make sense of any data purported to reflect condition because of variability.

The patterning in landscapes is initiated and maintained by a suite of processes such as erosion/deposition, runoff/runon, and harvest/concentration, a soil faunal-based process (Tongway, 1991; Tongway and Ludwig, 1994). These processes leave characteristic signs of their activity, signs that are easily observed and interpreted.

2. Evolution of the Methodology

The methodology for assessment was developed out of these concepts and models according to the following structure or stepwise advancement.

(1) The need to understand the landscape setting and geomorphic structure, e.g. the land system, land unit system (Christian, 1958) (landscape scale).
(2) The need to recognize within the whole landscape patterns of resource-rich and resource-poor zones and understand the reasons for their existence (patch scale).
(3) The assessment of the condition or relative quality of the soil in each of the pattern elements or patches (quadrat scale).

Landscape scale summaries of rangelands are commonly available and are typically mapped at scales of about 1:250,000. These summaries may originally have been devised as resource inventories and typically include data on geology, topography, and soil and vegetation types. In Australia, landscape scale summaries are mapped separately and as an integration called a land-system map (Christian, 1958). These compilations enable representative landforms to be selected and the monitoring site to be located in its landscape context (step 1).

Evaluating the size and spatial distribution of resource-rich and resource-poor patches within a landscape unit is the next step. This is the first data of the soil condition assessment proper because it focuses on the factors that affect the quality or condition of soil. The data collected at this scale may change over time due to climatic events, management, or some interaction. Usually this data can be quickly and easily captured if collected on an environmental gradient, such as slope using distance measurements, and analyzed by established methods. Some patterns are overt and self-evident (Tongway, 1991); others are more cryptic and need data analysis for confirmation (Ludwig and Tongway, 1995).

If repeated over time, such meso-scale assessment can give information about the dynamics of patches in a landscape. For example, a decrease in the type, size, number, and frequency of resource-rich zones indicates degradation. The development of the methodology, inspection, and analysis of several landscape types distributed across a wide biogeographical range found that the principle of patchy landscapes is a general one. Landscape types can in fact be characterized by the nature of their patchiness.

The Manual (Tongway, 1993) gives examples of this scale of observation in the field, both in methodology (page 10) and in recognizing a range of landscape patch types in Figures 7 to 20. Photographs of landscapes in differing condition are augmented by sketches that show the changing nature of processes.

At the fine or quadrat scale observations are focused on the effect of processes that impinge upon the surface of the soil. Having stratified the landscape into two or more patch types, researchers collect data reflecting the respective habitat quality factors of the soil. Eleven features need to be observed; each is addressed separately so that the observer can focus sharply on the criteria and the underlying processes. Features are assigned to a class representing the amount or severity of the factor in question. Each feature is closely described in the Manual, together with the process that it indicates or mediates. These features and processes are presented in the Appendix to this paper. In the Manual, photographs show a range of soil surfaces with the 11 features in all the different classes. The features are as follows:

- Soil cover (interception of raindrops)
- Soil cover (obstruction to overland flow)
- Crust broken-ness
- Cryptogam cover
- Erosion features
- Eroded materials
- Litter
- Soil microtopography
- Surface nature
- Slake test
- Surface texture

The data may be used as a simple index of soil condition in which a higher score means a better condition, or, alternatively, subsets of the data may be assembled to focus on soil behavior in terms of (1) stability, or resistance to erosion, (2) water capture and infiltration, and (3) nutrient cycling status. (See Appendix.)

Depending on soil type, this level of data can be quite dynamic over time if climatic events or management changes the processes that flow across the surface. Some features for a given range site will respond more to disturbance than others, so monitoring can be more sharply focused on these features after an initial period of testing.

3. Discussion

The methodology is built upon broadly based landscape ecology principles that stress the understanding of pattern recognition and process inference in landscapes. The methodology displaces notions of desertified soil as eroded or not eroded. These notions are clearly too simplistic for evaluating overtly complex systems with complex feedback loops. Specific responses of a given range site to the forces of desertification can be observed and scaled. Table II summarizes data from four sites in the eastern Australian woodlands. The data show that, although the overall score has a small dynamic range, stability and nutrients have wide ranges and hence pinpoint the specific soil problem in this example of desertification.

4. Summary

The proposed methodology is based on assessing the availability in space and time of scarce or limiting resources, as revealed by the status of processes that regulate those resources in the landscape. Clearly a nondesertified landscape will retain good control of water runoff and eroded or mobile solids (soil and litter) and will

TABLE II

An example of data from four sites in different stages of desertification. This data does not represent a wide range of desertification but is presented to show the utility of the method. The summary of this data into respectively overall, stability, infiltration and nutrient scores shows that the small range (5) in the overall score can be made more sensitive by looking at the stability score range (9). Infiltration scores do not vary widely and nutrient scores show low values with a useful range

	Site 1	Site 2	Site 3	Site 4
SCCC feature				
Interception of raindrops	1	1	3	5
Obstruction of overland flow	1	1	1	3
Crust broken-ness	4	2	3	4
Cryptogam cover	4	2	1	1
Erosion features	4	3	2e	1e
Eroded materials	4	3s,3g	3g	1r
Litter cover	1	1	1	1
Microtopography	1	2	1	1
Surface nature	3	2	4	4
Slake test	4	3	2	2
Soil texture	2	2	2	2
Summary of tabulated scores by category				
Overall score	27	22	23	25
Stability score	25	16	19	19
Infiltration score	8	8	9	11
Nutrient score	6	5	3	3

have sites or zones of high infiltration. Desertified landscapes will be seen from the data collected to have low resource control and be wasteful of resources, which will flow to waste.

This methodology leads to a predictive understanding, in a semi-quantitative sense, of landscape status and trajectory. The speed at which data can be collected allows this method to be a realistic monitoring tool. Typically, data for a single quadrat takes about 60 seconds of evaluation by a trained observer. The structure of the method means that the people who do the monitoring need to be constantly relating their observations to surface processes and overall landscape function and not to the mindless collection of numbers.

The notion of early warning signs is embedded in this approach as a whole. In principle, the warning signs are that mobile resources are being less retained and more lost from the landscape or the scale of resource redistribution patterns is favoring an undesired vegetative outcome (e.g. inedible shrub increase). Inevitably, specific critical points for landscapes must be identified on a landscape-by-landscape basis. The proposed method provides a common scale upon which a wide range of landscapes can be evaluated.

References

Bridge, B.J., J.J. Mott and R.J. Hartigan. 1983. The formation of degraded areas in the dry savanna woodlands of northern Australia. *Australian Journal of Soil Research* **21**: 91–104.

Christian, C.S. 1958. The concept of land units and land system. *9th Pacific Science Conference, 1957, Proceedings* **20**: 74–81.

Greene, R.S.B. and D.J. Tongway. 1989. The significance of (surface) physical and chemical properties in determining soil surface condition of red earths in rangelands. *Australian Journal of Soil Science* **27**: 213–25.

Ludwig, J.A., D.J. Tongway and S.G. Marsdenen. 1994. A flow-filter model for simulating the conservation of limited resources in spatially heterogeneous, semi-arid landscapes. *Pacific Conservation Biology* **1**(3): 209–213.

Ludwig, J.A. and D.J. Tongway. 1995. Spatial organisation of landscapes and its function in semi-arid woodlands, *Australia. Landscape Ecology* **10**(1): 51–63.

Mott, J.J., B.J. Bridge and W. Arndt. 1979. Soil seals in tropical tall grass pastures of northern Australia. *Australian Journal of Soil Research* **30**: 483–494.

Noy-Meir, I. 1973. Desert ecosystems: Environment and producers. *Annual Review of Ecological Systems* **4**: 25–52.

Noy-Meir, I. 1981. Spatial effects in modelling of arid ecosystems, in D. Goodall and R. Perry. (Eds.), *Arid land ecosystems*, Cambridge University Press, Cambridge, pp. 411–432.

Noy-Meir, I. 1985. Desert ecosystem structure and function, in I. Evanari, I. Noy-Meir and D. Goodall (Eds.), *Ecosystems of the World*, Elsevier, Amsterdam, pp. 93–103.

Soil Conservation Service, 1976. *National range handbook*, U.S. Department of Agriculture, Soil Conservation Service, Washington.

Tongway, D.J. 1990. Soil and landscape processes in the restoration of rangelands. *Australian Rangeland Journal* **12**: 54–57.

Tongway, D.J. 1991. Functional analysis of degraded rangelands as a means of defining appropriate restoration techniques. In *Proceedings of IVth International Rangeland Congress*, Association Française de Pastoralisme, Montpellier, France, pp. 166–168.

Tongway, D.J. 1993. Rangeland soil condition assessment manual. Commonwealth Scientific and Industrial Research Organisation, Melbourne. Available from author.

Tongway, D. T. and J.A. Ludwig. 1990. Vegetation and soil patterning in semi-arid mulga lands of eastern Australia. *Australian Journal of Ecology* **15**: 23–34.

Tongway, D.J. and J.A. Ludwig. 1994. Small-scale resource heterogeneity in semi-arid landscapes. *Pacific Conservation Biology* **1**(3): 201–208.

Tongway, D.J. and E.L. Smith. 1989. Soil surface features as indicators of rangeland productivity. *Australian Rangeland Journal* **11**(1): 15–20.

Tongway, D.J., J.A. Ludwig and W.G. Whitford. 1989. Mulga log mounds: Fertile patches in the semi-arid woodlands of eastern Australia. *Australian Journal of Ecology* **14**: 263–268.

Appendix

Soil Surface Condition Individual Features

This is a glossary of the nature, meaning and scope of each surface feature. In the assessment each feature on each quadrat will be assigned to a class.

TABLE A-I

Projected Cover		Class
nil	(<1%)	1
very low	(1–2%)	2
low	(2–5%)	3
moderate	(5–15%)	4
high	(15–50%)	5
very high	(>50%)	6

1. Soil cover

(A) Interception of Raindrops

The objective is to assess the degree to which surface cover resists rainsplash erosion: the projected percentage cover of perennial grasses and perennial shrubs to a height of 0.5 m, rocks, sticks and any relatively immovable and long-lived object that will intercept raindrops and protect the soil from rainsplash erosion. Soft annual herbage is excluded. "Gravity drops" falling from foliage drip-rings have a high erodibility when falling from heights of more than 0.5 m, so foliage at heights greater than 0.5 m are ignored in this section. Litter is assessed separately (see below). Soil with more protection has a lower potential for future erosion, even if the protection is in the form of a layer of stones caused by previous erosion.

(B) Interception of Overland Flow

The objective is to assess the cover of features that obstruct or divert overland flow. These are long-lived features that project at least 1 to 2 cm above the soil surface (e.g. plant tussocks or stems, rocks, wood) and obstruct and divert overland water flow so as to slow the overall flow rate, increase the depth of flow and thus reduce the transporting capacity of the flow for soil and organic matter. Soft annual herbage is excluded. Many small features can be more effective than a few large ones. The key question to ask is the following: To what extent will the observed features reduce the erosive power of overland flow?

2. Crust broken-ness

This observation assesses to what degree surface crust materials are broken or loosely attached and available for erosion. Crusts in this context consist of thin layers of fine-textured soil overlying the soil proper. Soils in good condition have a crust that is smooth and conforms to gentle undulations in the soil surface. The crust also commonly has polygonal cracks. But as long as they do not curve upwards at

TABLE A-II

Projected Cover		Class
nil		1
very low	(< 2%)	2
low	(2–5%)	3
moderate	(5–15%)	4
high	(15–50%)	5
very high	(>50%)	6

TABLE A-III

Crust Broken-ness	Class
extensively broken	1
moderately broken	2
slightly broken	3
intact	4

their edges, these cracks do not count as broken. Smooth, unbroken crusts will yield less material than crusts that are already partially broken into fragments and are mechanically fragile. Typically, as crust broken-ness increases, the crust becomes discontinuous, with sharp edges and sometimes loose crust fragments, marking zones where the crust material has been lost. Often drift material accumulates in the "pocket" where the crust has been lost.

Where a quadrat has more than 75% cover of eroded material (see item 5 below), crust broken-ness does not apply. Excluded are soils without natural crusts, such as self-mulching surfaces and loose sands on dunes and sandy banks.

Sandy textured soils that do have a thin cohesive crust are assessed under this heading.

Animal tracks and vehicle tracks are excluded from consideration.

The degree and extensiveness of the crust broken-ness needs to be considered. Quadrats that are judged to be outside the scope of this section should be scored as zero and the total score adjusted accordingly.

TABLE A-IV

Cryptogam Cover		Class
nil	(< 1%)	1
slight	(1–10%)	2
moderate	(10–50%)	3
extensive	(> 50%)	4

3. CRYPTOGAM COVER

The aim is to assess the cover of cryptogams. "Cryptogams" is a term that includes algae, fungi, mosses and liverworts, plants that can exist on stable (i.e. nonshifting) surfaces that have access to light. Cryptogams are often early colonizers of recovering soil and are positive indicators of surface stability (see also the notes on the slake test). They are useful in open situations (i.e. not under litter beds) in assessing surface soil stability and are assessed as percent cover.

Soil crusts with cryptogams are often flexible and resist pulverization (see also crust nature). If a cryptogam-crusted soil "plate" is detached from the surface by a paint scraper or chisel and gently broken, one can see across the break fine roots or hyphae, which contribute greatly to the robustness of the soil crust.

Analysis of soil beneath cryptogam crusts usually finds higher concentrations of nitrogen and carbon in the immediate soil surface, implying that these crusts help establish a healthy nutrient profile shape. Even where liverworts are growing on sandy deposits, their presence is a positive indication of the surface stability.

4. EROSION FEATURES

Erosion features are visible signs of active, current loss of soil material from the quadrat. Erosion features are used to assess how the surface soil responds to the erosive forces of wind and water. An erosion feature may be an incised channel or rill, a terracette, pedestalling around a plant, or sheet erosion. (A stony pavement on a smooth slope implies sheet erosion.) Observers must be careful not to confuse soil accumulated around grass butts as signs of pedestalling. In the case of accumulation hummocks, the soil in the hummock is unconsolidated and, if sectioned with a paint scraper, the hummock reveals layers of accumulated soil. Pedestals will be much more coherent and fine textured and have no sign of layering.

In the soil-surface method, these features will be small. Otherwise they would have been dealt with in Step 2. The severity of the feature is more important than percent cover per se. A feature may occupy a small area but be severe or serious. It is important to note the severity so as to flag a potential problem. Sometimes

TABLE A-V

Erosion Features	Class
extensive	1
moderate	2
slight	3
nil	4

TABLE A-VI

Eroded Materials		Class
extensive	(>50%)	1
moderate	(20–50%)	2
slight	(1–20%)	3
nil	(<1%)	4

spontaneous improvement in formerly active erosion scars will be evidenced by smooth, rounded edges and colonization by cryptogams. In such a case, reduce the assessment by one class.

5. Eroded Materials

The objective is to assess the degree to which materials are being eroded from one place in the landscape and deposited in another. These deposits may take the form of sand or gravel splays or hummocks. Because these materials are unconsolidated, they can be easily remobilized by an erosion event and carried away. The deposition of these materials around the landscape rather than in a streambed implies that, although materials are being eroded, there is potential for stabilization.

Sometimes erosion deposits can be quite productive. But until they have stabilized to the extent that they are no longer recognizable as erosion features, they need to be flagged as remaining potentially erodible.

Litter is treated separately (see below). Three size classes of materials can be seen: sand, less than 2 mm; gravel, 2 to 10 mm; and stone, greater than 10 mm. The percent cover is evaluated to classify this feature.

TABLE A-VII

Litter Cover		Class
nil	(<1%)	1
very low	(1–10%)	2
low	(10–25%)	3
moderate	(25–50%)	4
extensive	(50–100%)	5
very extensive	(100% but several cm thick)	6

6. Litter cover

The object is to assess the availability of organic materials for decomposition and nutrient cycling. Litter refers to such organic matter as detached leaves, stems, twigs, fruit and dung. The position of litter in the overall landscape also helps define fertile patches.

Litter may be local, i.e. accumulates and decomposes where it falls (e.g. under a tree or shrub), or it may be transported by wind or water and dispersed or concentrated depending on the nature of the landscape and the zones of potential accumulation.

Copious accumulation protects against rainsplash erosion and serves as a seed trap as well as source of organic matter for cycling. Transported litter is clearly more mobile and its value to the system more unsure.

Litter has three easily recognizable forms of decomposition value:

- loosely strewn on the surface (nil)
- in intimate contact with the surface (slight)
- partially or wholly covered with soil (extensive)

These categories reflect the likelihood of litter being incorporated in the soil *in situ*.

The cover is assessed first, then whether the litter is local (L) or transported (T) and finally whether the degree of incorporation is nil (N), slight (S) or extensive (E). A recording could therefore be 3LS.

These data are used in a calculation to modify the simple cover value *vis*.

- transported and nil incorporation, multiply by 1 (i.e. no change to simple cover)
- local and/or slight incorporation, multiply by 1.5
- extensive incorporation, multiply by 2.0.

Therefore, in the above example, 3LS, the score is $3 \times 1.5 \times 1.5 = 6.75$.

TABLE A-VIII

Surface Water Detention Capacity		Class
nil, smooth	(<3 mm)	1
slight, few, shallow depressions	(3–8 mm)	2
moderate, deeper depressions	(8–15 mm)	3
high, deep, extensive	(15–25 mm)	4
very high eg sink-holes	(>25 mm)	5

7. Soil microtopography

Soil microtopography refers to soil surface features that detain water on the quadrat between the time the surface becomes wet (ponded) and the time water actually runs off the quadrat area (runoff). Soils that can absorb and store rainfall will have higher biological activity than those from which this water runs off or is evaporated. Typically, soil microtopography consists of a series of unconnected micro-depressions or bowl-shaped structures several millimeters deep and about 5–50 cm across. Soils with deep cracks or sinkholes are also included if the scale of the feature is appropriate to the quadrat. Micro-depressions are often associated with cryptogam crusts or mats buckling at the edges so as to form embayments.

A second type of role for microtopographic features is in constituting safe seed sites because of their accumulation, retention and protection characteristics.

Macro structures such as Gilgai's and depressions larger than the quadrat are excluded and should be addressed in step 2. The assessment estimates the volume of the detention as shown by the depth criteria in the table.

Sheep or cattle foot impressions should not be included because their persistence is unknown and not necessarily related to the processes that naturally produce shallow depressions.

8. Surface nature

This observation assesses the robustness of the surface, or the degree to which the surface can withstand stress (e.g. trampling) or reform after rain. The features assessed are crust flexibility, brittleness and hardness and also the coherence of subcrust soil.

This test can be conducted only when the soil crust is air-dry. This is the only moisture content at which the test is informative because clearly all soils are soft and nonbrittle when wet or moist.

Hoof action easily breaks soils with weak or fragile crusts, releasing subcrust material for erosion. Flexible crusts indicate that fine roots or fungal hyphae are

TABLE A-IX

Surface Nature	Class
Crust shows some flexibility when pressed with pen or finger pressure, or surface is self-mulching clay. Sub-crust soil is coherent or strong crumb structure.	5
Crust is very hard (needs metal tool to break surface), but is brittle, breaking into amorphous fragments or powder. Sub-crust soil is hard and coherent.	4
Crust is moderately hard (needs plastic or metal tool to break), but brittle, breaking into amorphous fragments or powder. Sub-crust soil is coherent.	3
Crust is easily broken with finger pressure and is brittle. Sub-crust soil is noncoherent, e.g. sandy.	2
Surface is loose-sandy, over noncoherent sand.	1

holding the soil particles together (like a sticky string bag) and that biological activity is high on the surface.

Very hard soils resist detachment by mechanical means but indicate that infiltration rates may be very low and organic carbon levels depleted. Crust hardness is not a linear indicator of condition. Soil crusts are more likely to be reformed after disturbance in the medium hardness/flexible scale.

9. Slake Test

The slake test is a test for crust stability when wet. The ability of stable crusts to maintain their cohesion when wet enables them to resist the erosive effects of flowing water and to maintain a good physical structure after a wetting-drying sequence.

The test involves immersing fragments of surface soil (approximately 6 mm cubes) in a beaker of rainwater (saline waters are unsuitable) and observing the response of the fragments over several minutes (see the rating table for detailed criteria).

This test should be performed on each of the landscape strata soils, e.g. under trees and between trees. At least three fragments are used per test. One quadrat from each stratum is sufficient for testing unless soil crust broken-ness and/or cryptogam cover differs markedly along the transect.

10. Soil Texture

The objective of this test is to rate the permeability of the soil body as opposed to the crust. Permeability is rated with a pedologist's moist bolus test, a 0.5 cm sample and a simplified four-point scale.

TABLE A-X

Observed Behavior	Class
Fragment collapses completely in < 2 seconds with a myriad of air bubbles, into a shapeless mass. Very unstable.	1
Fragment substantially collapses over about 5 seconds. A thin surface crust remains, but $>50\%$ of the subcrust material slumps to an amorphous mass. Unstable.	2
Surface crust remains intact, some slumping of sub-crust material, but $<50\%$. Moderately stable.	3
Whole fragment remains intact over periods of 1 hr or more. Stable.	4

TABLE A-XI

Texture	Class
Silty clay to heavy clay (very slow infiltration rate).	1
Sandy clay loam to sandy clay (slow infiltration rate).	2
Sandy loam to silt loam (moderate infiltration rate).	3
Sand to clayey sand (high infiltration rate).	4

Rating Scales for Individual Features

See Table A-XI.

Soil Surface Condition

The individual observations of the soil surface are condensed into three categories, which have distinct significance for each monitoring site:

- Stability
  The ability of the soil to withstand erosive forces and to reform after disturbance.
- Infiltration/Runoff
  How the soil partitions rainfall into soil-water (water available for plants to use) and runoff water that is lost from the local system or may carry materials away.
- Nutrient cycling status
  How efficiently organic matter is cycled back into the soil.

These categories are useful at both the quadrat and site scale.

- At quadrat scale, because the locations of quadrats are fixed, changes in their assessed condition over time can be objectively used for assessing improvement or degradation.
- Soil condition can also be compared directly with plant data and so build up a cause and effect understanding.
- At site scale an overall frequency distribution of quadrats in different conditions gives an overall site value. Because there are three categories, the strengths and weakness of a given site will be determined more accurately.

The class values for each observation are summed for each category to give a value reflecting the status of a quadrat or site with respect to the category. The scales vary as follows:

Stability

- crust broken-ness (1–4)
- surface nature (1–4)
- slake test (1–4)
- erosion features (1–4)
- eroded materials (1–4)
- cryptogam cover (1–4)
- soil cover (rain) (1–6)
- litter cover (simple) (1–6)

(If all features are present, the scale ranges from 7 to 36.)

Note: simple litter cover only in this category.

Infiltration/Runoff

- microtopography (1–5)
- surface nature (1–4)
- litter cover (simple) (1–6)
- soil cover (flow) (1–6)
- soil texture (1–4)

(If all features are present, the scale ranges from 5 to 25.)

Note: simple litter cover only in this category.

Nutrient Cycling Status

- litter cover, origin and incorporation (1–18)
- cryptogam cover (1–4)
- microtopography (1–5)

(If all features are present, the scale ranges from 3 to 27.)

Note: full litter cover/origin/incorporation assessment in this category.

- If any feature does not apply to a given site, it should be recorded as zero and the scale of the category adjusted accordingly. For example, on a sandy site the soil may be loose and the slake test is impossible to perform. Score slake as 0 and adjust the category scale for stability to 6–26.
- The final assessment for each category should be converted to a percentage value.

PERSISTENCE OF DESERTIFIED ECOSYSTEMS: EXPLANATIONS AND IMPLICATIONS *

WALTER G. WHITFORD*, GUSTAVO MARTINEZ-TURANZAS** and ERNESTO MARTINEZ-MEZA***

Senior Research Ecologist, US-EPA, Environmental Monitoring Research Laboratory, PO Box 93478, Las Vegas, NV 89193, USA; **Department of Range Science, Colorado State University, Ft. Collins, CO 80523, USA; * U.S. Department of Agriculture, Agriculture Research Service Jornada Experimental Range, Box 30003, Dept. 3JER, New Mexico State University, Las Cruces, NM 88003, USA*

Abstract. Studies of rainfall partitioning by shrubs, responses of shrub-dominated ecosystems to herbicide treatment, and experiments using drought and supplemental rainfall were conducted to test the hypothesis that the shrub-dominated ecosystems that have replaced desert grasslands are resistant and resilient to disturbance. Between 16 and 25% of the intercepted rainfall is channelized to deep soil storage by stemflow and root channelization. Stemflow water is nutrient enriched and contributes to the "islands of fertility" that develop under desert shrubs. Drought and rainfall augmentation experiments during the growing season after 5 consecutive years of summer drought found that (1) growth of creosotebushes, *Larrea tridentata*, was not significantly affected, (2) perennial grasses and forbs disappeared on droughted plots, (3) nitrogen mineralization increased in the short term, and (4) densities and biomass of spring annual plants increased on the droughted plots. Doubling summer rainfall for 5 consecutive years had less-significant effects. Coppice dunes treated with herbicide in 1979 to kill mesquite (*Prosopis glandulosa*) had the same frequency of occurrence of the shrub as the untreated dunes when remeasured in 1993. These data indicate that the shrub-dominated ecosystems persist because they are resistant and resilient to climatic and anthropogenic stresses.

1. Introduction

Most examples of desertification in North America, Australia, and parts of Africa involve the replacement of grass-dominated ecosystems by shrub or small-tree-dominated ecosystems (Archer *et al.*, 1988; Buffington and Herbel, 1965; Gibbens and Beck, 1987; Grover and Musick, 1990; Hennessy *et al.*, 1983). Increases in woody plant cover and reduction or elimination of grass cover have greatly reduced the production of livestock in desertified ecosystems, but reducing or eliminating livestock grazing has resulted in neither the return of grasses nor the reduction of shrubs (Glendening, 1952; Hennessy *et al.*, 1983). Shrub-dominated ecosystems return even when challenged by herbicide treatment, root-plowing, or bulldozing (Roundy and Jordan, 1988; Rappole *et al.*, 1986). Shrub-dominated ecosystems have been suggested to represent alternative stable states replacing the predisturbance grassland or grass-dominated ecosystem (Westoby *et al.*, 1989).

* Notice: The U.S. Environmental Protection Agency (EPA), through its Office of Research and Development (ORD), partially funded and collaborated in the research described herein. This paper has been subjected to EPA's peer review and has been approved as an EPA publication.

Environmental Monitoring and Assessment **37:** 319–332, 1995.

2. Resistance and Resilience of Shrub-Dominated Ecosystems

From experimental and descriptive studies, evidence is accumulating that the shrub-dominated ecosystems that have replaced grasslands in many areas of the world are resistant to perturbation and resilient following perturbation (Roundy and Jordan, 1988; Rappole *et al.*, 1986). Resistance is a measure of the degree to which the ecosystem or ecosystem variable is changed from its equilibrium state following a disturbance. Resilience is the time required for a perturbed ecosystem or ecosystem variable to return to its equilibrium value. We hypothesize that the morphological and physiological characteristics of the shrub species that dominate many of the desertified ecosystems in the southwestern United States contribute to the formation of islands of resources centered on the shrubs. Water and nutrients essential for plant growth are concentrated under desert shrubs (Charley and West, 1975, 1977; Pressland, 1975, 1976; West and Klemmedson, 1978; Parker *et al.*, 1982; Virginia and Jarrell, 1983; Lajtha and Schlesinger, 1986; Nulsen *et al.*, 1986). Moreover, woody legumes may have root symbionts that fix atmospheric nitrogen, thereby enhancing the nitrogen fertility of soils around the plant (Virginia and Jarrell, 1983; Virginia, 1986; Virginia *et al.*, 1989).

Increased soil fertility is not limited to the contributions of nitrogen fixation. Mineralization rates are higher under shrub canopies where litter is trapped and soil organic matter content is higher than intershrub soils (Fisher and Whitford, 1995, Table I). The islands of fertile soils that develop under shrubs are partly a function of litter accumulation under shrub canopies due to wind redistribution of dead plant materials. Shrubs act like rocks in a stream with respect to wind movement, producing eddy currents that result in wind-entrained debris being deposited under shrubs or immediately on the downwind side of the shrub canopy (Reichman, 1984), but litter accumulation under shrub canopies is not the only contributor to the fertile island effect.

Rainfall intercepted by shrub canopies is redistributed to stemflow and throughfall in addition to the fraction that is evaporated from the leaf and stem surfaces (Martinez-Meza and Whitford, in press). In creosotebush (*Larrea tridentata*) stands stemflow that accounts for between 16 and 25% of the intercepted rainfall is nutrient enriched (Figure 1). There is a lesser enrichment of the throughfall water. Stemflow water enters the soil at the root crown. The soil immediately adjacent to the root crown has higher concentrations of many soil nutrients than the soil at mid-crown under a shrub (Figure 1). The root crown soil enrichment results mainly from stemflow water since in this shrub species the soil immediately adjacent to the root crown is generally litter free. The stemflow water enters the soil and follows preferential root pathways to deep storage. This movement has been documented in three of the dominant Chihuahuan Desert shrubs: honey mesquite (*Prosopis glandulosa*), creosotebush (*L. tridentata*), and tarbush (*Flourensia cernua*) (Martinez-Meza and Whitford, in press).

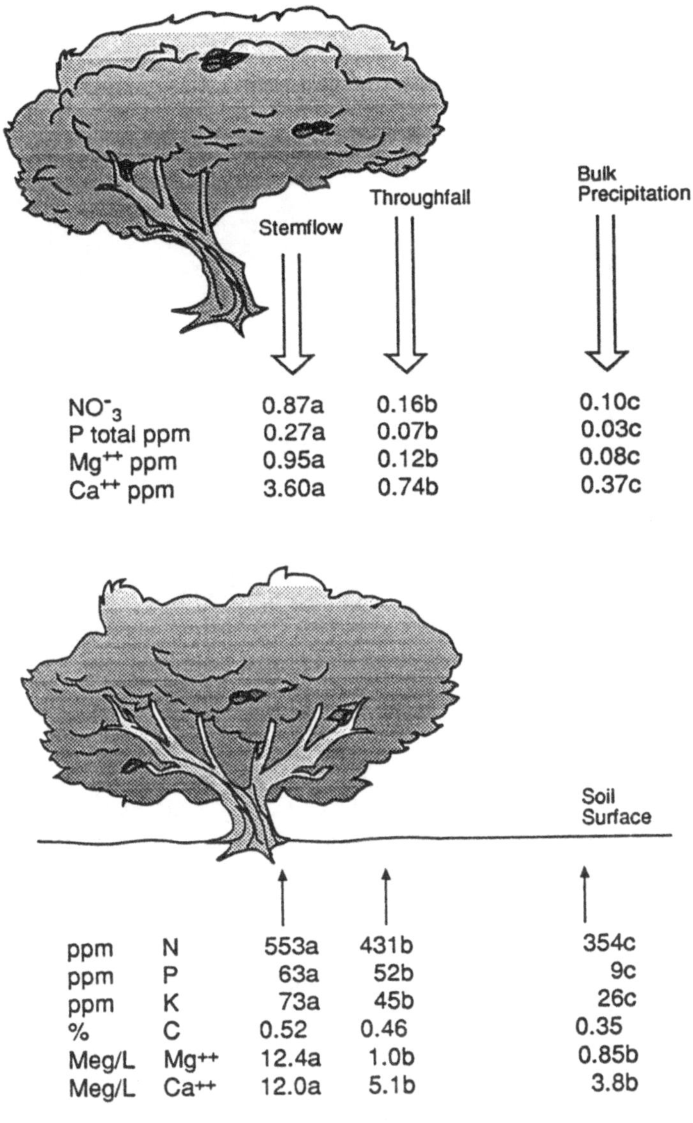

Fig. 1. Chemical composition of stemflow, throughfall, bulk precipitation, and soil chemistry affected by creosotebush (*Larrea tridentata*). Upper panel compares concentrations (ppm) of chemicals measured in stemflow, throughfall, and bulk precipitation. The bottom panel compares chemical concentrations in the soils adjacent to the stem of the shrub, soil under mid-canopy, and soil at the mid-point of the intershrub space. Letters following the numbers show significant differences in concentration of a chemical in different locations or sources at $p < 0.05$.

Water translocated to deep storage along root channels serves to initiate spring growth even in years with no effective winter-spring rainfall. The importance of this deep water source is evident in the phenologies of both creosotebush and mesquite and to a lesser extent tarbush. Mesquite and creosotebush initiate leaf production

Fig. 2. Rain-out shelter used to impose drought conditions on the shrub and surrounding soil in the enclosed, trenched plot.

and growth between mid-April and mid-May every year, even when winter-spring moisture is inadequate to initiate growth of herbaceous perennials or annuals. In drought years, tarbush will produce extremely small leaves in late spring and early summer. The ability to produce foliage despite inadequate rainfall to support that production is an important adaptation of these desert shrubs that allows them to dominate sites. Thus water redistribution, together with the enhanced mineral nutrient content of soils under shrub canopies, should provide the resources needed to make shrubs relatively independent of climatic constraints and contribute to the resistance and resilience of shrub-dominated ecosystems. This hypothesis was tested experimentally using repeated, consecutive summer drought and irrigation to examine the responses of a creosotebush (*Larrea tridentata*) ecosystem.

3. Drought and Rainfall Augmentation Experiments

One test of the resilience of an arid ecosystem is the recovery of that system from drought. We had an opportunity to examine the resistance and resilience of a creosotebush-dominated ecosystem in which we had 24, 3 m $\times$ 5 m plots assigned at random to drought, 2 X summer rainfall, and unmanipulated controls. Eight plots were irrigated at 2-week intervals with enough water to double the long-term monthly mean. Because of the variability of rainfall, the irrigated plots regularly received more than twice the long-term monthly average. Drought was achieved by "rain-out" shelters built of a steel frame (1.5 m high on the sides and 2.2 m high at the roof peak). The steel roof beams were covered with large-mesh chicken wire, which supported the ultraviolet-light-resistant greenhouse polyethylene covering the roof to exclude rain (Figure 2). In 1987 and 1988 the roof plastic sheets were rolled down at night and as storms approached. During that time period research focused on nitrogen mineralization patterns, and growth increments of the creosotebushes were not measured. From June 1989 through October 1991 rain-out shelter roof covers were installed in mid-June and removed in mid-September. With permanent roof coverings, the daytime air temperatures under the roofs were elevated 3–5 °C higher than in the control plots. This increase in temperature exacerbated the drought conditions of the plot, adding stress to the system and thereby more exactingly testing the resistance and resilience of the stressed system.

Growth was measured by painting a reference point on a terminal growth point 1 mm from the stem apex (approximately two leaf nodes) and measuring the total new-stem length added over a specified time period. Reference branches were tagged in late October-early November and were measured in late April-mid-May. Different branches were tagged in May and measured in September-late October, depending upon the year. Branch growth in creosotebushes exhibits an early summer growth spurt, a leveling off of growth in mid-summer, and a small late-summer growth surge (Fisher *et al.*, 1988).

Litter fall was collected in 30 cm × 30 cm trays built of aluminum window screen. Two trays were placed at random under the canopy of each shrub in a plot. Trays were emptied into plastic bags at the end of a sample period, returned to the laboratory, dried at 60 °C for 72 hours, and weighed.

The nitrogen mineralization potential and soil nitrogen content were measured on soil samples collected from under the canopy of shrubs in each plot. Nitrogen mineralization was measured by the batch incubation method of Fisher *et al.* (1987).

Residual effects of the experimental treatments were monitored by censusing all perennial plants in the plots in 1992 and 1993. Annual plants were censused on all plots in 1992. In 1993 densities of annual plants were estimated by a count of all annuals within five 40 cm × 40 cm quadrats placed at random within each plot.

Leaf litter decomposition was measured in 240 creosotebush leaf litter bags, 15 cm × 15 cm (fiberglass mesh 1.5 mm) filled with approximately 7.0 g (initial weight) of oven-dried creosotebush leaves. Sets of 10 litter bags were assigned at random to each plot of each treatment. Litter bags were placed under the shrub canopy at the mid-point of the canopy in July 1990. One leaf litter bag per plot was then retrieved at 6-month intervals and dried, cleaned, and weighed. Root decomposition was measured on 96 roots of the perennial herb, *Cassia bauhinioides*. Roots were washed and oven dried at 50 °C for 72 hours. Two roots per set were weighed and tagged with an aluminum tag, and two sets of roots were buried

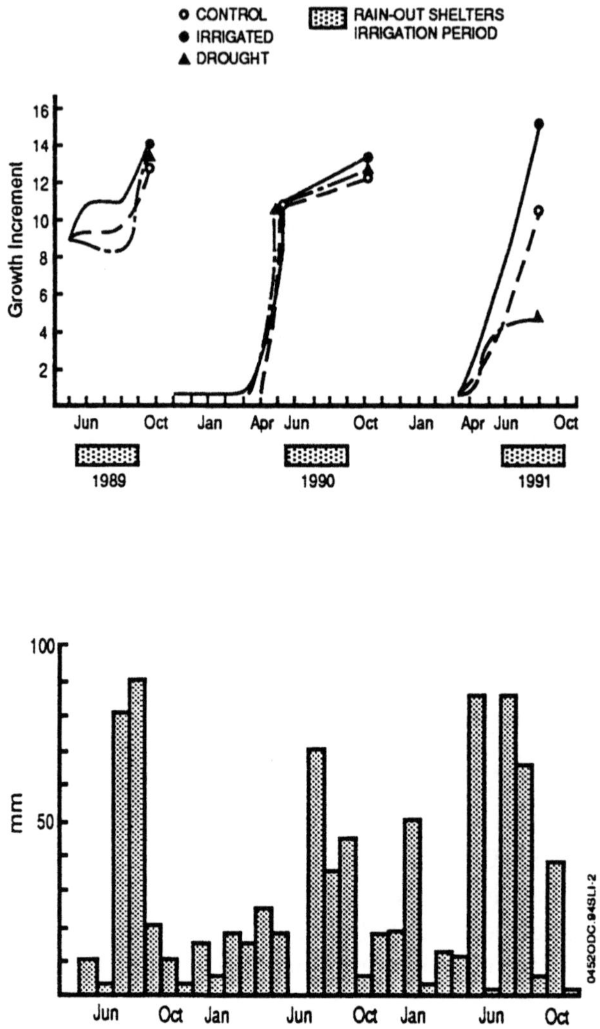

Fig. 3. The effects of drought and irrigation on incremental growth in *Larrea tridentata*. The bottom panel shows the monthly rainfall during the years of the experimental study. Symbols indicate actual measurements. Trajectory lines of growth have been inferred from data in Fisher *et al.*, 1988 and Lajtha and Whitford, 1989.

(parallel to the soil surface) at 5 cm depth in each plot. Sets of roots were retrieved at the end of October 1990 (end of the summer rainy season) and in March 1990. Root material remaining was brushed to remove adhering soil particles, oven dried at 50 °C for 72 hours, and weighed to estimate mass loss.

Data on growth measurements, annual and perennial plant densities, and decomposition were analyzed by ANOVA using the General Linear Model Procedure (PROC GLM) of SAS (SAS Institute Inc., 1985). Statistically significant treatment

TABLE I

Effects of rain exclusion (dry) and irrigation (wet) on soil inorganic N ($NH_4 + NO_3 + NO_2$) and on mineralization potentials (inorganic N accumulated in a 2-week incubation). Data are for soil samples collected approximately 2 weeks after the end of treatments. Numbers followed by different letters significantly differ at $p < 0.05$

	Soil Inorganic N mg N/kg soil	Mineralization potential mg N/kg soil
Control	2.1 a	9.6 a
Dry	3.7 b	10.7 b
Wet	2.3 a	9.2 a

effects were identified with the Least Significant Difference means comparison test.

4. Results and Discussion of Drought-Rainfall Augmentation Experiments

Summer rainfall during the years of the experiment was above average for July through September, the period of drought application and irrigation (Figure 3). In 1989 and 1990 growth increments were measured more than 30 days after the rainout shelter roofs were removed. Exposure of the droughted plots to even a small amount of rainfall following the "drought" period resulted in growth increments that did not differ among treatments (Figure 3). Since the pattern of measurement produced no difference due to drought, growth increments were measured in 1991 immediately after the roof covers were removed and before rainfall affected the uncovered plots. Treatments significantly differed at that time. Growth increments, however, were not further measured in 1991 to determine if the plants on the control and drought plots reached the growth increment of the shrubs on the irrigated plots. Because of the large coefficients of variation (30% irrigated, 85% control, and 50% drought), the growth increments significantly differed at $p < 0.05$ between the drought and irrigated plot shrubs, but the growth of shrubs in the control plots did not significantly differ from shrubs receiving treatments.

The growth increment data demonstrate that creosotebush (*L. tridentata*) rapidly recovers following drought and that irrigation does not benefit this species in the long term. The resilience of this dominant species following drought is an important factor contributing to the persistence of creosotebush-dominated ecosystems in areas that were historically grass dominated. The failure of irrigation to provide a long-term differential in stem elongation and biomass production is the result of the depletion of available nitrogen and the nitrogen requirements of this species (Fisher

TABLE II

Effects of rain exclusion (dry) and irrigation (wet) on creosotebush litter fall collected in 30 cm × 30 cm litter traps. Data are average g/litter trap. Rainfall was excluded for periods marked with *. Dates represent beginning of a sampling period, which ran until the next date listed. Letters show significant difference at $p < 0.10$

	7/7/87*	10/28/87	3/30/88	6/30/88–10/7/88*
Control	82 a	50	48	98 ab
Dry	45 b	53	46	72 b
Wet	84 a	58	54	126 a

TABLE III

Effects of rain exclusion (dry) and irrigation (wet) on biomass (g/m^2) of annual plants collected on the dates shown. U indicates under shrubs, O indicates intershrub space, and OU indicates data combined for both locations because there was no location effect. Letters show significant difference at $p < 0.10$

	May 88 U	May 88 O	Oct 88 OU
Control	16.3	3.9 a	6.9 a
Dry	14.6	14.9 b	0.8 b
Wet	15.9	7.0 b	2.9 ab

et al., 1988). In earlier studies of drought and irrigation effects on available nitrogen and nitrogen mineralization potential, the increases in available nitrogen and the rate of mineralization approximately 2 weeks after termination of treatments were higher in the drought plot soils than in the control or irrigated plot soils (Table I). These data suggest that the rapid recovery in growth following drought can be attributed to the increase in available nitrogen in the droughted plots.

Another feature of the growth dynamics of creosotebush in response to drought and irrigation is change in the dynamics of litter fall. During drought litter fall significantly decreased but was not significantly affected by irrigation (Table II). When treatments were terminated, litter fall amounts were the same on all plots (Table II). This feature of the growth of creosotebush exhibited the same resilience pattern that stem growth showed in later years of the experiment.

Litter fall patterns are related to leaf production and stem growth, with leaves being retained during periods with little or no growth (Lajtha and Whitford, 1989). This pattern of leaf demography allows for rapid carbon gain as soon as water is

TABLE IV

Average densities (no./m^2) of the perennials: fluff grass (*Erioneuron pulchellum*) and the forb *Bahia absinthifolia* on plots subjected to drought and irrigation in the growing season for 6 consecutive years (1987–1992). Data were collected in May 1992 and May 1993. No perennial grasses or forbs were in the drought plots. Data for species that occurred in fewer than four plots are reported as present. Numbers followed by different letters are significantly different at $p < 0.05$

	1992		1993	
	Control	Irrigated	Control	Irrigated
Erioneuron pulchellum	10.3 ± 12.7	18.9 ± 10.5	3.0 ± 3.6a	22.3 ± 10.2b
Bahia absinthifolia	4.4 ± 4.3	6.9 ± 4.5	3.8 ± 3.9	6.6 ± 7.1
Number of plots with species present				
Muhlenbergia porteri	2	3	3	2
Opuntia phaeacantha	2	5	2	3
Zinnia acerosa	2	1	2	2
Gutierrezia sarothrae	1	2	1	2
Aristida longiseta	0	1	0	1
Average cover: cm^2/plant				
Erioneuron pulchellum	23.4 ± 3.2a	5.1 ± 0.7b	No data	
Bahia absinthifolia	23.2 ± 9.0a	6.9 ± 1.5b	No data	

available following a drought. This pattern of leaf demography contributes to the resilience of the system by allowing for immediate response to rainfall.

By changing the timing and quantities of available water and nutrients, drought and irrigation changed the production patterns of annual plants on the experimental plots. Drought during the previous growing season resulted in increased biomass production of spring annual plants and marked differences in the biomass production of annual plants in the intershrub space compared to controls (Table III). But the drought treatment resulted in a significant decrease in the biomass of summer annuals, and the irrigation also reduced the biomass of summer annuals compared to controls (Table III).

After 5 consecutive years of growing season drought and irrigation, perennial grasses and forbs had disappeared on the drought plots. Repeated growing-season irrigation significantly affected the density and size of the dominant perennials (Table IV). The higher densities of fluff grass and *Bahia absinthifolia* in the irrigated plots were offset by the smaller size of the plants in comparison to those in the control plots. The decrease in fluff grass on the control plots in 1992 was not matched on the irrigated plots (Table IV). The changes in fluff grass between 1992 and 1993 may reflect demographic differences in the fluff grass populations in the treatment and control plots. Numbers of other species of perennials on the irrigated plots slightly increased in comparison to the controls (Table IV). The other

TABLE V

Densities of spring annual plants (number/m^2 $\pm$ standard deviation) on plots subjected to repetitive drought and/or irrigation from 1987–1991. Numbers in parentheses are numbers of plots on which a species was found

	1992			1993		
	Control	Wet	Dry	Control	Wet	Dry
Erdi	3.8 ± 1.9	1.9 ± 1.0	20.3 ± 4.8	8.3 ± 5.8	7.1 ± 5.7	39.2 ± 34.7
Depi	0.6 ± 0.2	0.2 ± 0.2	0.6 ± 0.3	15.7 ± 19.0	13.0 ± 7.3	33.2 ± 8.8
Lela	0.9(2)	0.1(2)	1.6(6)	0.4(2)	0	2.6(5)
Ertr	0.6(3)	0.3(3)	1.6(3)	0.2(2)	0.3(2)	0.4(2)
Erab	0.2(2)	0.3(3)	0.3(3)	0	0	0
Cran	0	0.4(4)	2.1(5)	0.1(1)	0.5(3)	7.5(5)
Crmi	2.3(4)	5.3(4)	8.0(7)	4.1 ± 3	8.7 ± 10	7.6 ± 9
Total	6.5 ± 2.7	2.9 ± 1.9	22.5 ± 18.4	29.5 ± 6.2	28.9 ± 11.4	104.0 ± 62.0

Erdi – *Eriastrum diffusum*, Depi – *Descurainia pinnata*, Lela – *Lepidium lasiocarpum*, Ertr – *Eriogonum trichopes*, Cran – *Crypthantha angustifolia*, Crmi – *Crypthantha micrantha*.

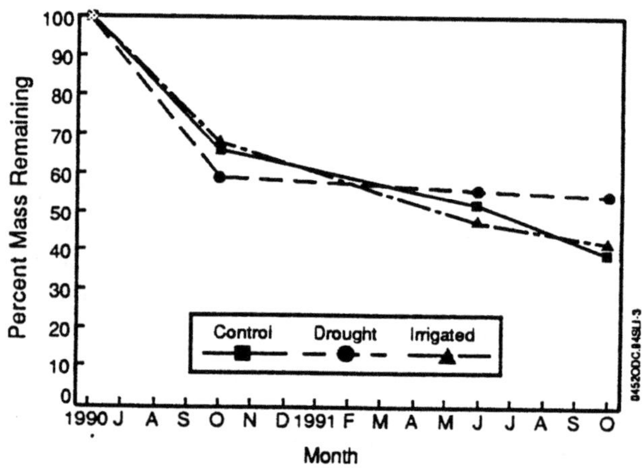

Fig. 4. The effects of drought and irrigation on the decomposition of creosotebush (*Larrea tridentata*) leaf litter.

species were present on fewer than half of the plots, and no statistical analysis was performed.

The absence of perennial grasses and forbs on the drought plots may be the primary reason for the increased densities of spring annuals on the drought plots in 1992 and 1993 (Table V). The differences in spring annuals appear not to be related to soil nutrient differences among treatments because the most responsive species are those shown not to respond to fertilization (Gutierrez and Whitford, 1987).

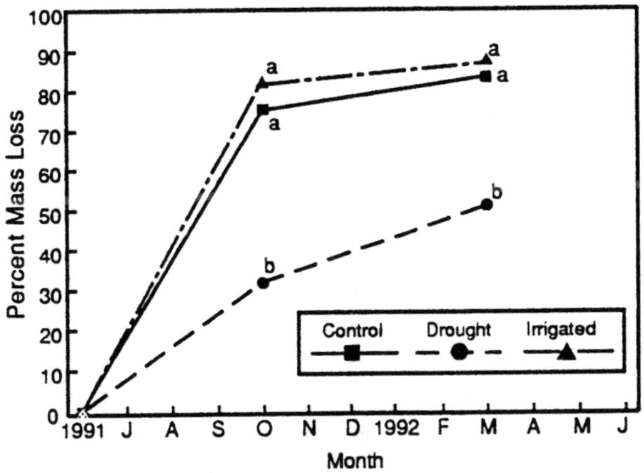

Fig. 5. The effects of drought and irrigation on the decomposition of roots of the perennial herb, *Cassia bauhinoides*. Points with different letters significantly differ at $p < 0.01$.

Decomposition rates of surface litter were slightly affected by drought and not at all by irrigation (Figure 4). Root decomposition was not affected by irrigation but was affected by drought (Figure 5). In roots most of the mass was lost within the first 4 months in the control and dry plots. During the next 6 months the roots in irrigated and control plots lost approximately 8% of their initial mass, while the roots in the drought plots lost 37% of their initial mass (Figure 5). Extrapolating the results of these measurements to a second full year would be unlikely to reveal large differences in the mass remaining among treatments, suggesting that in this ecosystem, recovery from drought and/or above-average rainfall has little long-term effect on decomposition and release of nutrients from dead plant material.

These studies confirm the hypothesis that islands of resources develop under creosotebush (*L. tridentata*) and that the concentration of these resources results in an ecosystem that is extremely resilient to drought stress. The rates of recovery of components and processes within this ecosystem differ, but all reach predisturbance levels within 2 years. The experiments were conducted over five consecutive growing seasons during which summer drought was not found to measurably harm the shrubs. Drought and supplemental rainfall, however, did affect the density and biomass of perennial herbaceous plants (Whitford, 1994). Drought reduced the abundance and biomass of one perennial forb, *Bahia absinthifolia*, and one grass, *Erioneuron pulchellum*. Irrigation affected only the fluff grass, *E. pulchellum*, (Whitford, 1994).

Although drought affected the decomposition of roots, it had little effect on the decomposition of leaf litter. The decomposition of roots and the effects of drought on the microbial biomass were translated into temporal changes in the pulses of available nutrients. The increased availability of nitrogen in the droughted plots contributed to the rapid recovery of the shrubs and to the increased biomass production of spring annual plants.

4.1. Shrub Removal and Recovery

One measure of the resilience of shrub-dominated ecosystems is the recovery of those systems after disturbance. We examined the recovery of plots treated in 1979 with the herbicide tebuthiuron applied at rates varying between 1.3 kg/ha and 2.6 kg/ha. The treatment resulted in nearly a complete kill of the coppice dune mesquite (*Prosopis glandulosa*). In 1993, we ran ten 100 m line transects across the treated area and ten 100 m line transects across an adjacent, untreated, mesquite coppice dune area. The average grass cover was 2.3% on the herbicide-treated coppice dune area and 10.2% on the untreated area. Mesa dropseed (*Sporobolus flexuosus*) and fluff grass (*Erioneuron pulchellum*) were the dominant grasses on both sites. The frequency of *P. glandulosa* on the treated and untreated plots in 1993 did not differ: 15.7$\pm$7.1/line treated; 14.3$\pm$3.7/line untreated reference area. The average canopy cover per plant was 0.28 m^2 on the treated area and 9.8 m^2 for the coppice dune plants. Thus, only 14 years after shrub removal, mesquite had re-established on a site that was mesquite free in 1979. Most of the young mesquites in the herbicide-treated area were growing on the residual mounds of the treated coppice dunes. Despite the smaller size of the shrubs in the herbicide-treated area, these shrubs will eventually reach the size of those on the untreated coppice dunes.

5. Conclusions

The historical trend of desertification in North America has been the replacement of perennial grassland ecosystems with shrub-dominated ecosystems (Grover and Musick, 1990). This experimental study demonstrates that the creosotebush ecosystem remains relatively unchanged after 5 consecutive years of growing season drought. The persistence of this ecosystem when subjected to long-term stress suggests that there is little probability that the shrub-dominated ecosystem will revert to grassland or that the ecosystem will undergo a transition to another state. Applying herbicide to kill shrubs often results in a short-term shrub-free ecosystem that reverts to a shrub-dominated ecosystem. The permanence and resilience of the "desertified" ecosystems implies that these ecosystems should be managed for the values that can be derived from them in their present state. The failure of shrub-control measures and other manipulations to change structure and function of the desertified ecosystems in the long term is further evidence that management should focus on the values that can be derived from the degraded systems or on improving one ecosystem service such as enhanced water storage rather than attempting large-scale reconstruction.

References

Archer, S., C. Scifres, C.R. Bassham and R. Maggio. 1988. Autogenic succession in a subtropical savanna: Conversion of grassland to thorn woodland. *Ecological Monographs* **58**: 111–127.

Buffington, L.C. and C.H. Herbel. 1965. Vegetational changes on a semi-desert grassland range from 1858 to 1963. *Ecological Monographs* **35**: 139–164.

Charley, J.L. and N.E. West. 1975. Plant-induced soil chemical patterns in some shrub-dominated semi-desert ecosystems of Utah. *Journal of Ecology* **63**: 945–964.

Charley, J.L. and N.E. West. 1977. Micro-patterns of nitrogen mineralization activity in soils of some shrub-dominated semi-desert ecosystems of Utah. *Soil Biology and Biochemistry* **9**: 357–365.

Fisher, F.M., L.W. Parker, J.P. Anderson and W.G. Whitford. 1987. Nitrogen mineralization in a desert soil: Interacting effects of soil moisture and nitrogen fertilizer. *Soil Science Society of America Journal* **51**: 1033–1041.

Fisher, F.M. and W.G. Whitford. 1995. Field simulation of wet and dry years in the Chihuahuan Desert: Soil moisture, N mineralization and ion-exchange resin bags. *Biology and Fertility of Soils* **20:**: 137–146.

Fisher, F.M., J.C. Zak, G.L. Cunningham and W.G. Whitford. 1988. Water and nitrogen effects on growth and allocation patterns of creosotebush in the northern Chihuahuan Desert. *Journal of Range Management* **41**: 387–391.

Gibbens, R.P. and R.F. Beck. 1987. Increase in number of dominant plants and dominance classes on a grassland in the northern Chihuahuan Desert. *Journal of Range Management* **40**: 136–139.

Glendening, G.E. 1952. Some quantitative data on the increase of mesquite and cactus on desert range in Southern Arizona. *Ecology* **33**: 319–328.

Grover, H.D. and H.B. Musick. 1990. Shrubland encroachment in southern New Mexico, U.S.A.: An analysis of desertification processes in the American southwest. *Climatic Change* **17**: 305–330.

Gutierrez, J.R. and W.G. Whitford. 1987. Chihuahuan Desert annuals: Importance of water and nitrogen. *Ecology* **68**: 2032–2045.

Hennessy, J.T., R.P. Gibbens, J.M. Tromble and M. Cardenas. 1983. Vegetation changes from 1935 to 1980 in mesquite dunelands and former grasslands of southern New Mexico. *Journal of Range Management* **36**: 370–374.

Lajtha, K. and W.H. Schlesinger. 1986. Plant response to variations in nitrogen availability in a desert shrubland community. *Biogeochemistry* **2**: 29–37.

Lajtha, K. and W.G. Whitford. 1989. The effect of water and nitrogen amendments on photosynthesis, leaf demography, and resource-use efficiency in *Larrea tridentata*, a desert evergreen shrub. *Oecologia* **80**: 341–348.

Martinez-Meza, E. and W.G. Whitford. In press. Stemflow, throughfall, and root water channelization by three Chihuahuan Desert shrubs. *Journal of Arid Environments*.

Nulsen, R.A., K.J. Bligh, I.N. Baxter, E.J. Solin and D.H. Imrie. 1986. The fate of rainfall in a mallee and heath vegetated catchment in southern Western Australia. *Australian Journal of Ecology* **11**: 361–371.

Parker, L.W., H.G. Fowler, G. Ettershank and W.G. Whitford. 1982. The effects of subterranean termite removal on desert soil nitrogen and ephemeral flora. *Journal of Arid Environments* **5**: 53–59.

Pressland, A.J. 1975. Rainfall partitioning by an arid woodland (*Acacia aneura* F. Muell.) in south-western Queensland. *Australian Journal of Botany* **21**: 235–245.

Pressland, A.J. 1976. Soil moisture redistribution as affected by throughfall and stemflow in an arid zone shrub community. *Australian Journal of Botany* **24**: 641–649.

Rappole, J.H., C.E. Russell, J.R. Norwine and T.E. Fulbright. 1986. Anthropogenic pressures and impacts on marginal, neotropical, semiarid ecosystems: The case of south Texas. *Science of the Total Environment* **55**: 91–99.

Reichman, O.J. 1984. Spatial and temporal variation of seed distribution in Sonoran Desert soils. *Journal of Biogeography* **11**: 1–12.

Roundy, B.A. and G.L. Jordan. 1988. Vegetation changes in relation to livestock exclusion and rootplowing in southeastern Arizona. *Southwestern Naturalist* **33**: 425–436.

SAS Institute Inc. 1985. *SAS user's guide: statistics, version 5*, SAS Institute Inc., Cary, North Carolina.

Virginia, R.A. 1986. Soil development under legume tree canopies. *Forest Ecology and Management* **16**: 69–79.

Virginia, R.A. and W.M. Jarrell. 1983. Soil properties in a mesquite dominated Sonoran Desert ecosystem. *Soil Science Society of America Journal* **47**: 138–144.

Virginia, R.A., W.M. Jarrell, P.W. Rundel, D.H. Kohl and G. Shearer. 1989. The use of variation in the natural abundance of ^{15}N to assess symbiotic nitrogen fixation in woody plants, in P.W. Rundel, J.R. Ehleringer and K. Nagy (Eds.), *Stable isotopes in ecological research*, Ecological Studies Series Volume 68, Springer-Verlag, Heidelberg and New York, pp. 375–394.

West, N.E. and J.O. Klemmedson. 1978. Structural distribution of nitrogen in desert ecosystems, in N.E. West and J.J. Skujins (Eds.), *Nitrogen in desert ecosystems*, Dowden, Hutchinson and Ross, Stroudsburg, Pennsylvania, pp. 1–16.

Westoby, M., B. Walker and I. Noy-Meir. 1989. Opportunistic management for rangelands not at equilibrium. *Journal of Range Management* **42**: 266–274.

Whitford, W. G. 1994. Desertification: Implications and limitations of the ecosystem health metaphor, in D. J. Rapport, C. Gaudet and P. Calow (Eds.), *Evaluating and monitoring the health of large scale ecosystems*, Springer-Verlag, Heidelberg and New York.

A FIELD STUDY OF THE INFLUENCE OF LAND MANAGEMENT AND SOIL PROPERTIES ON RUNOFF AND SOIL LOSS IN CENTRAL SPAIN

ANDREW G. WILLIAMS*, J. LES TERNAN*, ANDY ELMES*,
MARTA GONZALEZ DEL TANAGO** and RAOUL BLANCO**
*Department of Geographical Sciences, University of Plymouth, Plymouth, Devon, PL4 8AA, United Kingdom; **Escuela Superior de Ingenerios de Montes, Universidad Politecnica de Madrid, Madrid, Spain

Abstract. Since the 1950s afforestation of degraded land has been the principal means of combatting erosion in a seasonally arid area of central Spain. In the 1970s tree planting of steep hillsides and gully sides was preceded by bench terracing. Experimental sites have been established to monitor runoff and soil losses under mature *Pinus* forest, 12-year-old *Pinus* forest, and *Cistus* matorral. The experiment is being conducted at three scales: large gully or small watershed (c. 3.5 ha), runoff erosion plot (10–21.5 m^2), and rainfall simulation plot (1 m^2). Monitoring began in October 1992. Discharge was recorded continuously, while sediment loss and soil moisture content were measured on a storm basis. The paper presents summary data on runoff and soil erosion for the three scales and comments on relationships between land management, site characteristics, and these losses. We stress the crucial role of vegetation and its interrelationship with soil properties such as structure and aggregate stability. Matorral was effective in combatting water and soil loss, but we question the practice of afforesting seasonally arid, steeply sided areas that have highly erodible soils.

1. Introduction

Substantial land has degraded in central Spain as a result of past and present land use. For many years the major approaches in dealing with the problem of accelerated soil loss were to afforest hillslopes in headwater areas and to build structures such as check dams to arrest sediment removal from the gullies. The objective of afforestation in Spain, as previously done, for example, in Britain, was to protect soil rather than produce timber, but the ecological effects were often overlooked. Research in Britain has shown that the impact of upland afforestation may frequently be deleterious to the environment at the watershed scale (Calder and Newson, 1979; Newson, 1992; Peterken, 1993). But in central Spain few experiments to date have examined either the hydrological or ecological impact of afforestation, its effectiveness at erosion control, or indeed the impacts of different forestry practices. At a time when the forestry sector plans to invest almost 4 billion pesetas to afforest about 8.4 million ha, such studies are of crucial importance.

The object of this paper is to describe how an experiment was set up to collect baseline data to determine the influence of the different types of land management – mature *Pinus* forest, 12-year-old *Pinus* forest, and *Cistus* matorral – on runoff and soil loss at different scales. We will discuss methodological aspects of evaluating

Environmental Monitoring and Assessment **37:** 333–345, 1995.

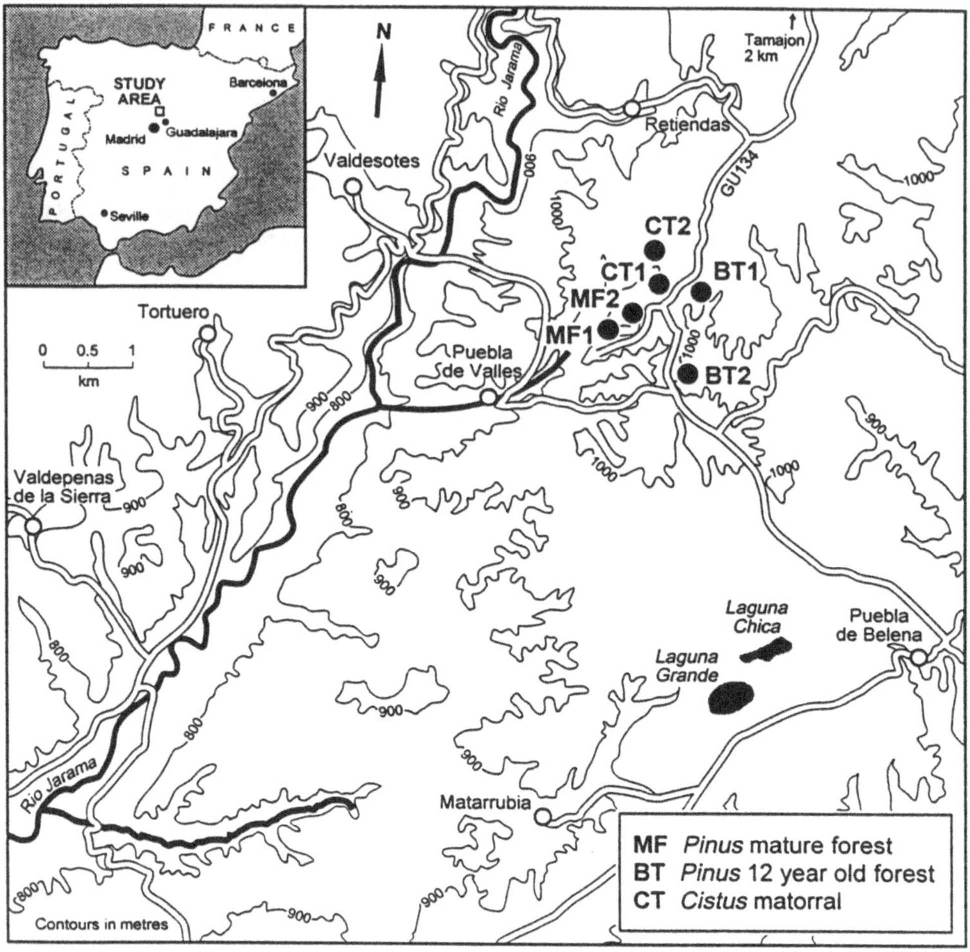

Fig. 1. Runoff-erosion study area showing location of gully catchments.

the impact of afforestation at the gully, erosion plot, and point scale of rainfall simulation plots. The experimental sites were established under the European Community R & D Programme in the Field of the Environment (Desertification in the Mediterranean Area).

2. The Study Area and Land Use

The study was conducted in the Puebla de Belena – Puebla Vallees – Retiendas area, some 70 km northeast of Madrid (Figure 1). The locality receives 700–800 mm of rain per year (Munoz *et al.*, 1989), with an average maximum daily intensity of 20 mm/day. Summer temperatures frequently exceed 35 °C, there is a precipitation deficit from May to September, and the area may be described as seasonally arid.

The dominant sediments, locally referred to as alluvial deposits of the Rana formation, are Tertiary to early Quaternary. The landscape has a series of wide depositional and planation surfaces with steeply incised valleys and badlands developed on the margins. Soils are classified as alfisols and have a distinctive argillic horizon (Espejo Serrano, 1985). The fine earth fraction ranges from silty loam to sandy silt loam with a highly variable stone content.

In common with many such Mediterranean or seasonally arid areas, the indigenous mixed oak woodland has been replaced almost entirely on the hillslopes by *Cistus*-dominated matorral and by cultivated dryland farming on the plateaus. The dominant vegetation type is matorral, which is found over substantial tracts of gently sloping land. Common species include *Cistus landanifer*, *Cistus laurifolius*, and *Rosmarinus officinalis*. The *Cistus* plants are often 1.5 m high, and where not degraded, the ground surface is completely covered. Limited nomadic sheep grazing continues on these areas of *Cistus* scrub, and occasionally the vegetation has been scraped away to provide young, less woody vegetation for grazing.

Soil erosion has caused the loss of productive topsoil and has also been responsible for reservoir sedimentation in the area. A policy of afforestation of degraded land has been pursued since the 1950s as the main means of combatting erosion to protect the reservoirs, often without regard to the environmental costs (Ternan *et al.*, 1992). A conifer plantation was established on former cultivated land along the spine of one of the plateaus. Principal species are *Pinus nigra* and *Pinus pinaster*. Planted in the 1950s, the trees have reached maturity. Litter cover of the ground surface is usually 100%, and there is no ground layer of herbs or grasses. The steeper hillslopes and gully walls were afforested in the 1970s. Ground preparation was substantial: bench terraces were built with bulldozers, and their treads were subsoiled. In general the treads and risers were about 2 m across with a vertical distance of 1 m between treads. These operations exposed substantial areas of subsoil, including the clay-rich Bt horizon. The treads were planted with *Pinus pinaster*. This afforestation has had limited success: the trees have a low survival rate, many show signs of stress, and their heights are mostly between 1.5 and 5 m.

3. Experimental Design and Techniques

Developing a methodological approach to study the influence of land management on runoff and erosion in an area must consider the complexity of the ecosystem together with the interrelationships between components. Other points for consideration include scale of impact and the most suitable temporal and spatial scale for the investigation. Ideally, one should be able to include the work within a framework of future research as further needs and policies develop. In addition, this work must all be conducted within a finite budget.

One of the major dilemmas faced by field scientists is how to investigate the underlying processes of land degradation at the level at which policies can be applied. In many circumstances the Newtonian approach offers a useful framework within which to study natural systems. This approach is based on breaking down the system into its components in an effort to understand how they function, eventually leading to the discovery of universal laws (Perez-Trejo, 1989). In some circumstances the complexity of systems, together with Newtonian assumptions about equilibrium and fixed structures, may make applying this approach inappropriate (Perez-Trejo, 1989). But the so-called top-down approach has many advantages for field scientists trying to grapple with problems of scale-dependent variability as determined by physical processes.

From a pragmatic point of view, following Newson (1992), much can be gained from a nested approach to scales of study. One can consider a hierarchy of scales ranging from that of a watershed to plots within that watershed and to points within that level. At the landscape scale the collective effects of individual processes are masked as they are integrated. For example, surface runoff produced on areas of soils of low hydraulic conductivity may be lost on adjacent high conductivity areas. Likewise, sediment transported from some areas may be deposited when the water infiltrates. At the more-detailed erosion plot scale, different response units as determined by slope, soil, and vegetation types can be defined, and the relationships between runoff and soil loss with controlling factors investigated. By isolating key parameters, this approach facilitates an understanding of the way the degradation processes function to control the watershed response.

One of the fundamental parameters controlling runoff and erosion is vegetation, which influences erosivity and erodibility. Vegetation protects the soil surface from raindrop impact but more particularly has a crucial indirect role in determining soil structure and structural stability. Soil structure affects the retention, transport, and redistribution of soil water in a profile, all of major importance in generating surface runoff and sediment transport. Stability of aggregates also resists erosive forces and the detachment of soil particles. The influence of vegetation can be studied at a variety of scales. Imeson *et al.* (1992), for example, demonstrated that one useful approach may be to consider the fundamental significance of individual plants to the erosion system and to set this information into the wider landscape context.

The gully scale is both the primary unit of hydrological research and also the minimum scale at which land use policies are implemented. At this scale inputs and outputs can be simply measured. Following from these measurements, one can calculate discharge and erosion rates per unit time, rainfall-runoff ratios, and water balances. Considerable information regarding water pathways, flow routes, and sediment production and transport can, therefore, be derived. At this gully scale the runoff and soil loss of smaller interlinked response units are combined to produce an averaged watershed response.

The object of the gully-scale investigation was to monitor the runoff and erosion loss from catchments under different land managements. Ideally, all other factors such as soil type, topography, and aspect would be similar, but inevitably compromises were made. Sites were selected to ensure that land use was uniform,

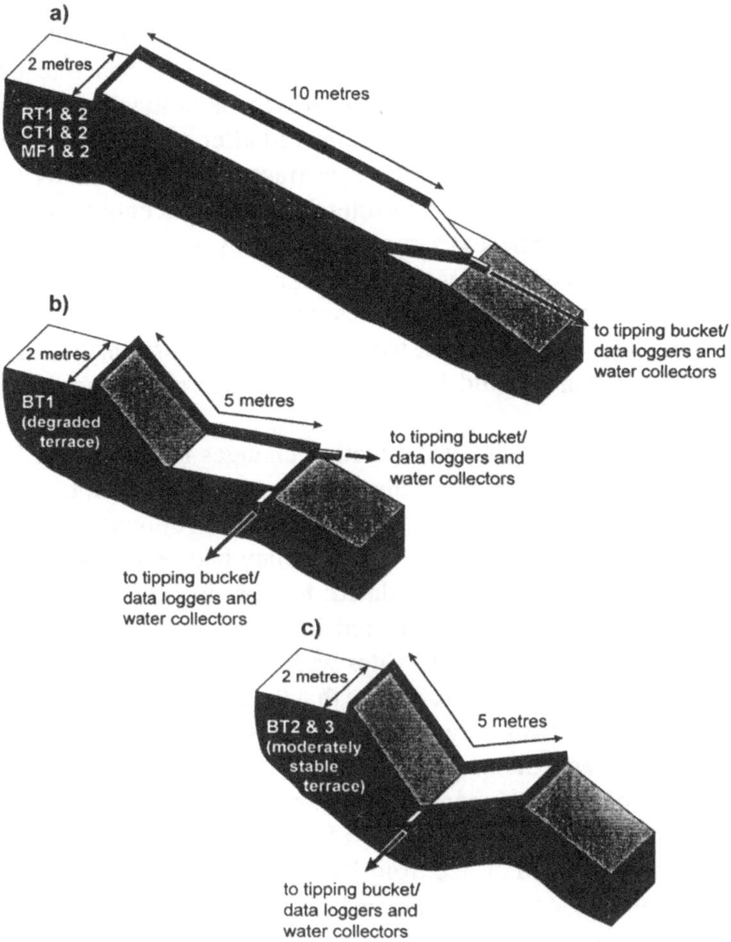

Fig. 2. Design of runoff erosion plots in a) *Cistus* matorral and *Pinus* mature forest b) & c) *Pinus* 12 year old, bench terraced site.

channel orders were the same, the divide was clearly defined, and there were no check dams.

Six gully catchments were instrumented, two being set up in each of the management units, i.e. mature forest, 12-year-old forest, and matorral. These catchments ranged from 2.4 to 5.8 ha. Channels are deeply incised at the head of the gullies, and slope forms are complex convexo-concave in shape, with steep slopes (15–20°) adjacent to the channel and gentle slopes (5°) toward the divide.

Runoff and sediment loss from the ephemeral channels were monitored at the exit of the gully catchments. A particular problem was that local factors, including microrelief, variable source areas, and channel roughness at the catchment exit, could significantly affect the measured response. The gauging stations were therefore set up in relatively uniform locations where the channel was relatively straight and incised. Conventional stage discharge recorders were used, and a stable rated section was provided by thin weir plates installed in the bed of the channel. Sediment loss was monitored by collecting sediment on the concrete apron behind the weir. Suspended sediment concentration was determined using rising stage samplers. Where possible, bedload was removed after every storm. In the laboratory this bedload was weighed and the size fraction determined by wet seiving. Similarly, the suspended sediment was filtered and its concentration calculated.

At the next level in the nested sampling hierarchy, erosion plots were set up to measure runoff and soil loss under the three management types (Figure 2). Annual and storm-based runoff and sediment losses per unit area could be compared because the physical characteristics were reasonably uniform. As with the gully scale, the storm response could be monitored to determine runoff patterns and coefficients.

Furthermore, at this level of investigation changes in controlling factors such as soil moisture and crusting could also be monitored. The plots were bounded and therefore isolated from the rest of the hillslope following the MEDALUS recommendations (MEDALUS, 1991). This policy had several advantages: water and sediment balances could be calculated; topographic, soil physical characteristics, and plant cover could be measured in some detail; and the results could be compared with MEDALUS plots. Sites were selected using a stratified random approach. The main criterion was to ensure that the slope angles of the erosion plots in the mature forest and *Cistus* areas were similar. Replicate plots were established in the mature forest and *Cistus* plots, while in the 12-year-old forest three plots were established to reflect the complex form of the bench terraces (Figure 2) and the variety of cover.

The erosion plots ranged in size from 10 to 21.5 m^2 and were bounded by sheet aluminium inserted 15 cm below the surface and protruding 15 cm above. The area at the plot exit was concreted to ensure that the lip did not erode. Flow was routed via a tipping bucket system plus switching device and datalogger to a large tank that collected all the runoff and sediment. Suspended sediment samples were collected after every storm and concentrations measured as before. Soil moisture was monitored at 10 points per plot on a frequency determined by the storms.

A series of 1 m^2 rainfall erosion plots were also established where possible within the sampling grids to examine the runoff and sediment losses under controlled conditions. The influence of land management and vegetation cover was investigated: two sites within mature forest, 100% leaf litter; three sites within the 12-year-old forest, on a vegetated riser, vegetated tread, and unvegetated tread; and two sites in the matorral, one with 100% cover and the other with 50% cover. A rainfall simulator designed at Silsoe College, U.K., was used for the experiment. A rotating boom, 2m above the ground surface was set to deliver 45 mm/hr. Runoff was monitored using a tipping bucket and data logger system, and a water sample was collected every 10 minutes to determine sediment loss. Lasting 1 hour, each simulation was divided into periods of 10 minutes spraying and 3 minutes without

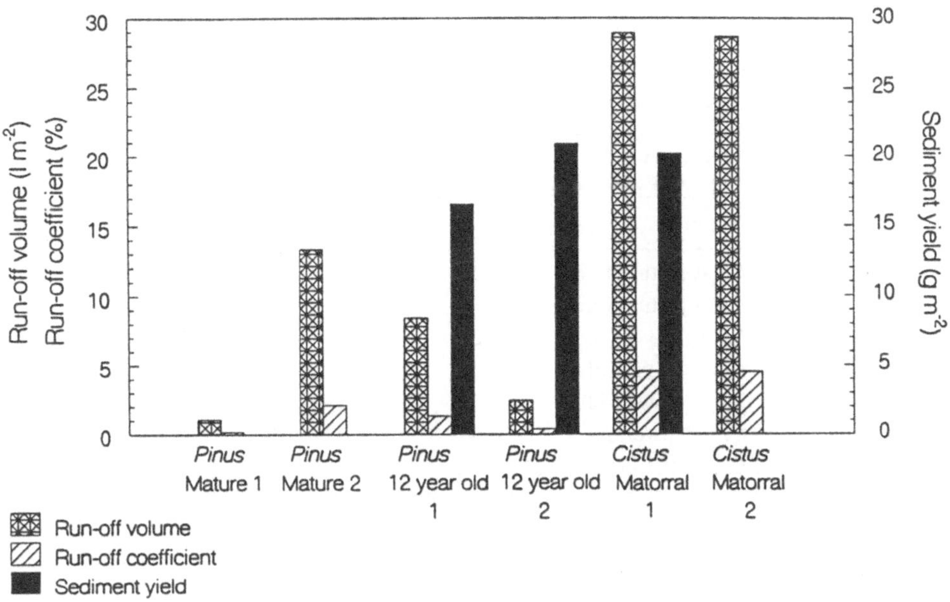

Fig. 3. Runoff and sediment yield from the gullies (30 Nov. 1992–7 Nov. 1993).

TABLE I

Runoff and sediment yield from the gullies (30 Nov. 1992 to 7 Nov. 1993, rainfall = 634.1 mm)

Vegetation and management	Area (ha)	Runoff (l/m^2)	Runoff volume coeff (%)	Sediment yield ($g\ m^{-2}$)
Pinus mature forest	2.4	13.3	2.1	0.0
Pinus mature forest	2.6	1.1	0.2	0.0
Pinus 12-year-old	2.4	8.4	1.3	16.6
Pinus 12-year-old	3.2	2.5	0.4	20.9
Cistus matorral	5.8	28.9	4.6	20.2
Cistus matorral	4.2	28.7	4.5	0.0

spray. During this time volumetric soil moisture contents were measured. At the end of each rainfall simulation experiment the biomass was weighed and the percent stone cover recorded. Saturated conductivity was measured in the field using a ring permeameter technique.

Rainfall was measured at a point using a standard "Rimco" rain gauge with a tipping bucket connected to a logger. Recorded values of incident rain were assumed to be representative of the larger gully scale.

TABLE II

Runoff and sediment yield from erosion plots (30 Nov. 1992 to 7 Nov. 1993, rainfall = 634.1 mm).

Vegetation and management	Runoff (l/m^2)	Runoff volume coeff (%)	Sediment yield ($g\ m^{-2}$)
Pinus mature forest	0.3	0.1	0.3
Pinus mature forest	0.2	0.1	0.2
Pinus 12-year-old	149.2	23.5	318.3
Pinus 12-year-old	43.2	6.8	82.8
Pinus 12-year-old	32.1	5.1	5.1
Cistus matorral	0.6	0.1	0.7
Cistus matorral	0.2	0.1	0.4

4. Results and Discussion

Gully scale: runoff and sediment yield data for the six instrumented gullies are presented in Table I and Figure 3. Runoff losses per unit area were in the order *Cistus* > mature *Pinus* > 12-year-old *Pinus*. Similarly, average rainfall-runoff coefficients were highest in the matorral. These results are probably more of a function of the local site conditions than of the gully catchments as a whole. The stony floor of the gully beds, i.e. thalweg, encourages infiltration. Runoff, therefore, strongly reflects the proximity of bare gully slopes generating stormflow to the gauging station rather than land use management.

Table I shows that runoff coefficients were very low for both mature *Pinus* (0–2%) and bench terrace sites (0–1%), in contrast to the higher values recorded for the *Cistus* gullies (4%). The low runoff coefficients in the mature *Pinus* forest and bench-terraced area are partly due to the many storages available, including tree canopies, litter retention, and ground surface depressions.

On average, the 12-year-old *Pinus* forest recorded the highest sediment yield (Table I, Figure 3). Given the relatively low runoff per unit area, erosion was very active at the immature forest sites. Field observation showed that gullies were extending along the line of the bench terrace treads and that considerable material was being transported and deposited at the back of the treads.

In summary, probably the main reason for the higher runoff losses in the *Cistus* matorral at the gully scale was simply a reflection of runoff-generating areas being located adjacent to the gauging station. Conversely, high sediment losses from the 12-year-old forest were due to the influence of management practices and in particular the construction of bench terraces.

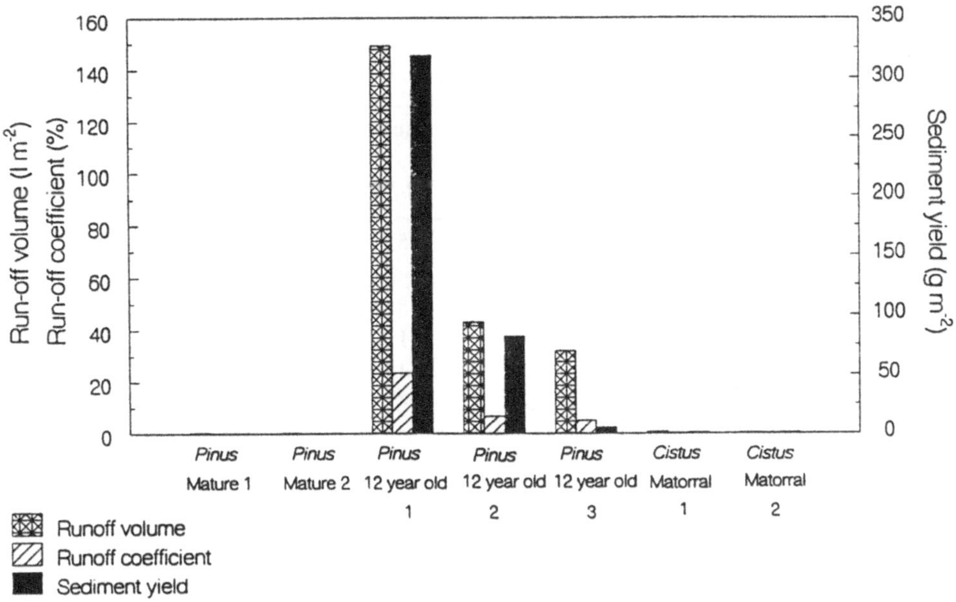

Fig. 4. Runoff and sediment yield from the erosion plots (30 Nov. 1992–7 Nov. 1993).

The gully-scale investigations alone may not be the most appropriate for investigating the influence of management on runoff and erosion since the measurements may reflect local conditions at the monitoring point rather than the larger gully system. A number of problems regarding exactly how the gully systems operate are difficult to resolve for two reasons:

1. Collective effects are averaged at the landscape scale.
2. Point measurements may not be suitable for representing whole areas.

Such problems include a consideration of whether the matorral soil surface sheds water but not sediment and the major controls on runoff generation in the bench terrace area. Questions about geomorphic processes can be answered only at a more-detailed plot scale.

Results presented in Table II and Figure 4 for the erosion plots show that runoff at the bench terrace sites was two orders of magnitude greater than either the mature *Pinus* or *Cistus*. Similarly, the sediment yield was also substantially higher. Runoff and sediment losses per unit area were as follows: 12-year-old *Pinus* >> *Cistus* = mature *Pinus*. The erosion plot data in Table II, therefore, presents a rather different runoff-erosion picture than the data for the gully scale in Table I.

Runoff coefficients were about 0.1% in the mature *Pinus* and matorral sites. The thick layer of pine needles and the lichen cover in the matorral plots seem to be controlling runoff. Walsh and Voight (1978) discuss the crucial role of pine needles retaining and releasing surface water. They found that substantial amounts of water can be held and transmitted without infiltrating to the soil surface. Soil moisture contents were least in the mature pine because of interception and transpiration

TABLE III
Runoff and sediment yield from rainfall simulation plots

Vegetation and management	Biomass kg/m^2	Runoff (l/m^2)	Runoff volume coeff (%)	Sediment yield (g m^{-2})
Pinus mature forest	0.9	25.6	48.0	0.8
Pinus 12-year-old, bare grnd	0.0	34.5	66.0	51.0
Pinus 12-year-old, riser	2.6	18.5	42.0	1.2
Cistus matorral bare	0.0	41.2	56.0	18.9
Cistus matorral int.	1.7	4.9	11.0	0.4
Cistus matorral mature	6.6	'6.1	19.0	1.2

losses in the forest. Volumetric soil moisture contents were highest in the bench-terraced areas due to the predominantly bare nature of the ground surface. The higher clay content, due to the ground disturbance and exposure of the subsoil, may also be important. Detailed soil moisture content measurements across the bench terraces showed that the soil moisture was low on the riser (6%) but was greatest at the lower riser/tread boundary (31%). Runoff coefficients were highest (23%) for the 12-year-old bench-terraced *Pinus* area because for much of the study period the soil was close to saturation (27% compared to 36–44% at saturation).

In a study of the role of land management and aggregate stability for the area, Ternan *et al.* (in press) found that the water stable-aggregate index (WSA) >3.0 for the *Cistus* matorral exceeded 80% (range 84–91%, mean 87%, n = 5), and the WSA of the mature forest ranged from 33–81% (mean 61%, n = 7). Later research found WSA for the bench terrace area to be only 1–55% (mean 25%, n = 9). Sediment losses from the mature *Pinus* area and matorral were minimal due to their high to moderate aggregate stability. These areas, therefore, had a lower potential for detachment. Furthermore, the soil losses of up to 318 g/m^2 under 12-year-old forest were probably exacerbated by the high runoff (149 l/m^2) (Table II, Figure 4).

Runoff and erosion data for the rainfall simulation plots are presented in Table III and Figure 5. Substantial amounts of runoff were recorded from the bare *Cistus*, bench-terraced *Pinus,* and mature *Pinus* plots compared with the relatively small volumes from the intermediate and mature *Cistus* plots. Runoff volumes were consistently higher in plots with high soil moisture contents.

Saturated areas, limited in extent, were observed in the bare areas of the unvegetated *Cistus* and the bench-terraced immature forest rainfall simulation plots. Such saturated areas may have arisen from development of a surface crust or possibly because microtopography controlled the flow pattern. Mualem *et al.* (1993), for example, reported that in a sealed profile the wetting front is shallower and the water content lower, except for the upper surface soil layer, which becomes saturat-

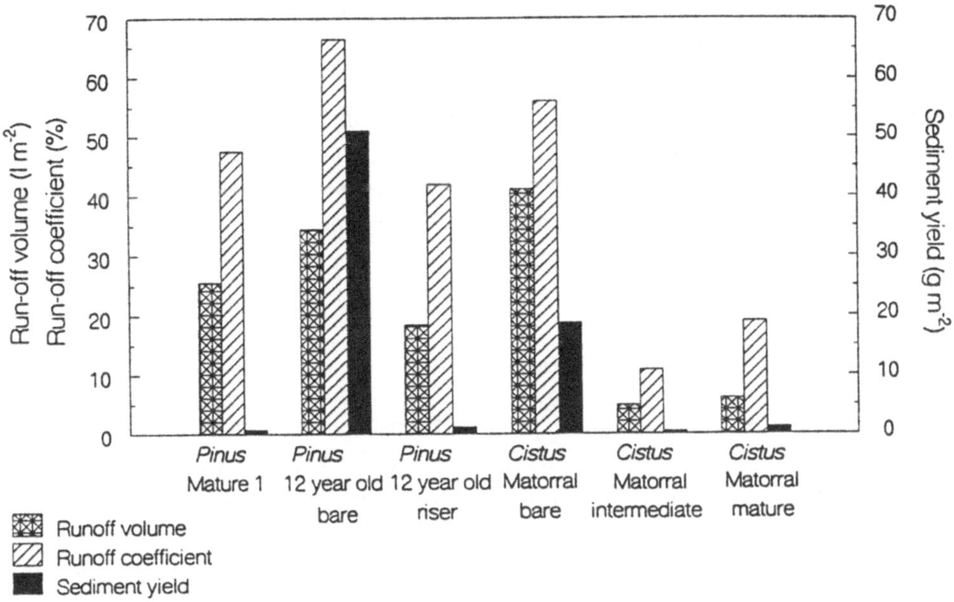

Fig. 5. Runoff and sediment yield from the rainfall simulation plots.

ed very quickly, leading to overland flow generation. Detailed examination of the soil moisture patterns showed that crusting probably exercised a dominant control on runoff only on the bare plots. Soil moisture contents rose relatively little there, and much of the incident rain was transmitted overland. On vegetated plots, however, the steady rise in moisture content during the experiment, the higher values at the lower edge of the plots, and the higher values at depth all suggest that water is transmitted rapidly through the soil and that a crust did not form on these soils.

Sediment yield was substantial from the bare plots of the bench terrace and *Cistus* areas, and significent volumes of runoff were derived from both these and the two other *Pinus* sites (Table III). Erosion loss is therefore controlled mainly by the detachment of soil particles by splash or previous mechanical action rather than the transporting ability of the overland flow. This result was confirmed by analysis of the sediment concentrations through time: in all cases the sediment concentration declined rapidly through time as the supply of loose material became exhausted.

The bare *Cistus* site showed no evidence of features usually associated with matorral, such as organic matter in the soil, the presence of lag gravels with good lichen cover, or indeed a well-developed lichen crust. High values of sediment loss recorded from the bare plot in the *Cistus* area (Table III) are therefore considered to be atypical.

Results from the rainfall simulations on the tread of the bench-terraced sites suggest that the high runoff and sediment yield were due to the poor physical characteristics of the soil, such as poor structure, low hydraulic conductivity, low aggregate stability, and tendency to crusting. The soil characteristics are due in part to the design of the bench terrace in which the topsoil is buried and subsoil is exposed at the back of the tread. The poor physical properties at the junction of the tread and riser plus the relatively high soil moisture content there due to the terrace profile combine to lead to substantial degradation. Forest management, which desired to prevent erosion by afforesting the area, has exacerbated the problem by building bench terraces. Although soil erosion may decline as the bench terraces become vegetated, it is uncertain how much the system will degrade before that goal is reached.

5. Conclusion

The hierarchical sampling scheme provided a useful framework for investigating runoff and soil losses. Summary data presented for the erosion plot and rainfall simulation plot scale showed the importance of the role of land management and its influence on vegetation and key soil properties such as structure, aggregate stability, and hydraulic conductivity. Preliminary findings from this experiment suggest that, while the effectiveness of matorral in combatting water and soil loss can be clearly demonstrated, the effects of afforesting seasonally arid, steeply sided areas with highly erodible soils must be carefully considered.

Acknowledgements

Financial support for the research was provided by the European Union R & D Programme in the Field of the Environment (Land Management Practice and Erosion Limitation in Contrasting Wildfire and Gullied Locations in the Iberian Peninsula), Contract EV5V-0041. We gratefully acknowledge assistance in the field by Kevin Solman and Richard Hartley.

References

Calder, I.R. and M.D. Newson. 1979. Land use and upland water resources in Britain – A strategic look. *Water Resources Bulletin* **15**(6): 1628–1639.

Espejo Serrano, R. 1985. The ages of soils of two levels of 'rana' surfaces in central Spain. *Geoderma* **35**: 223–239.

Imeson, A.C. 1992. The effects of fire and water repellancy on infiltration and runoff under Mediterranean type forest. *Catena* **19**(3–4): 345–361.

MEDALUS. 1991. *Mediterranean desertification and land use field manual* (compiled by L.H. Cammeraat). Amsterdam.

Mualem, Y., S. Assouline and D. Eltahan. 1993. Effect of rainfall induced soil seals on soil water regime: Wetting processes. *Water Resources Research* **19**(6): 1651–1659.

Munoz, M.J., R.A. Aldeanueva and J.M.R. Arnaiz. 1989. El clima de la Provincia de Guadalajara paralelo 37°. *Revista de Estudios Geographicos.* **13**: 227–252.

Newson, M.D. 1992. *Land, water and development: River basin systems and their sustainable management*, Routledge, London.

Perez-Trejo, F. 1989. Impact assessment methodologies for complex natural systems, in P.M. van der Staal and F.A. van Vught (Eds.), *Impact forecasting and assessment: Methods, results, experiences*, Delft University Press, Delft, The Netherlands, pp. 143–157.

Peterken, G.F. 1993. *Woodland conservation and management*, 2nd edition, Chapman and Hall, London and New York.

Ternan, J.L., A.G. Williams, A. Elmes and R. Hartley. In press. Aggregate stability of soils in central Spain and the role of land management. *Earth Surface Processes and Landforms.*

Ternan, J.L., A.G. Williams and M. Gonzalez del Tanago. 1992. Soil properties and gully erosion in the Guadalajara Province, central Spain, in J. Rickson (Ed.), *Proceedings of the First International European Society for Soil Conservation Congress*, Silsoe, United Kingdom.

Walsh, R.P.D. and P.J. Voight. 1978. Vegetation litter: An underestimated variable in hydrology and geomorphology. *Journal of Biogeography* **4**: 253–274.

DESERTIFICATION IN RUSSIA: PROBLEMS AND SOLUTIONS (AN EXAMPLE IN THE REPUBLIC OF KALMYKIA-KHALMG TANGCH)

IGOR S. ZONN
First Vice-President, Russian National Committee for the United Nations Environment Programme, Moscow, Russia

Abstract. During the second half of the 20th century Kalmykia has undergone severe desertification. Under Soviet rule, rangelands were increasingly devoted to animal production, and pastures were converted to cropland in a campaign to increase crops. Pastures were grazed two to three times their sustainable production, saiga populations and habitat greatly decreased, more than 17 million ha were subjected to wind erosion, 380,000 ha were transformed into moving sands, and 106,000 ha were ruined by secondary salinization and waterlogging. By the 1990s almost 80% of the Republic had undergone desertification, and 13% had been transformed into a true desert. In 1986 the General Scheme of Desertification Control was formulated. The scheme called for rotating pastures, reclaiming blown sand using silviculture, tilling overgrazed pastures and sowing fodder plants, and developing water supplies for pastures. In its early years the scheme has been successful. But the management of restored pastures usually reverts to the same farms responsible for the poor conditions, and there is great apprehension that degradation could reoccur. This case study concludes that the general cattle and agriculture development in Kalmykia is unviable for ecological and economic reasons, that Kalmykia should implement an adaptive policy oriented toward conservation and accommodating the interrelation and variability of land resources, that the desertification problem can be solved only by changing agrarian policy as a whole, and that a desertification control program must become an integral part of economic and social development of the Republic.

1. Introduction

The Republic of Kalmykia – Khalmg Tangch – occupies a peculiar place on the map of the world. Kalmykia is a nation isolated in race (Mongoloid), language (Altaian family, Mongolian branch), religion (Buddhist), and socioeconomics (long-term development of extensive animal husbandry). A geographically extensive and ecologically isolated part of the Russian Federation, Kalmykia belongs to a pastoral regional economic complex of the Caspian Sea basin. The natural resource potential of the Caspian area is limited to horse breeding, sheep and cattle production for meat, and maintenance of wildlife habitat. Natural semishrub and grass pastures, long warm growing seasons, and established traditions of cattle production favor this type of natural resource use.

Previously the Republic was burdened with a centrally administered command decisionmaking system. This political system completely ignored the traditions, customs, and established natural resource management systems of the native population. The spontaneous development of agriculture, industry, and transportation over the past 70 years has disregarded possible ecological consequences of development, contributing to the overexploitation of natural resources and degradation

Environmental Monitoring and Assessment **37:** 347–363, 1995.

of pasture ecosystems. This degradation has resulted in the sustained decline in productivity known as desertification. One might argue that these actions led to the only "desert" in Europe created by human activity.

The problem of desertification in Russia was raised only after the disintegration of the Former Soviet Union (FSU), when arid territories that had formerly been part of the FSU found themselves outside Russian boundaries but suffering from ecological problems that were a legacy of FSU rule. The State Report (1992) *On the State of the Environment in the Russian Federation in 1991* noted, "The condition of arid areas in the Russian Federation is impaired in general; their degradation and desertification is of considerable scale. The most prominent example of desertification is in the region of Black Lands (Kalmykia) and Kizlyar pastures where a desert was formed and is progressively increasing now." Thus, Kalmykia is one of 13 regions of Russia in a very grave ecological situation, and on August 1, 1993, the President of Kalmykia pronounced a state of ecological emergency in the Republic.

World and FSU maps do not show Kalmykia as a region of desertification because it is a small, easily overlooked area with a limited natural resource base that can support only free-ranging animal husbandry and thus has limited economic potential. But within the Russian Federation this region has acquired importance as an all-Russian and European zone of desertification.

Originally, desertification in Kalmykia was recognized in the UNEP/USSR Project "Desertification Control by Integrated Approach" (Zonn, 1986). Later, when UNEP assessed global desertification and revised its original concept, Kalmykia was featured as an example of a FSU region that had experienced severe desertification (Rozanov, 1990; Dregne *et al.*, 1991).

2. Geography and Natural Resources

The Republic of Kalmykia is situated in the south of European Russia on the western shore of the Caspian Sea (the shoreline length is 110 km within the Republic). Kalmykia's total area is 76,100 km^2. Its three distinct areas reflect differences in geological past and current natural conditions. The Caspian Lowland in the east, divided into Black Lands and the Sarpa Lowland in the north, is a plain, the highest parts of which extend 15–20 m above sea level but the greater part of which lies below sea level. The Yergeni Upland in the southwest, the oldest part of Kalmykia, forms the watershed between the Caspian Lowland and the basin of the Lower Don. The Kumo-Manych Valley is a depression in the southwest. From the mid-Tertiary it was repeatedly flooded by sea water that formed the strait connecting the Caspian and Black seas. When these waters retreated, the Kuma, Western, and Eastern Manych river valleys were formed, and in these valleys were built the Proletarian and Chorai reservoirs (Figure 1).

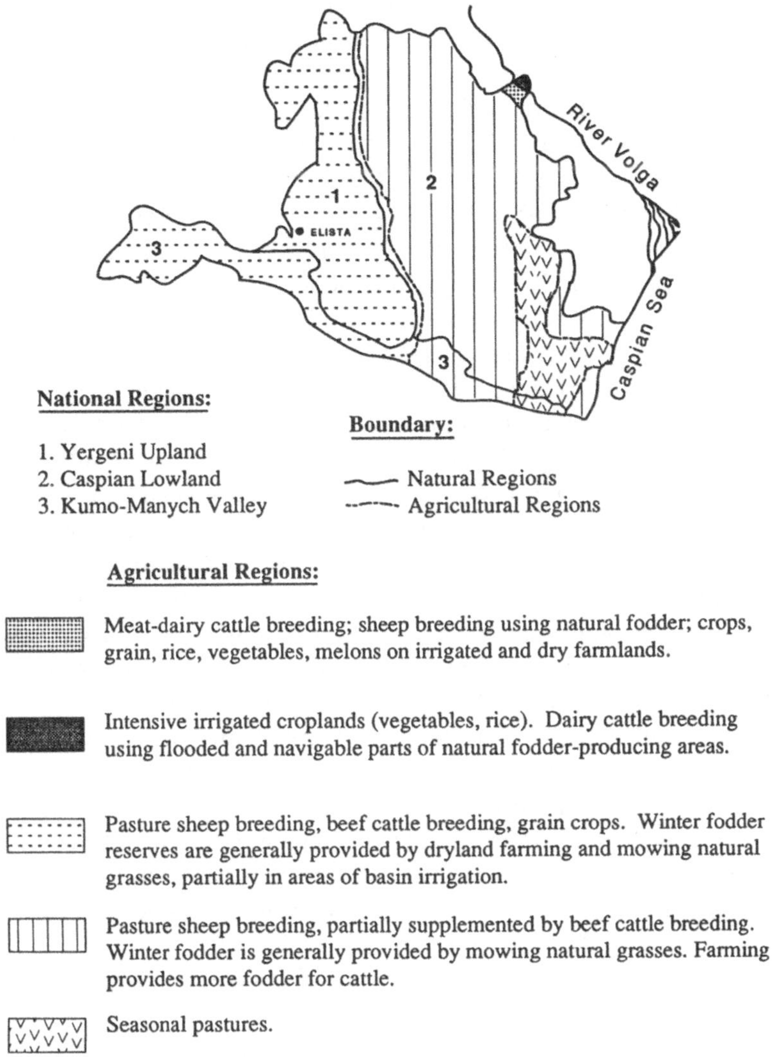

Fig. 1. Natural agricultural regions of Kalmykia.

Kalmykia's climate is strongly continental: generally warm and dry with long hot summers and rather severe but stable winters. Average temperatures in January range from −8 to −4.5 °C, and in July from 23 to 26 °C. The growing season is about 213 days. Average annual precipitation ranges between 200–350 mm/year, with 75 percent occurring during the warm part of the year. Winters typically have little if any snow.

High air temperatures and low humidity often lead to the formation of dry winds. During this period wind velocity reaches 15–20 m/sec, air temperature rises to +35 °C, and relative air humidity falls to 10–20%, resulting in dust storms. In

1969 a dust storm that originated on the Black Lands reached Poland. According to space survey data, in 1984 a storm from the same region reached France. The probability of dry and arid years in the region is 88–95%. In aridity, Kalmykia is exceeded only by the deserts of Central Asia.

The Republic is water deficient. Rivers flowing to the Manych Depression dry up during summer, while water in the tributaries to the Yergeni – the Elista, Sarpa, Yashkul – is more abundant. In both cases, the rivers flow to lakes or dry inland deltas (Sarpa Lake in the north, Sosta lakes in the south, and several lakes in the central part of the Kumo-Manych Depression). In the geologic past, these lakes were closely connected to the Caspian Sea basin, a connection reflected by groundwater pattern. The depth of groundwater and salinity changes towards the coastal strip from 6 to 10 m with 15–30 g/l on the Yergeni and up to 2–3 m with 40 g/l near bays. Fresh water lenses (1–4 m below the surface) are generally related to sandy soil and subsoil of the Black Lands. When tapped by wells for watering cattle, shallow groundwater is rapidly depleted and replaced by salt water.

The soils of Kalmykia are of either the brown or chestnut types. Brown soils developed in the south and southeast parts of the Caspian area, and chestnut soils developed in the northern part of this region on the Yergeni and over floodplain terraces of the Manych Valley. Chestnut soils occur in complexes with solonetz and meadow chestnut soils in microdepressions. Among other soils, meadow, solonchaks, takyrs, and wind-blown sands occur in the region.

Vegetation includes Eurasian steppe and Afroasian desert steppe types. Steppe is found in the most arid subzone-Zavolzhsko-Kazakhstanian desert steppe, and desert in the less arid Northern Touranian steppe desert. Desert steppes are found in the Yergeni Upland, the eastern part of the Manych Valley, and the south and southeast of the Caspian area. Presently, true diversified grass-soddy cereal and soddy-cereal steppes are almost completely given over to cultivation. Desert steppes are extensive, occupying 40% of the Republic's area.

Depending on their texture, soils are further divided into four types. Xerophyte desert steppe is developed on loamy light-chestnut soils of the Yergeni and eastern part of the Manych Valley and has a white wormwood-tipchak-feather-grass stand. *Agropyron desertorum* is an indicator of these soils. Plant canopy cover of these steppes is 35–40%, and productivity amounts to 300–400 kg/ha. The semi-sandy desert steppe (hemipsammophyte) lies in the southeast of the Caspian area with loamy sand and light loamy soils. Its grass vegetation is the white wormwood-tipchak-tyr type. *Agropyron fragile* is an indicator of these soils. Plant canopy cover is between 30 and 35%. The sand desert type (psammophyte) is developed on brown sands of the same part of the Caspian area, where dominant plants are *Agropyron fragile* and white wormwood. The halopsammophyte desert steppe is related to saline sandy soils of littoral area. Apart from psammophytes and grasses, there are also halophytes and thickets of *Tamarix ramosissima.*

In the North-Touranian deserts in Kalmykia low shurb communities occupy 31.8% of the area. These communities are related to solonetz, and their composition therefore changes according to variations in soil. Most typical desert communities are developed on crusted solonetz. Stands are composed of haloxerophile low shurbs, including *Camophorosma monospeliaca* and black wormwood (*Artemisia panciflora*).

According to its subsoil, vegetative cover, climate features, and topography, the territory of Kalmykia is considered a semiarid zone of steppe and semidesert.

3. Status of Desertification

In 1991, out of a total area of 7,610,000 ha, agricultural lands occupied 5,880,000 ha, including 904,000 ha of cultivated lands, 131,000 ha of hay lands, and 4,845,000 ha in unimproved pastures.

Historically, the Kalmyks pursued nomadic livestock breeding on pastures, similar to their predecessors, the Sarmatians, Huns, and Scythians. The Kalmyks had learned that the earth had its own laws to be observed in order to maintain a stable equilibrium. In spite of climate continentality, they had no need to store fodder because in winter the herds were driven to the Black Lands (Chernye Zemli), which were green all year round.

"Chernye Zemli" or "Khar Gazr" is one of the more unusual places in the south Russian steppes. For centuries this region has served as seasonal pasture. Kalmyks, who brought their cattle here as early as the 12th century, are said to understand the true value of these pastures where horses, "could graze almost without bending their heads." The steppe provided forage for thousands of fat-rumped sheep, horses, local breeds of Kalmykian cattle, and camels without endangering the million-head herd of relict sheeplike antelope known as saigas.

Nonetheless, the steppe is easily damaged. Within several centimeters of the surface lay quicksand that had been deposited on the former bottom of the Caspian Sea. But the trampling of comparatively flat hooves of fat-rumped sheep had little effect on grass stands. Also, in the southeast part of the steppe, where snow is rare and is seldom deeper than 20 cm, the grass stand was 0.5 m high, and long-term grazing was forbidden. To preserve pastures for winter, they were rested for the whole summer, and sheep were excluded.

Active development of the steppe began in the middle of the 19th century in the Lower Volga and Northern Caucasus, considerably increasing pressure on the Black Lands. To control this pressure, a special service existed to regulate grazing. In the autumn of each year, steppe congresses for grazing control were convened with representatives of the steppe administration and cattle owners. At the congresses Kalmyk pastures on the Black Lands were allotted for the coming winter, and trustees were elected to control the observance of grazing boundaries set up by the congress for the winter.

In 1913 about 300,000 humans inhabited Kalmykia, and about 1 million sheep, 300,000 cattle, 200,000 horses, and 20,000 camels were kept on the pastures. This mix of animals was considered optimal because it yielded the highest animal production per unit area while preserving the long-term stability of pasture productivity (Vinogradov, 1993).

Crop farming in Kalmykia was always a very risky enterprise because of the extremely dry climate and low crop yields. Therefore crops were cultivated on small plots and only in some settlements of the western steppe near the crop-farming regions of Povolzhye and North Caucasus. In 1913 agricultural production on the Kalmykian steppe occupied a little more than 110,000 ha, where spring wheat, winter wheat, oats, and millet were cultivated and yields ranged from 380–455 kg/ha. Crops often failed during years of famine. During dry years, crop yield dropped disastrously to 100–200 kg/ha. Under these conditions, the land could not return enough seed for sowing the following year's crop.

During Soviet rule, Kalmykia rangelands were increasingly devoted to animal production, mainly fine-fleece Karakul sheep and beef cattle on large-scale farms. More importantly and aside form this general policy, after alleged collaboration with the Germans, the Kalmykian people were collectively exiled to Siberia in 1943 and were not allowed to return to their lands until 1958. The Russian population that remained during the intervening period damaged pasture condition. Changes in pasture vegetation harmed animal populations, especially the saiga. The ill-conceived conversion of the Kalmykian economy was begun by sacrificing the Kalmykian people and their profound understanding of how to manage this delicate rangeland ecosystem.

By 1950 forage crop yields were inadequate for sustaining the 1.8 million head of sheep kept there. Desertification in the form of pasture degradation had begun but covered only 5% of the pasture area.

In the beginning of the 1960s a widespread campaign to increase agricultural production was begun. When forced to adopt uniform government policies that recognized no differences between areas, local authorities neglected or ignored the unique ecological character of Kalmykian steppes and their susceptibility to human impacts (salinization of subsoil, compaction of fine textured-soils, etc.). These attitudes and actions fully unleashed the mechanisms of desertification. The delicate balance between nature and man was quickly upset by a dramatic growth in animal population – mainly sheep – whose sharp hooves destroyed the plant cover. Within the last 70 years, the proportion of large animals in the animal population declined from 40% to 6.7% (Bananova, 1993).

In 1970 the sheep population reached 2.3 million head; in 1975, 2.9 million head; in 1980, 3.0 million head; and in the mid-1980s, 3.4 million head. According to official statistics, in 1913, 1 million sheep yielded more than 25 kg of meat per head, whereas in 1985, 3.4 million sheep yielded only 7–9 kg of meat per head. Thus, total production of meat remained relatively constant, but the number of animals increased by a factor of 3.5. Further, this estimate of sheep numbers is undoubtedly low because it does not include about 0.5 million sheep kept for individual use or another 1.5 million head kept illegally. In total, the number of

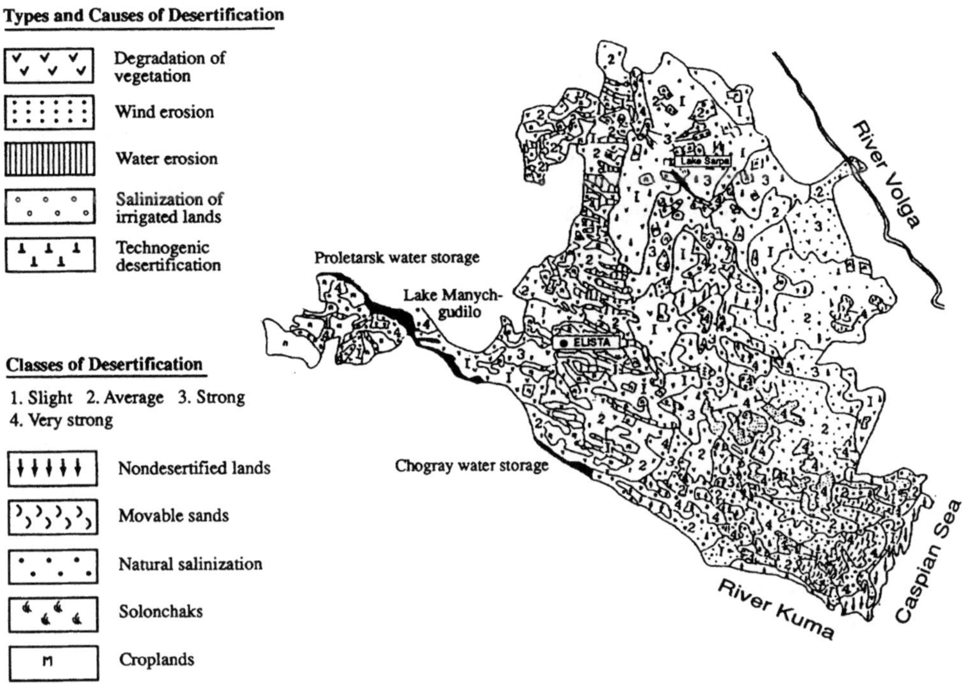

Fig. 2. Antropogenic desertification of the arids lands of Kalmykia (Bananova, 1993).

sheep could be 5 million head or more, suggesting overgrazing by a factor of 2–3 times sustainable production. Despite recommendations to bring the number of livestock into some balance with grazing capacity, the total herd in 1990 numbered nearly 3.2 million sheep and 348,000 cattle (Vinogradov, 1993).

"Chernye Zemli" sustained particularly heavy impact. The number of sheep that might be grazed there should probably not exceed 750,000. Now, however, there are 1,613,000 sheep. Today, the number of saigas ranges from 130,000 to 200,000, depending on the source of data. In 1978, their number was 780,000. In 1960, saigas ranged over an area of 100,000–110,000 km^2, but their range now has been reduced to 23,000 km^2. (The FSU holds the monopoly on saiga meat. The cost of three pair of saiga horns at foreign exchange rates is equivalent to the cost of 10–12 mink skins.)

Apart from the growth in livestock numbers, from the 1970s onward a policy was pursued to increase crop production. This policy included converting pasture to cereal and melon production as well as increasing the amount of cropped area under irrigation. Combined with other miscalculations in planning, mismanagement, and law violations, these activities resulted in sharp degradation of land resources, which manifested itself in accelerated desertification. During the 1970s, 37% of the Republic had experienced some degree of desertification (much of this accompanied the recession of the Caspian Sea level and consequent drying of

TABLE I
Present status of desertification in Kalmykia (1000 ha/%)

Desertification type	Desertification stages				Background	Total
	slight	average	strong	very strong	level	area
I. Zoogenic desertification Pasture degradation						
1. Overgrazing	359.2	1806.6	1556.9	175.0	339.2	4236.9
%	4.7	23.7	20.5	2.3	4.5	55.7
2. Wind erosion	–	43.1	–	298.3	–	341.4
%	–	0.6	–	3.9	–	4.5
3. Water erosion	196.3	51.0	–	40.5	–	287.8
%	2.6	0.7	–	0.5	–	3.8
Total:	555.5	1900.7	1556.9	513.8	339.2	4866.1
%	7.3	25.0	20.5	6.7	4.5	64.0
II. Zoogenic-technogenic desertification						
1. Grazing and extraction of mineral resources	14.6	106.9	298.7	495.6	–	915.8
	0.2	1.4	3.9	6.6	–	12.1
Total:	14.6	106.9	298.7	495.6	–	915.8
	0.2	1.4	3.9	6.6	–	12.1
III. Technogenic desertification						
1. Settlements, industry transport	–	–	–	345.7	–	345.7
%	–	–	–	4.5	–	4.5
2. Wind erosion, (plowing under dry land farming)	–	–	–	62.8	–	62.8
%	–	–	–	0.8	–	0.8
3. Salinization under flooding (during irrigation)	–	–	17.0	58.1	30.7	105.8
%	–	–	0.2	0.8	0.4	1.4
Total:	–	–	17.0	58.1	439.2	514.3
%	–	–	0.2	0.8	5.7	6.7
Total for anthropogenically desertified area	–	570.1	2024.6	1913.7	1448.6	6017.9
	–	8.5	26.6	25.2	19.0	79.1

coastal reed brushwood habitat, formation of maritime solonchaks, etc.). Figure 2 and Table I show the present status of desertification in Kalmykia. B.Rozanov (1991) noted that almost 13% of the whole territory has been transformed into a

TABLE II

Evolution and dynamics of agricultural development and related ecological changes in Kalmykia

Years	Political situation	Agricultural situation	Animal population		Ecological effects
1900–1917	Private property, noncollective natural resource management.	Livestock breeding on pastures (fat-rump sheep; mutton and wool)	Sheep Horses Cattle Camels	972,000 200,000 300,000 20,000	Stable equilibrium, maintenance of biological productivity. No evidence of desertification.
1920–1930	Acquisition of autonomy in Russian Federation.	Transition from semi-nomadic to sedentary livestock production.	Sheep Horses Cattle Camels Saigas	< 1,000,000 100,000 220,000 15,000 100	Desertification caused by extreme annual climate events. Spontaneous recovery of pasture vegetation and wild animals.
1931–1957	Socialist transformation of national economy; establishing collective and state farms; deportation of Kalmyks; economic ruin and restoration, 1941–45.	Sedentary livestock production; transition to fine-fleece sheep production (Karakul).	Sheep Horses Cattle Camels Saigas	1,800,000 10,000 137,000 10,000 2,000	Disruption of seasonal pasture rotation. Trampling of pastures (40–50%) and compaction of clay soils (20–30%). Increase in roads. Wind and water erosion. Desertification of 5% of territory.
1958–1970	Socioeconomic reorganization.	Replacement of fat-rumped sheep by Karakul.	Sheep Horses Cattle Camels Saigas	2,200,000 to 2,800,000 20,000 137,000 7,000 550,000	Desertification initiated due to increased numbers of animals and change in sheep breeds. Desertification of 37% of territory.

Table II. Continued.

Years	Political situation	Agricultural situation	Animal population	Ecological effects
1971–1990	Econonomic intensification.	Mechanization; increased livestock production; extension of irrigation and dry farming.	Sheep 3,100,000 to 3,500,000 Sheep (total, including illegal) 5,000,000 Horses 35,000 Cattle 348,000 Camels 4,000	Overgrazing by 2 × to 2.5 ×. Conversion of pasture to cropland. Increase in xerophytes and halophytes. Intensive wind erosion. Salinization and water-logging. Desertification reaches 80%.
1991–	Mythical autonomy and transition to market economy.	First steps toward ecologically proper agriculture: balanced use, reconstruction of pastures and forests.	Sheep 1,500,000 Horses 20,000 Cattle 250,000 Saigas 130,000 to 200,000	Grave conditions. Anthropogenic desert area of 1 million ha.

true desert, all within the past 20 years. The evolution and dynamics of agricultural development and related ecological changes are summarized in Table II.

4. Degradation of Pastures

The dominant type of desertification in Kalmykia consists of pasture degradation. If the peanut proved fatal to the Sahel, and cotton to Central Asia, then sheep proved to be the fatal factor in Kalmykia. The following were the main causes of degradation: (1) the number of livestock were increased beyond pasture carrying capacity, (2) farms that included grazing were established during a period of sustained drought (the 1960s), and (3) the Caspian Sea was retreating (1970s).

According to Vinogradov (1993), the yield of forage crops in 1954–1961 was good. Loamy tipchak-white wormwood pastures produced 600–800 kg/ha; sandy-rye grass-sandy-wormwood pastures produced 100–120 kg/ha; and bekmannian produced 4000 kg/ha. During the past 20–25 years the yield of pasture forage crops has declined 2.2–2.1 times, but the number of sheep has increased 1.6–1.8 times. Including animals that are not officially reported, the number jumped 2.2–2.5 times and for the semibush deserts, 2.2 times. On loamy pastures the forage yield declined to 200–300 kg/ha and to 100–200 kg/ha on sandy pastures. As a result, summer fodder deficits became a problem in some regions. Although formerly able to export fodder, these regions now import from 40,000 to 80,000 fodder units annually. According to different sources, from 100,000 to 600,000 head of livestock starved during the winter of 1986/1987 due to fodder shortages.

An increase in the number of sheep pastured during the winter without permission in the Black Lands has resulted in excessive overgrazing and degradation of pastures. In 1959 overgrazed pastures made up 32% of the area. By 1972 the figure was 59%, and by 1986 it had risen to an impressive 76% (Reznikov, 1993). Out of 3 million ha of pasture here, more than two-thirds is overgrazed (1,119,500 ha), strongly overgrazed or very strongly overgrazed, (534,000 ha), or suffering from blown sands and barren solonchaks (665,000 ha) (Kazakov, 1985). Annually the desert claims 50,000 ha of pastures from the Black Lands. If farming continues as it is, one can argue that this unique place of Russia will vanish, and 3 million ha will be transformed into desert.

Pastures are especially overgrazed around settlements and any other place where livestock concentrate (e.g. sheep sheds, wells). Most farms are overgrazed not only because of an increase in load per unit area of pasture but also because of "unsystematic" grazing in which early and continuous stocking results in the trampling of vegetation and thus the absence of favorable conditions for normal plant reproduction and growth. Unsystematic grazing became such a major factor that it practically liquidated seasonal variability of pasture crop yields. Crop yield indices became completely dependent on the degree of degradation. Overgrazing changes vegetation composition. The abundance of perennial fodder plants in all arid ecosystems declines as desertification progresses. In addition, overgrazing simplifies the horizontal and vertical structure of vegetation communities. Well-marked three-five member diversification of a background level disappears under conditions of very strong desertification. Similarly, multiple canopies, canopy cover, yield, and forage value decline, and noxious plants increase. Thus, native vegetation degenerates and is replaced to the degree that it is ultimately lost as a source of fodder (State Report, 1992).

Between 1950 and 1985, the area of steppe pastures in Kalmykia was reduced 1.2 times due to the increase of xerophyte semibush and halophytes. The true diversified grass-tipchak-feather-grass steppe vegetation type has essentially disappeared from the natural pasture areas of Kalmykia (Bananova, 1993).

5. Wind Erosion

Kalmykia is one of the most erosion-susceptible regions of southeast European Russia. Out of 5,880,000 ha of agricultural lands 5,400,000 ha present deflation hazards. Out of this amount, 474,100 ha are subjected to water erosion and 1,742,900 ha to wind erosion. A total of 380,000 ha have been transformed into moving sands. Under natural conditions soil erosion is essentially unknown.

Wind erosion is especially threatening in parts of the Black Lands that are underlain by sand. The impact of livestock on these pastures has extended centers of sandy soil deflation and transformed them into mobile sands.

Even in areas of finer textured soils wind erosion is a problem. Animals compact soils, destroy surface structure, and loosen soil through trampling. These actions permit the movement of fine particles through wind action, often forming barchan dunes.

When winds come out of the southeast, humidity may drop to 20–25%, temperatures may reach 30–40 °C, and velocities may exceed 20 m/sec. Under such conditions, blowing sand may sever plant stems through abrasion or even uproot plants by exposing and undermining root systems. As a result of these mechanics, plowed fields become the centers of wind erosion, especially on fine-textured soils. Such erosion is the main cause of the development of "sand sores," spawning areas of sand hillocks and deflation depressions. In 1970 such areas constituted 25,000 ha of the Republic, and in 1980 they had swollen to 560,000 ha (Vinogradov, 1993).

Concurrent with the deterioration of Kalmykian pastures, farms began plowing to convert pastures to fodder crops (maize for silage, rye for hay). This cultivation created new nodes for expanding desertification processes. As a consequence, in 1956–1959 about 3.5% of the Black Lands were embraced by wind erosion. By 1971–1972 this percentage had risen to 37.2%, and by 1984–1986 it had extended to 94.6%.

The attempt to fill the fodder deficiency by increasing fodder crops reduced pasture area, thus increasing the concentration of livestock on remaining pastures.

The cost of fodder produced on agricultural fields is higher than that produced on natural pasture.

The intensification of livestock production also results in increased demands for labor and services. These demands, in turn, place more demand on the environment. As a result of these converging forces, wells are drying up, and settlements are covered with dunes. Thus, barchans, abandoned farms, and sand-covered villages are the legacy of plowing in the cattle-breeding Republic.

6. Salinization of Lands

By 1970 irrigation in Kalmykia covered 124,000 ha. Since the soils of Kalmykia are developed on saline Caspian deposits with poor drainage, the result has been salinization. For the past 20 years irrigation-watering systems were built in different regions of the Republic. During their operation and maintenance secondary salinization and solonetzification embraced from 40% of the irrigated lands in the Pravo-Egorlyk's system to 90% in the Caspian. The development of rice – in the Sarpa system particularly – led to an increase of secondary salinization and waterlogging up to an area of 67%. As a whole, about 106,000 ha of lands in the Republic have been ruined (Bananova, 1986).

No statistical information exists on those lands outside state-engineered systems or "initiative" irrigation systems supplied from local sources. Taking everything into account, we might argue that 100% of irrigated lands are salinized to some degree.

Other processes of salinization, among other problems, originated during the construction of the Volga-Chograi Canal. Almost 80 kilometers of this 20 m deep, 100 m wide canal were built without lining and are presently filled with saline water. The rise in groundwater levels that followed canal construction has increased salinization along its course because of the uncontrolled outflow of saline waters from artesian wells.

7. Desertification Control

As the most important component of the biosphere, vegetative cover of the extensive pasture of Kalmykia determines not only fodder for animals but also the quality of human habitat. Favorable conditions for pasture ecosystems are determined by their resilience to exploitation, in their ability to produce and reproduce.

For the most efficient use of pastures the number of livestock must be regulated in accordance with grazing capacity as it varies through time and in space. Livestock can be regulated by adjusting the number of livestock and rotating their use of pastures throughout the season to spread and minimize their impact.

But if the pasture has been degraded, it must first be restored. During primary pasture restoration livestock populations must be reduced below what might be considered optimum. In Kalmykia livestock numbers would have to be reduced by 1.6 million animals/year, and herd composition would have to be changed. Instead of single-breed sheep for wool only, other animals such as meat sheep, horses, or camels should be introduced. In the future, perhaps by the turn of the century, the herd could be increased to the optimal level – 1.8–2.2 million head (Vinogradov, 1993).

One of the main methods of desertification control of natural fodder-producing areas is rotational grazing. The essence of this system is subdividing pastures and allocating them to certain flocks for seasonal use. Typically a pasture is subdivided into 15–16 small sites that would contain fodder reserves for 5–6 days. During the grazing season livestock would visit each site twice (in the beginning and towards the end). In semiarid conditions, this technique allows grass to recover.

But in Kalmykia, where a considerable area is occupied by heavily grazed pastures that have experienced severe desertification, special technologies are required for pasture improvement. At present, natural fodder-producing areas are being improved in two ways. The first way involves stabilizing open sands by planting brushwood (juzgun, teresken), sowing sand-fixing crops (small oats), and later introducing such perennial plants into the grass stand as rough grasses, wormwood, and rye grass. The second way is "radical" improvement of heavily grazed pastures where grass stands cannot be restored, even with long-term rest. Radical improvement involves physically preparing a seedbed through tilling and then sowing fodder plants.

Recognizing the importance of the Black Lands as the main fodder base of the Kalmykia, the "General Scheme of Desertification Control" was formulated for this region in 1986. The scheme envisages implementing works between 1993–2000 on reclamation of blown sands using silviculture on 661,000 ha, radical improvement of 440,600 ha, and development of water supplies for pastures over the entire 3,375,100 ha area. Scientifically justified pasture rotations must be introduced to all preserved natural or restored pastures, and the most erosion-prone pastures exceeding 600,000 ha are recommended for use only as winter pasture (October-April).

The results of the first year's work on the General Scheme show that where a complex of measures have been implemented, within 2–3 years the processes of desertification are slowed or completely abated, and that fodder reserves have increased several fold. Within 3–5 years, introducing existing technologies will allow pasture communities (with a dry weight yield of 2–2.5 t/ha) to become established on barren sands. In a number of regions average yields of gigantic dune grass on fixed sands amounted to 2.4 t/ha and 1.7 t/ha of dry weight of teresken. These amounts exceed the fodder reserves of natural pastures by a factor of 6–8 (Reznikov, 1993).

It is difficult, however, to hold much optimism because management of these restored pastures usually reverts to the same farms that brought them to such a state in the first place. There is great apprehension that such degradation could be repeated. The pasture could once again be overgrazed within one or two seasons.

Implementing a complex of measures on pasture restoration favors preserving and developing saiga. Saiga do not compete with domestic animals for fodder and may coexist or be combined with other herds in ecologically organized pasture farming. Saiga should be involved in pasture rotation, which will allow their numbers to increase to 0.4–0.75 million head by 1995 and to 1.2–1.5 million head by 2000. These numbers would be sufficient for a stable European population.

Regardless of its efficiency, pasture restoration is a means of overcoming the aftereffects of unsystematic, nonadaptive use of natural resources. Land degradation is a product of the incompatibility of technologies applied to brittle natural and reconstructed territories. Pastures restored or established on sands may be again be irrevocably lost if the ways they are exploited do not change.

Land users have yet to tap scientific developments in the efficient use of pastures through a system of rotation and seasonal use of a part of the region's pastures. In our opinion this is a result of the lack of any linkage between economic interest and the responsibility of land users for the state of fodder croplands.

Because of the uncertainties that surround the future economic prospects for Kalmykia and the possibility of climatic change, it is difficult to evaluate the long-term effectiveness of measures to combat desertification.

Realization of the actions suggested here depends on many factors. Among them are the general structure of power (and control) existing in the Republic and locally and the conflict of interest among forces, bureaucratic departments, local authorities, and public ecological organizations. Alternative decisions and ecological forecasts are often trumps in the political game of force and interests. They could be ignored in whole or in part, depending on the interests of administrators (Zonn, 1994).

In a variety of respects, the territory of Kalmykia assumes importance as a model for investigating desertification aimed at developing the systems of control and preventing its advancement.

8. Conclusions

1. Kalmykia is a vivid example of P. Erlich's description of an "economic system, embraced by delusions of growth and nature conservation, which are opposed to each other fundamentally." For a long time the agricultural sector of the Republic pursued an ecologically destructive cattle-breeding type of monoculture development. Its objective was to increase cattle production in a manner that was often incompatible with the functioning of natural resources. The dimensions of herds considerably exceeded the opportunities of existing pasture resources, which resulted in intensive development of desertification processes that eventually led to the formation of a true desert.

2. This general type of cattle and agricultural development is unviable for both ecological and purely economic reasons. Evidence points to an increasing imbalance between productive output and economic input. Deterioration in the quality of cattle breeding and agricultural production is a dangerous consequence of such development, which will have an immediate effect on the viability and health of the local population.
3. Until now the rehabilitation policy of desertification control considers desertification a cumulative phenomenon rather than a process. Such a perspective assumes that the process cannot be resumed. Instead, we should implement an adaptive policy based on "planning from what has been achieved." Such a policy should be oriented toward conservation and should accommodate the interrelation and variability of land resources.
4. The desertification problem will be solved only by changing agrarian policy as a whole. As a first step we need to establish a conservative regime of pasture exploitation that is defined by ecological limits and will lead to a stable, ecologically balanced, adaptive type of farming.
5. We cannot control and combat the process of desertification until a program of desertification control becomes an integral part of the program of economic and social development of the Republic.

Acknowledgements

The author expresses his gratitude to the following people: Dr. N. Reznikov, Head of Directorate Agrolesomeliostroill of the General Department of Black Lands and Kizlyar Pastures, who provided useful advice and information; Dr. V. Bananova, Head of Department of Desert of the Institute of Sociology, the Kalmykian Branch of the Russian Academy of Sciences; and Professor S. Zonn, Consultant, Institute of Geography, Russian Academy of Science, for cartographic materials and valuable advice.

References

Bananova, V.A. 1986. *Methodical directions for investigation of desertification processes of arid areas in the Kalmyk ASSR*. Kalmyk University, Elista, p. 38.

Bananova, V.A. 1993. Anthropogenic desertification of arid areas of Kalmykia. Synopsis of thesis for Doctor's Degree of Geographical Sciences. Ilym, Ashgabad, p. 44.

Bananova, V.A., G.M. Borikov and J.E. Khahulov. 1986. Towards investigation of present state of desertification processes in Kalmykia. In *Ecological problems of desert development and wilderness conservation*, Ylym, Ashgabad, pp. 92–94.

Dregne, H., M. Kassas and B. Rozanov. 1991. A new assessment of the world status of desertification. United Nations Environment Programme *Desertification Control Bulletin* **20**: 6–18.

Kazakov, B.G. 1989. Chernye Zemli. In *The Caspian Sea area for us and our children*, Moscow, pp. 358–383.

Reznikov, N.I. 1993. *First results of the realization of the general scheme of desertification control of Black Lands and Kizlyar pastures*, Agrolesomeliostroy, Elista, p. 12.

Rozanov, B. 1990. Assessment of global desertification. In *Desertification revisited.* United Nations Environment Programme, Nairobi, pp. 45–122.

State Report. 1992. *On the state of the environment of the Russian Federation in 1991.* Moscow.

Vinogradov, B.V. 1993. The Present dynamics and ecological forecasting of natural environment in Kalmykia. *Problems of Desert Development* N1: 29–37.

Zonn, S.V. 1986. "Black Lands" of Kalmykia. In *Arid land development and the combat against desertification: An integrated approach,* USSR Commission for UNEP, Moscow, pp 124–127.

Zonn, I.S. 1994. A creeping environmental phenomenon in Russia: Desertification in Kalmykia, in M. Glantz (Ed.), *Workshop Report on Creeping Environmental Phenomena,* National Center for Atmospheric Research, Boulder, Colorado, pp. 169–174.

GPSR Compliance

The European Union's (EU) General Product Safety Regulation (GPSR) is a set of rules that requires consumer products to be safe and our obligations to ensure this.

If you have any concerns about our products, you can contact us on ProductSafety@springernature.com

In case Publisher is established outside the EU, the EU authorized representative is:

Springer Nature Customer Service Center GmbH
Europaplatz 3
69115 Heidelberg, Germany

Batch number: 10607320

Printed by Printforce, the Netherlands